The Dobsonian Telescope

A Practical Manual for Building Large Aperture Telescopes

David Kriege • Richard Berry

published by

T.M.
Willmann-Bell, Inc.

Publishers and Booksellers Serving Astronomers Worldwide Since 1973
P.O. Box 35025   Richmond, Virginia 23235

Published by Willmann-Bell, Inc.
P.O. Box 35025, Richmond, Virginia 23235

Copyright © 1997 by David Kriege and Richard Berry
First English Edition
    First Printing , January 1998
    Second Printing, June 1998

Printed in the United States of America

Library of Congress Cataloging-in-Publication Data.
Kriege, David
   The Dobsonian telescope : a practical manual for building large
   aperture telescopes / David Kriege, Richard Berry. -- 1st English
   ed.
        p.   cm.
   - Includes bibliographical references and index.
      ISBN 0-943396-55-7
      1. Dobsonian telescopes --Design and contruction--Amateurs'
   manuals. I. Berry, Richard, 1946-   . II. Title.
   QB103.K75   1997
   681'.4123--dc21   97-43034
                                                    CIP

The following are acknowledged as trademarks: Campmor, Carborundum, Chivas Regal, Dawn Dishwasing Detergent, DIG, Inc., Dob-Driver, Ebony Star, Elmer's Glue, Ferrari, FinnPly, Ford Motor Co., Hex-Tek, Ivory Soap, Jell-o, JoAnn Fabrics, Kydex, Lumicon, MicroGrit, Mylar, Nagler, Obsession Telescopes, Oreo, Panoptic, Porta-Pak, Post-it, Q-tip, Rip-Stop, Saran Wrap, Sequentia, Sonotube, Stardust, Structoglas, Styrofoam, Tangent Instruments, Tech 2000, Thinsulate, Thompson's Water Seal, Touchstone, Vaseline, Windex, Yellow Pages and Ziploc

97 98 99 00 01 02 03 04   9 8 7 6 5 4 3 2 1

For my dad, who helped me build my first telescope a long time ago.

David Kriege

Foreword
Give credit to those who helped you so much

Unlike the "Big Bang," this book didn't just happen. It came into being because I have a debt to repay. I owe the many telescope makers before me who shared their ideas and made astronomy and telescope making one of the highlights of my life. In fact, telescope making has become an obsession for me. I want to acknowledge these talented people, starting with Sir Isaac Newton, by giving them the credit they deserve. More importantly, I want to pass along their ideas with mine so you too can build your own truss tube telescope and have a lifelong affair with astronomy.

Some of you already know me from the articles I wrote for *Telescope Making #35, #37, #41*, and *#44*. Those "how I built it" stories were extremely popular. In 1990, I became a professional telescope manufacturer because people were willing to pay me to build them a telescope like mine. Before long I had made over a hundred telescopes for others. At the same time amateurs and a few professional telescope makers were building telescopes based on my "Obsession." Within five years, the telescope I built and wrote about had become the most imitated design in the hobby.

This success is obviously a great source of personal satisfaction for me. Yet my contribution has been built on the ideas of others before me. As you read in later chapters how today's amateur telescope designs came about, you will realize why things are done they way they are. You will also realize that we are very lucky to be living in a time when so much is available to the amateur telescope maker.

This book shows you how to construct a high-performance, large-aperture Dobsonian telescope. As many amateur astronomers already know, Dobson's original design is fundamentally excellent. What I have done is taken Dobson's concepts and tried to realize their full potential.

Now there are nearly as many variations of the Dobsonian design as there are amateur telescope makers. But as different as they are, all share an important similarity: building them gave their makers a deep sense of accomplishment and pride. All of them are fun to use. Their very variety should encourage aspiring telescope makers. I have seen a great many Obsession clones at various star parties. Each of these telescopes reflects the desires, knowledge and skills of its builder.

So, if you don't want to build the Obsession design, feel free to try other designs and to develop your own ideas. Think openly. Don't feel guilty if you want to alter things; that is your right as a telescope builder.

However, before you rush out and start sawing sheets of plywood, I want to encourage you to *actually read the text of this book*. Although you could probably make a successful telescope by copying what you see in the photos, you would miss out on the opportunity to understand fully what you are doing. Once you understand why the design works, you're equipped to build a telescope that is a top-notch performer.

Why do I say this? Since the designs appeared in *Telescope Making*, Richard and I have had the chance to examine hundreds of telescopes based on those articles. Although all of these telescopes seem to work acceptably well, some of them do not perform nearly as well as they could have. Read and learn. Think and study. Maybe this will sound old-fashioned, but it doesn't take much more time and effort to build a telescope that works really well, and you'll be a lot happier with it.

The design described in this book doesn't have any secrets. Everything is out in the open. Understanding the design boils down to basic high school algebra and physics. If those two words scare you, don't worry. Just knowing why different materials behave the way they do is a major step, and probably more important than the math. Once you develop a feel for the "whys" of telescope design, you'll be better equipped to manage the "hows" of their construction.

Besides, it's much more fun to do something when you know what you are doing. Those of you with serious math-phobia can get all the dimensions you need from the tables and plans. Don't let your math abilities—or any lack thereof—deter you from building your dream telescope.

But please be forewarned. Reading this book might—no, almost certainly *will*—affect your mind. Aperture fever is a serious illness, and building a big telescope often evokes obsessive-compulsive behavior. Because you are spending all your free time working on "the scope," you may discover that you are neglecting your loved ones. Chasing all over town for obscure fasteners. Devoting long hours to doodling out new ideas on paper. Spending too much money on equipment you feel you just "gotta have." Scrounging parts. Driving the countryside searching for dark sites. Becoming a recluse at New Moon. These are obsessive behaviors. If you develop these symptoms, try to remind yourself that the point of it all is to have some fun. And never forget that the stars will still be there tomorrow.

Next, I want to address a somewhat touchy subject: mirror making. Frankly, I don't recommend it. But I feel that it's only fair to recognize how important mirror making has been in amateur astronomy. It all goes back to 1921, when an arctic explorer named Russell Porter began writing articles for *Popular Astronomy* on how to make reflecting telescopes. Porter's practical advice soon earned him an eager following of amateur telescope makers in his native Vermont. Then *Scientific American* invited him to write two articles on the subject. Reader interest was so strong that in 1926, Albert "Unk" Ingalls at *Scientific American* encouraged Porter to help write what became the definitive work on the subject, *Amateur Telescope Making*.

By titling one of his first articles "The Poor Man's Telescope," Porter showed that he was a man in touch with his time. The article showed the would-be amateur astronomer how to make inexpensive mirrors for a reflecting telescope. During the Great Depression, folks had little disposable income but plenty of time. Almost everybody knew somebody who could run a lathe—what they lacked were the skills and know-how for making mirrors. No instructions were available, and even if you had the cash, nobody had mirrors for sale. The bottom line was: if you could make a mirror, you could own a telescope. For the next 50 years, mirror making was an integral part of amateur astronomy.

In 1951 a great French optician, Jean Texereau, wrote what would become a classic work, *How to Make a Telescope*. This book is still the bible of mirror making, so if you want to make your own mirrors, get this book.

Porter and Texereau deserve a standing ovation for their contributions to amateur astronomy. From 1921 through the 1970s, thousands of amateurs made reflectors, some as large as 12 inches in aperture, and a scattering made even larger mirrors. When someone said he had built his own telescope, it almost always meant he had made everything—including the primary mirror.

But mirror making is tedious, exacting, and slow. And homemade mirrors tend to be small—6 to 10 inches in diameter—which is a lot smaller than the mirror that you want for your large-aperture Dobsonian. For small mirrors, times have changed and the economic incentive has vanished: you can now buy a finished mirror for about the same price as you can buy the raw materials.

Of course, I can hear the dyed-in-the-wool mirror makers out there moaning and groaning. Yes, making a mirror is a great learning experience and I'll concede that you'll probably appreciate your telescope more if you grind and figure its mirror. No argument from me. But I don't do it and most other telescope makers don't grind their own mirrors any more either.

With all that said, you're probably wondering why we included a lengthy Appendix by optician Bob Kestner on making big, thin telescope mirrors. I suppose that I could make the excuse that Richard put me up to it, which is true, but the real reason is that we both want you to understand everything that goes into making a high-quality telescope. Furthermore, as this book travels around the world, it will reach places where big, thin, commercially made telescope mirrors are not available locally and are too expensive to import. Even in the United States, you can save a bundle if you grind, polish and figure a large telescope mirror. If those reasons make sense to you, then grasp the nettle and grind the mirror.

Before concluding, I want to be 100 percent upfront about my little company, Obsession Telescopes. We make and sell complete telescopes. Some of the components in the commercial telescope have been custom designed, and others are hard to locate. Rest assured you can make everything for an Obsession-style telescope by following the instructions in this book. Sources for all of the materials and off-the-shelf parts are provided in an Appendix in the back of the book.

This book has benefited from Harold R. Suiter, Mel Bartels, Thomas Firebaugh, John Koester, David Groski, Chris Bechtler and Robert Lombardi who read and commented on the manuscript prior to publication. Richard and I appreciate their assistance which has made this a better book because of their efforts.

Well, that concludes my introduction. As you read, once in a while remember this: you have fallen heir to knowledge that took thousands of amateur telescope makers seven decades of building telescopes and two decades of building Dobsonians to learn. Consider how fortunate you are to be living in a time when you can build a huge telescope, drive to a remote site where the sky is black, and get set up for observing in a few minutes. You can see every nebula, cluster, and galaxy in the *New General Catalog* and thousands more besides. Have you died and gone to heaven, or what?

David Kriege
Lake Mills, Wisconsin

Table of Contents

Chapter 1
Large-Aperture Dobsonians

Be honest. You want to observe through the biggest telescope you can get your hands on. It's every amateur astronomer's fantasy. And, if you think we're not telling the truth, go to any star party and notice which telescopes attract long lines of people. Do people line up behind 6-inch, or 8-inch, or 12-inch telescopes? No. They line up to look through the big Dobsonians.

Why? Well, when it comes to astronomy, nothing beats aperture. And Dobsonians are the poor man's way to own a very big telescope. They are low-tech instruments; ordinary mortals can build their own. Easy to set up and a joy to use, Dobsonians can be taken to desert or mountain skies for truly spectacular observing. Most important, big Dobsonians give big, bright images. For visual astronomers, they are the ultimate observing experience.

For the Dobsonian, we owe thanks to John Dobson. Dobson is a visionary. With his home-built telescopes, he smashed traditional "small" expectations for amateur instruments. He said, "Big telescopes are easy to build," and he proved it. John Dobson pointed the way to today's dream telescopes.

To appreciate Dobson's impact on the hobby, consider the following: Before 1980, 90% of all homemade telescopes were equatorially mounted reflectors under 12.5 inches aperture. Since 1990, 90% of all homemade telescopes have been Dobsonians, and many of them are much larger than 12.5 inches aperture.

What could have prompted such a drastic change in amateur astronomy? How did the Dobsonian-mounted Newtonian telescope, commonly called the "Dobsonian telescope," capture the interest of so many amateurs in only a decade? Obviously, the Dobsonian had a lot going for it. Let's look at the features that amateur telescope makers found attractive in the *classic* Dobsonian.

- It had a thin, large-aperture primary mirror. This meant the mirror was light in weight, but offered enormous light gathering power. In the early days, 16 inches was considered a large telescope. Today's big Dobsonians have apertures as large as 40 inches.

- Plywood triangles and a sling supported the thin primary mirror. Although this mirror cell was easy to build, it provided support so the mirror delivered outstanding images. Today's

Fig. 1.1 *The classic Dobsonian "Sidewalk Telescope" had a cardboard tube, a massive rocker box, and a thin porthole-glass mirror. Here, observers have set their telescope at Glacier Point in Yosemite National Park. Photo by Robert Kestner.*

cells are open to the air, so the mirror also cools rapidly and gives even better images.

- The mount was a simple altitude-azimuth ("alt-az") design made from flat pieces of plywood. It was easy to construct and held the telescope rock-steady on the celestial object of your choice. The same is true today, but the mount is lower, lighter, and stiffer for better all-around performance.

- The bearing surfaces of the mounting used low-friction Teflon plastic on smooth countertop laminate. This provided enough friction to hold the telescope in place, but the observer could easily point the telescope at any object in the sky. Today we have found even better laminates, so bearings are bigger, more stable, and give smoother motion.

- The bottom of the telescope tube was a simple wooden box, very easy to construct, that opened so the mirror could be taken out for storage in a separate box. Today we've reduced the weight of the box to a minimum, and we store the mirror safely inside it.

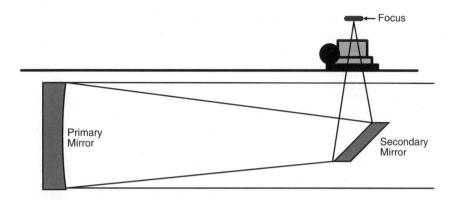

Fig. 1.2 The Newtonian telescope is a simple instrument with only two optical surfaces—a curved primary mirror that focuses the incoming light and a flat diagonal mirror that directs the focused light beam to the eyepiece. Dobsonians are Newtonians.

- The top half of the telescope was a big cardboard tube. Concrete form tube was inexpensive, light in weight, and worked great. This made the classic Dobsonian easy to build. Today we use a featherweight aluminum truss tube for a telescope that is even lighter and more compact.

Imagine the delight of the telescope makers who built the first Dobsonians in one-tenth the time and for one-tenth the cost of a conventional equatorial Newtonian. The telescope was big and it was portable. It needed no machine-shop work. For a remarkably small investment, you could build an enormous "light bucket" that gave sharp, crisp views of the sky. When lightweight, affordable, large-aperture primary mirrors hit the market, the Dobsonian revolution *really* took off.

Today, a few decades after the revolution began, tens of thousands of Dobsonian telescopes have been built. The greatest number are "small" ones (that is, telescopes under 12 inches aperture) but a huge number lie in the "classic" size range between 14 and 25 inches aperture. Finally, a tiny but highly visible minority belong to the group we call "The Big Ones," with apertures of 30 inches and more. As you read this book, you will be tempted to build the biggest telescope possible. Right up front, we want you to know that is not our goal in writing this book. We feel that you will be happiest if you build a telescope that's "the right size" for you, and that's not necessarily the biggest. In **Chapter 13** we describe an 8-inch Dobsonian. If this is not the right time for you to build a large telescope, then consider a small one.

1.1 Why Dobsonians Are Newtonians

Newtonians are the best telescopes going. They offer the largest aperture for the lowest price. They have simple optical systems that perform well. Nowadays you don't see much about Newtonians in telescope advertising, but that's because

Newtonians provide the manufacturers relatively small profit margins—and naturally enough, they prefer to sell telescopes with lots of fancy bells and whistles. Is that what you really want? This book will show you how you can own more telescope for fewer dollars than you ever dreamed possible.

Before the 1930s almost all amateur telescopes in America were refractors. The lens in a refractor gathers light and focuses it to an image that is then magnified and observed with an eyepiece at the bottom end of the telescope tube. Although refractors offer wonderful images, optical glass is expensive and a fine lens is difficult to make, so refractors tend to be small and costly. Today, one factor that limits the popularity of the refractor is tube length. Refractors traditionally needed long tubes to hold the main lens the proper distance away from the eyepiece. The classic 6-inch achromatic refractor has a tube 8 feet long! For these reasons, refractors over 8 inches aperture are rare. Great as they are, refractors yield good crisp images that are relatively small and dim. But let's face it: we astronomers want good crisp images that are *big and bright*.

Historically, the breakthrough arrived in 1678, when Isaac Newton invented a new type of telescope. Newton was among the first scientists to try to understand optics, and to study how the refracting telescopes of his day worked. After carrying out a long series of experiments with prisms, he declared that it would remain forever impossible to eliminate chromatic aberration from refractors, and proposed using a mirror instead of a lens to form an image. Although Newton was not right about refractors, the reflecting telescope design he proposed had some major advantages.

Newton suggested that a concave mirror could gather and focus starlight better than a lens. In fact, his design calls for two mirrors. The primary mirror at the bottom of the tube gathers light and reflects it toward a second, small mirror located near the top of the tube. Why the second mirror? It reflects the beam of light to the side of the tube where you view it with an eyepiece; without the secondary mirror, your head would block some of the incoming light. Newton made a small reflecting telescope which he presented to the Royal Society of London. Although it took several centuries to be perfected, Newton's new telescope design eventually revolutionized astronomy.

Reflecting telescopes produce wonderful images, totally free of chromatic aberration, just as Newton said they would. From our modern-day perspective, the best feature of the Newtonian is economic: the primary mirror can be made from large disks of inexpensive glass, so the images it forms are big and bright.

In addition, you can make a reflector much shorter than you can make a refractor of the same aperture. The tube for a typical 6-inch reflector is only one-third to one-half as long as the tube of a 6-inch refractor. Furthermore, the same number of dollars you could spend on a 6-inch lens for a refractor buys you a 16-inch mirror for a reflector. Catadioptric telescopes, those with both lenses and mirrors, are quite popular today, but they are expensive compared to reflectors of the same aperture. Dollar for dollar, money spent on a reflector will buy you more aperture and better performance than you can get from any other type of telescope.

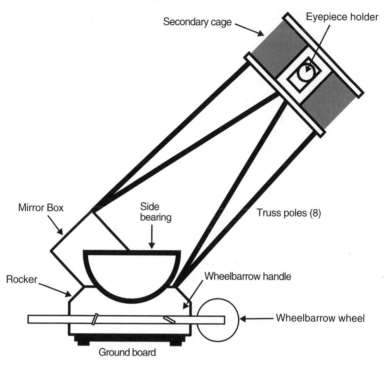

Fig. 1.3 *Twenty years of steady evolution have made the modern-style Dobsonian lighter and easier to transport than the classic Dobsonian, yet the names and functions of the major components remain the same.*

In short, Isaac Newton's bright idea meant better images from a package that was less expensive and easier to use. The combination of image quality, ease of construction and affordability makes reflectors the design of choice for large telescopes.

1.2 How Dobsonians Work

You now see that Dobsonians are also Newtonians, but what is it about these Newtonians that earns the Dobsonian label? John Dobson probably summarized it best when he said: "A Dobsonian telescope is just a pile of stuff."

Dobsonian telescopes, he says, are held together by gravity and run on yogurt power. Gravity holds the primary mirror in its cell, keeps the rocker box in place on the bearings, and provides just enough friction so the telescope stays where you point it. As for yogurt power—well, you eat the yogurt. There are no batteries, no wires, no motors. Just you in communion with the stars.

Let's take a look at the modern Dobsonian starting from the top, with the *secondary cage*. This lightweight structure has three jobs: it holds the diagonal mirror in place, it carries the eyepiece holder (focuser), and it blocks stray light. It consists of two rings of plywood and a cylinder of thin Kydex plastic—the classic monocoque construction used in the aircraft industry. The focuser rides on the *focuser board* between the two plywood rings. The whole upper tube assembly on a 25-inch telescope weighs only 15 pounds.

The secondary cage rides on a *truss* of eight lightweight aluminum poles. Their job is to support the upper tube assembly rigidly above the primary mirror below. In early Dobsonians, a length of spiral-wound concrete form tube ("Sonotube") did this job, but the truss is superior. Not only does it transmit fingertip pressures to the bearings below, but it is also extremely stiff and you can you take the truss apart for easy transport.

At the base of the truss is the *mirror box*. This is the nexus of the Dobsonian telescope. The truss tubes are clamped firmly to the upper end of the mirror box. The primary mirror rides in the mirror cell mounted at the bottom end of the mirror box, and attached to its sides are the *altitude bearings*. The function of the mirror box is to keep these essential components in precise alignment.

The *primary mirror* is, of course, the real heart of any telescope. The job of the *mirror cell* is to pamper the mirror on a system of levers, "floating" the glass so that it holds its figure against the pull of gravity. The wide-open design of the cell permits air to circulate around the mirror, cooling it to ambient temperature quickly so you can get the best possible views of the heavens.

Together, the mirror box carrying the primary mirror, the truss tube, and the secondary cage constitute the *optical tube assembly*, that is, the telescope proper. The optical tube assembly of a Dobsonian weighs but a fraction of that of an equatorially mounted telescope. A 20-inch modern-style Dobsonian, including the mirror, weighs a mere 150 pounds.

The job of a *telescope mounting* is to hold the telescope solidly in the desired position and yet allow an observer to point the telescope easily toward any part of the sky. The Dobsonian alt-azimuth mounting excels in both of these functions. The rocker box supports the telescope from both sides with widely spaced supports so all the weight is centered over the bearings. This minimizes the sagging and vibration that plague the under-engineered mountings on other types of telescopes, and keeps the total weight of the instrument very reasonable. You need a fork-lift and a couple of strong helpers to set up a conventionally designed 25-inch equatorial telescope, yet with the Dobsonian design, one person can handle the entire task.

On each side of the mirror box you will see a prominent *altitude bearing*. Although the tube assembly rests entirely on these two large semicircles, light pressure suffices to move it. A *plastic laminate* such as *Formica* covers the outer surface of each bearing. These bearings are just another example of Dobson's fantastic yogurt-powered design; because Formica slides easily on Teflon plastic, the motion has that unique "buttery" feel of a well designed and well made Dobsonian.

So, under each altitude bearing you'll see two *Teflon pads* mounted on the solid structure of the *rocker box*, so called because the telescope rocks up and down in it. For maximum stability, the rocker box should be very low and stubby.

On the bottom of the rocker box is the *azimuth bearing*, a circle of plastic laminate riding on three Teflon pads. A *center pivot*, usually a large machine bolt, keeps the rocker box from sliding sideways, so when you're observing and you press lightly against the telescope, the whole tube and mounting rotates smoothly, with that "buttery" feel of a good Dobsonian. The base of the telescope is the *ground board*; its sole job is to hold the Teflon pads in place under the azimuth bearing. Finally, at the very bottom, the ground board rests on *feet* directly under the Teflon pads.

A design as sophisticated as the Dobsonian didn't just happen. It became what it is through a long series of inventions and experiments, systematic engineering design, and just plain luck. It took many inventive telescope makers many years to come up with a telescope this good—and you are the beneficiary of that tradition.

1.3 The History of the Dobsonian Telescope

In 1784, the German-born English amateur astronomer William Herschel climbed to the eyepiece of a large reflecting telescope to continue his comprehensive survey of the heavens. Herschel's instrument, at that time the most powerful in the world, is the ancestor of today's big Dobsonians.

His big reflector did not resemble a sleek, modern Dobsonian. Instead, the 20-foot-long tube hung among the poles of a massive rotating scaffold. Herschel stood on an elevated gallery and peered into an eyepiece mounted at the front of the tube. With groans and rumbles, the whole bulky contrivance turned in azimuth; with the creak of ropes and the squeal of pulleys, Herschel could raise and lower the heavy tube. Herschel's 20-foot telescope shared two key features with today's Dobsonians: it had an alt-azimuth mounting, and because he observed with an exceptionally large aperture, Herschel could see thousands of nebulae.

The young musician and choirmaster had become interested in astronomy in 1773; he was an amateur astronomer. Because commercially made telescopes were prohibitively expensive, Herschel set out to build his own. He pursued his new hobby with tremendous diligence: by 1776, he had mastered the art of casting, grinding, polishing, and figuring mirrors up to 9 inches diameter from hard, shiny speculum metal. Fame came in 1781, when Herschel, who was systematically sweeping the sky with a handcrafted 6.2-inch reflector, discovered the planet Uranus.

The discovery of a new planet brought Herschel to the notice of King George III, who granted Herschel the title "Royal Astronomer" with a salary large enough to allow him to devote his full energies to the study of the heavens. Pushing the technology, Herschel constructed the most powerful telescope in the world, an alt-azimuth reflector boasting an 18.8-inch primary mirror of 20-foot focal length, and began a systematic and unprecedented exploration of the heavens. The nebulae, clusters, and galaxies Herschel found in his sweeps became the basis for the *Gen-*

Fig. 1.4 *Although not a Dobsonian, William Herschel's giant 40-foot reflector rode in an alt-azimuth mounting, as modern Dobsonians do. When he completed it in 1787, Herschel's 48-inch aperture telescope was the largest in the world.*

eral Catalogue of Nebulae and Clusters of Stars, a list of 4,500 deep-sky objects that is the forerunner of the *New General Catalogue* (the NGC objects) in use today.

Herschel's 20-foot telescope was so successful that, with the aid of the royal treasury, he undertook building an enormous one: a 40-foot long reflector with a 48-inch mirror. Although this telescope remained the world's largest for over 60 years, it was too large for the technology of his era, and after 25 years of declining use, Herschel's great telescope fell into disrepair.

The next large alt-azimuth telescopes were built in the 1840s, by William Parsons, third Earl of Rosse. As any new builder must, Parsons worked his way through a series of increasingly large telescopes—with 15-inch, 24-inch, and 36-inch mirrors—mounted on alt-azimuth platforms that closely followed the plan of Herschel's. Using a more reflective alloy for his mirrors with better designed and

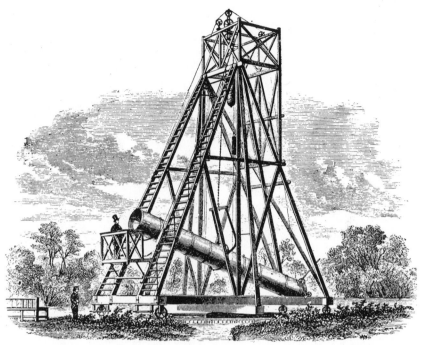

Fig. 1.5 By 1840, William Parsons, third Earl of Rosse, built a successful 36-inch alt-azimuth telescope. Parsons then caught aperture fever and built a 72-inch telescope in which the tube was suspended between heavy masonry walls. It did not meet his expectations.

smoother mountings the performance of Parsons' 36-inch alt-azimuth undoubtedly surpassed that of Herschel's 48-inch telescope.

Yet Parsons could not be satisfied: he decided to build a 72-inch reflector. The "Leviathan of Parsonstown," as it became known, was only a partial success. Supported by chains between two massive walls of masonry, the telescope could be raised to any altitude, but moved over only a short range of azimuth. Poor seeing, its limited pointing ability, and the terrible weather of central Ireland severely limited its use. Despite these limitations, however, the immense mirror revealed spiral structure in nebulae for the first time.

The last of the great ancestors of the Dobsonian was a 20-inch Cassegrain reflector constructed in the 1840s by the English engineer James Nasmyth. In silhouette, this telescope closely resembled a Dobsonian. The tube rode in trunnions mounted on a turntable, and could be pointed to any spot in the sky. To use the telescope, an observer sat on a chair in front of the eyepiece mounted on one trunnion and turned two large handwheels that, through gearing, moved the telescope. Nasmyth's telescope established the potential of the alt-azimuth mounting, although his innovation was quickly forgotten. Large alt-azimuth telescopes were soon thereafter eclipsed by far smaller equatorials, and more than a century would pass before large alt-azimuth telescopes reëmerged.

Fig 1.6 *James Nasmyth, a English engineer, built this 20-inch Cassegrain reflector in the 1840s which in silhouette looks a lot like a modern Dobsonian.*

Fig 1.7 *"I'm the one who has always said, 'Maksutov, Schmaksutov, Cassegrain, Schmassegrain, Schmidt, Schmidt. Now it's Dobsonian, Schmobsonian.'" This photograph shows John Dobson shortly before he spoke to an enthusiastic crowd at the 1987 Stellafane convention.*

1.3.1 John Dobson Invents the Dobsonian

Despite their alt-azimuth mountings, the telescopes of Herschel, Lord Rosse, and Nasmyth were decidedly not Dobsonians. The true Dobsonian—the telescope held by friction alone—was invented by John Lowry Dobson. Born in China in 1915 to missionary parents, it fell to Dobson to reduce the alt-azimuth telescope to its essentials.

When his family returned to San Francisco in 1927, he attended Lowell High School and later the University of California at Berkeley to study biochemistry. An exceptionally able and original thinker, Dobson soon rejected conventional institutions and joined the Carol Beals dance group. However, a lecture by Swami Ashokananda changed the direction of young Dobson's life, sending him on a quest for "the reality behind the universe" under the Swami's instruction. The Swami advised returning to school, and in 1943, Dobson graduated with degrees in chemistry and mathematics. He immediately found work at Berkeley, later transferring to Caltech and then to the Berkeley Radiation Laboratory.

In 1944, Dobson quit his job and entered a monastery as a monk of the Ramakrishna order. At the monastery, Swami Ashokananda assigned him the task of uniting the ancient thinking of India with atomic physics and astronomy, the sciences that deal most closely with the "first cause" of the universe.

In 1956, Dobson built his first telescope. The mirror was made from a 12-inch disk of porthole glass using the instructions found in Allyn Thompson's *Making Your Own Telescope*. The sight of the moon through this instrument helped him decide that everyone in the world had to see the heavens through a telescope.

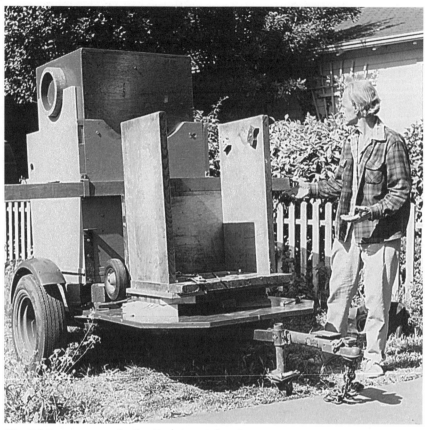

Fig. 1.8 *Every summer the San Francisco Sidewalk Astronomers trailered their huge telescopes to national parks in the West. In this 1980 photograph John Dobson prepares the rocker boxes of two large instruments for the annual tour.*

In 1958, Dobson was transferred to the Vedanta Society's monastery in Sacramento, where he served as the monastery's gardener and surreptitiously built telescopes with cardboard hose-reel barrels and porthole-glass mirrors.

At night Dobson trundled his reflectors on wagon wheels around the monastery neighborhood and taught local children how to build telescopes. But monastery rules forbade leaving the monastery grounds without permission, and in 1967, after 23 years as a monk, Dobson was expelled. At that time, he had constructed fifteen 12-inch and two 18-inch telescopes from scavenged junk.

On his own again, Dobson returned to San Francisco. On every clear night, he rolled his 12-inch *Stellatrope* to the corner of Jackson and Broderick Streets and showed the heavens to anyone who would look. One of the thousands of passersby realized that Dobson could teach others how to make telescopes, and arranged for him to begin teaching telescope making and astronomy at the Jewish Community Center, and later at the Lawrence Hall of Science and the California

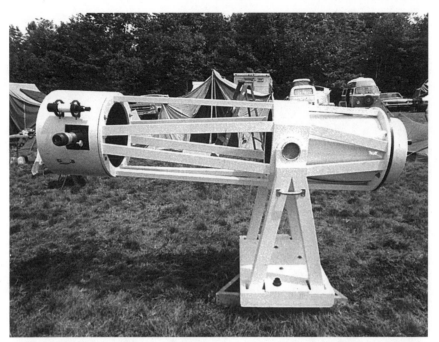

Fig. 1.9 Shown at Stellafane in 1978, Dennis di Cicco's 16-inch reflector was among the first large alt-azimuth telescopes built east of the Mississippi. Although not a true Dobsonian, it demonstrated the potential of large alt-azimuth telescopes.

Academy of Sciences.

The following year, two of Dobson's friends insisted that he join them in forming a club to be named the San Francisco Sidewalk Astronomers. This club met at Jackson and Broderick Streets, on the sidewalk, and brought telescopes for public stargazing. Among the telescopes were *Cyclops*, a 16-incher, and *The Little One*, one of Dobson's original 18-inchers from Sacramento. During the 70s and 80s, the Sidewalk Astronomers toured national parks in the West, showing tens of thousands of people how their universe looks.

Big, thin mirrors, the sling support, Teflon-on-Formica bearings, and the practical alt-azimuth mount are Dobson's contributions. If it weren't for John Dobson, you probably wouldn't be reading this book. Yet amateur astronomers were slow to see the value of these innovations. In that era, amateurs yearned for precisely machined equatorials crafted in basement and garage workshops. As Charles A. Federer, Jr., the publisher of *Sky and Telescope*, noted in a letter to Dobson, "While your shortcuts undoubtedly help to demonstrate large amateur telescopes, they can hardly lead to satisfactory instruments of the kind most amateurs want in these large sizes." Federer may have been right: the time was not yet ripe for the Dobsonian.

Another decade passed before the Dobsonian finally received nationwide exposure with two short articles in the newly-founded magazine *Telescope Making*.

Fig. 1.10 *The performance of Douglas Berger's 16-inch "Edelweiss" inspired Richard Berry to feature Dobsonian telescopes in TM#4 and TM#5, so that thousands of amateur telescope makers began to dream of building a large aperture telescope.*

Dennis di Cicco had seen Dobson's large telescopes at the 1978 Riverside Telescope Makers Conference, and had been sufficiently impressed that he based a 16-inch telescope on Dobson's ideas and exhibited it at Stellafane. Richard Berry, the editor of *Telescope Making* and one of the authors of the present book, saw di Cicco's 16-inch and that autumn remounted his own 12½-inch *f*/7 Newtonian as an alt-azimuth. Although neither of these telescopes was a true Dobsonian, they confirmed to Berry that Dobson's ideas worked for large telescopes.

1.3.2 The Dobsonian Revolution

The following spring, at the 1979 Riverside Telescope Makers Conference, Douglas Berger and Earle Watts, members of the Sidewalk Astronomers, brought their 16-inch Dobsonian telescopes. Berry, very much on the alert for Dobsonians,

Fig. 1.11 *Earle Watts demonstrates the simplicity of the azimuth bearing of his home-built Dobsonian. At the time, most amateur telescope makers found it difficult to believe that a bearing so simple would work so well.*

spent an evening observing with Berger's telescope, felt how smoothly the mount moved and saw fine images from a properly made 1-inch-thick mirror. He subsequently persuaded Robert Kestner, another Sidewalk Astronomer and a talented optician, to write two articles about Dobsonian telescopes for *Telescope Making*. The articles appeared in *TM#4* (Summer 1979) and *TM#5* (Fall 1979).

Kestner's first article, "It's Stability that Counts," explained the thinking behind the Dobsonian that John Dobson had taught him, and underscored the importance of a stable mounting. The second, "Building a Dobsonian Telescope—All the Details," gave instructions for constructing Dobsonians from 6 inches to 20 inches aperture.

That fall and winter Berry, now fully aware of the details of Dobson's innovations, completed a 20-inch Dobsonian of his own and described his experiences building it in *TM#7*, published in April 1980. Although the resulting telescope was bulky by later standards, it was nonetheless considerably smaller and lighter than many Dobsonians of the same aperture.

Fig. 1.12 Robert Kestner's June, July and August 1980 articles in Astronomy *described building a classic 10-inch Dobsonian, alerting the magazine's readers that home-built telescopes could be both large and inexpensive.*

That same April, *Sky and Telescope* carried a lengthy story, written by Dobson, entitled, "Have Telescope, Will Travel," and *Astronomy*'s Forum department published an interview with Dobson written by Ben Mayer. Dobson was a colorful figure and his appeal was strong—not only was he dedicated to helping others see the universe, but he wowed amateurs as he spoke of building amateur telescopes with unheard-of apertures of 16, 18, and 24 inches! When a detailed story describing how to build a basic 10-inch Dobsonian, coauthored by Kestner and Berry, appeared in the June, July, and August issues of *Astronomy*, a significant number of that magazine's 100,000 readers realized how easy it would be to build one of these simple, large-aperture telescopes.

Meanwhile, other events were leading toward a Dobsonian "revolution." By 1979, most telescope makers purchased rather than made their mirrors, and yet there was no commercial source for large mirrors suitable for Dobsonians. However, the owner of the Coulter Optical Company, Jim Jacobson, had recently expanded his production facilities in anticipation of a major contract from another manufacturer. When the contract failed to materialize, he was stuck with an expensive investment and far more production capacity than he needed. Jacobson decided to put this overcapacity to work making mirrors for large Dobsonians.

1.13 *Featured in TM#7, Richard Berry's massive 20-inch f/5 Dobsonian served as a model for many of the Dobsonians built in the next few years. For portability, the tube disassembled into three sections.*

Although the Coulter Optical Company had long been known for its inexpensive optics, Jacobson asked the extremely attractive price of $595 for his 17½-inch *f*/4.5 mirrors and galvanized the amateur community. Because *Telescope Making* was a new magazine with a very small circulation, Jacobson approached Robert Maas, the magazine's publisher, for 2,500 copies of a reprint booklet including the *Telescope Making* articles on Dobsonians to ship with his mirrors. Maas asked Berry to whip up something quick, and the resulting booklet, *How to Build a Dobsonian Telescope*, was published in July 1980. Jacobson mailed a copy of the booklet to everyone who requested information about Coulter mirrors, ultimately placing Kestner's articles in the hands of over 10,000 amateur astronomers and telescope builders.

The Dobsonian had an immediate impact on amateur astronomy. In an August 1980 "Behind the Scenes" editorial in *Astronomy*, Berry predicted that be-

Fig. 1.14 *All across the United States, amateurs rushed to build Dobsonians. Here, members of the Bowie Astronomical Society in Maryland proudly show off the society's new 14-inch f/5 Dobsonian. Photo by Walter Hamler.*

cause of John Dobson, "a lot of people who might never have gotten into observational astronomy will give it a try." This prediction became a reality. Amateur astronomers who had formerly thought of their hobby as a casual interest built telescopes and became serious observers.

In particular, many of these newly minted observers were attracted to deep-sky viewing. Although interest in deep-sky objects had grown slowly through the 60s and 70s, spurred largely by Walter Scott Houston's "Deep-Sky Wonders" column in *Sky and Telescope*, large-aperture Dobsonians drove the subsequent explosive growth in the subject. David Eicher's *Deep Sky Monthly* (later renamed *Deep Sky* magazine) provided observers with extensive lists of objects to observe and a rallying point for their enthusiasm through the 1980s. Although it carried a column for users of "small" telescopes, *Deep Sky* catered primarily to amateurs with large-aperture telescopes. This drove observers to search for the tens of thousands of deep-sky objects now within reach of their large Dobsonians.

But deep-sky observing was just the tip of the iceberg. During the early 1980s, amateur astronomy expanded prodigiously. Although factors such as the success of the Voyager missions to Jupiter and Saturn contributed, Dobsonians undoubtedly played a key role in the expansion of the hobby. By lowering the cost of acquiring a telescope, Dobsonians made amateur astronomy more attractive to more people. In 1979, the circulation of *Astronomy* passed that of *Sky and Telescope* and rose toward the all-time record of 240,000 that it attained in 1985, at the height of the Comet Halley craze. Amateur astronomy grew to four times the size it had been a decade earlier; new companies were founded, advertising revenues swelled, astronomy clubs appeared all over, star parties multiplied, and amateur astronomy appeared to be a growth industry.

But, inevitably, there were casualties.

The first was the venerable home-built pipe-mounted Newtonian. Like dinosaurs at the end of the Cretaceous, an entire class of home-built telescopes abruptly went extinct. By 1981, no one who realized what had happened the summer before would even consider building a pipe mount. Not only were Dobsonians easier and less expensive to build, but they also offered superior performance.

Another casualty may have been commercially made, equatorially mounted Newtonians. Despite the rapidly growing astronomy market, sales of Newtonian telescopes on German equatorial mounts declined rapidly after Coulter Optical introduced its Odyssey line of low-cost Dobsonians. Coulter's 13.1-inch Dobsonian sold for $399, about the same price as a 6-inch equatorially mounted Newtonian. As a result, several firms abandoned Newtonians and switched to new product lines during 1980 and 1981. By 1982, equatorial Newtonians had all but disappeared from telescope advertising.

The Dobsonian "revolution" also coincided with the beginning of a long-term decline in interest in amateur astrophotography. Although a cause-and-effect relationship remains unproved, it seems likely because observers who used large Dobsonians were satisfied with what they could see in the eyepiece and therefore did not feel the need to take pictures; and also because Dobsonians do not follow the motion of the sky, they are not well suited to astrophotography. Before Dobsonians, astrophotography represented the pinnacle of achievement for an amateur astronomer; after Dobsonians, visual observing with a large-aperture telescope came to be regarded as the new ultimate experience.

In the March 1981 issue of *Astronomy*, Canadian amateur astronomer Leo Enright summed up the Dobsonian revolution as "a new philosophy of building telescopes," noting that "10, 12, and even 14-inch telescopes seem to be everywhere." (At the time, these apertures were regarded as surpassingly large.) According to Enright, the Dobsonian revolution meant that "faint and far-off vistas of our universe are open to our wondering eyes."

1.3.3 Optimizing the Dobsonian

As the Dobsonian revolution swept across America in 1980 and 1981, the majority of builders closely followed the archetypal design set forth in Kestner's article in *Telescope Making #5*. The tube was made from a length of concrete form tube inserted into a close-fitting plywood box. A hinged "tailgate" on the tube box held the mirror. On opposite sides of the tube box were altitude bearings supported on tabs of Teflon plastic nailed to cradle boards attached to a rotating rocker box faced on the bottom with kitchen countertop Formica. The whole instrument turned on Teflon pads mounted on a square ground board. As Kestner pointed out in his article, the design was hard to improve because it had been optimized by several decades of Sidewalk Astronomers. But amateur telescope makers, being

Fig. 1.15 Optician Robert Kestner described how to grind, polish, and figure large, thin mirrors in a series of articles published in Telescope Making #12, #13, and #16. Now revised and brought up to date, Kestner's instructions appear as **Appendix B** of this book.

what they are, immediately set out to improve the design.

As in any evolutionary process, many early mutations were dead-end paths of development. But others survived, got noticed, and were copied in the next generation of telescopes. Significant variations included using a full-length round tube with a box clamped over it near the balance point, building the tube as a square plywood box, building the tube as separable components, cutting away unneeded parts of the tube, and substituting a variety of materials.

One of the first amateurs to examine the performance of the Dobsonian analytically was Richard Berry. In an effort to understand the kinematics of his own

Fig. 1.16 *Dennis Donnelly was among the first to find celestial objects with alt-azimuth setting circles and a pocket calculator. In 1980, he could locate stars in the daytime sky using plastic-coated measuring tape refills on the bearings to read his telescope's position.*

20-inch Dobsonian, Berry derived equations that enabled a telescope maker to calculate the force required to move a Dobsonian in altitude and azimuth. Berry measured the static and kinetic coefficients of friction for a variety of materials at a variety of different relative velocities.

In the course of his measurements, he discovered that because its coefficient of friction rises with increasing relative velocity, the Teflon-on-Formica combination acts as a speed regulator. He also pointed out the crucial relationships between the diameter of the bearings, the materials used in the bearings, and the weight of the telescope. An updated version of this analysis is presented in **Chapter 3**.

Yet another contribution was made by Bob Kestner with a three-part series "Grinding, Polishing, and Figuring Thin Telescope Mirrors." The series, which appeared in *Telescope Making #12, #13,* and *#16,* described how to make large mirrors using very simple methods. By combining what he had learned as a young Sidewalk Astronomer from John Dobson with his early experience as a professional optician, Kestner codified the word-of-mouth craft of making big mirrors. Rather than advocating porthole glass as the Sidewalk Astronomers had, Kestner recommended annealed and diamond-generated Pyrex blanks. However, he also showed that an amateur could grind a 25-inch mirror on the living-room carpet. The series not only helped hands-on amateur astronomers, but also served as the bible for workers in the optical shops that were springing up all over to supply the growing demand for large Dobsonian mirrors.

Across the country and around the world, telescope builders sought to add enhancements such as setting circles and equatorial tracking to the Dobsonian. With newly available handheld scientific calculators, advanced observers realized that they could convert right ascension, declination, date, and local time into precise altitude and azimuth coordinates. With a compass and protractor, the average amateur could make giant "setting circles" and attach them to the large side and bottom bearings of a Dobsonian. Once aligned and calibrated, the Dobsonian could even find stars in the daytime sky. Although the majority of observers continued to star-hop, the availability of goodies like alt-azimuth setting circles began to give a distinctly high-tech aura to the humble Dobsonian.

Equatorial tracking was an even more desirable feature. Although Adrian Poncet's groundbreaking article in the January 1977 issue of *Sky and Telescope* set forth the concept of the "equatorial table," the design was a theorist's delight, elegant in its geometric perfection, but unable to bear the weight of a large telescope. An update on Poncet mountings published in the February 1980 issue of *Sky and Telescope* focused on tables for 10-inch and smaller telescopes. Variations such as the "nested corner" and "bending pier" attracted some attention, but the low, sturdy platform built by Tom Martinez and described in *Telescope Making #19*, finally established that Poncet-style equatorial platforms could carry loads typical of large Dobsonian telescopes. In the same issue, Wally Pursell proposed an alternate geometry that was first realized as a practical platform by Tom Osypowski in early 1984, carrying a 16-inch *f*/5 Dobsonian.

Osypowski, deluged with requests from observers to "build me a platform just like yours," started a small manufacturing company called Equatorial Platforms. This enabled observers who were not themselves telescope makers to buy a low-cost "equatorial mounting" and "clock drive" for their large Dobsonian. Osypowski has used large Dobsonians for deep-sky astrophotography and CCD imaging with exposure times of many minutes.

Other products spurred interest in the Dobsonian, and the Dobsonian in turn made these products successful. When amateur astronomer Steve Kufeld introduced the inexpensive "unit power" Telrad finder, he made alt-azimuth star-hopping easier and also cut the cost of equipping a Dobsonian with a finder. Without the Dobsonian revolution to drive sales, the Telrad might not have been such a tremendous success.

Light-pollution filters and high-performance eyepieces also emerged as

Fig. 1.17 *Inventive telescope builders tried many novel ways to make their Dobsonians track. In this picture, Tom Fangrow pumps up an inner tube that will provide a uniform motion in right ascension as it deflates. Photo by Dana Huntly.*

Fig. 1.18 *Amateur telescope maker Thomas Martinez of Kansas City demonstrated the first equatorial tracking platform sturdy enough to carry the weight of a large Dobsonian and still track accurately.*

Fig. 1.19 *The 1980s was a time of experimentation as builders sought to lighten and simplify the too-heavy classic Dobsonian design. Two 1 x 4 strips of laminated redwood replace the tube in Mike Hill's elegant 17.5-inch Dobsonian. Photo by Robert E. Cox.*

products of the Dobsonian revolution. As deep-sky observing with large exit pupils became common, observers were desperate for filters to remove the harmful glare of mercury and sodium streetlights and for eyepieces that gave sharp images over wide fields of view. Among the first to offer an effective light-pollution filter was physicist-turned-entrepreneur Jack Marling, the owner of Lumicon. Lumicon now offers a line of light-pollution filters, each designed for a different type of observing.

Modern, high-performance oculars have been a passion of optical designer Al Nagler. Nagler ran a small company called TeleVue, and he entered the market with an optimized version of the venerable Plössl eyepiece. Nagler later introduced his own design, the Nagler ocular, epitomized by a well-corrected 13 mm model with an 82° field of view. Nagler eyepieces outperformed the classic Erfle by a wide margin, and observers found the high-power 82° field so addictive that they were willing to pay premium prices for them. Nagler's eyepieces helped the Dobsonian's popularity, and the popularity of the Dobsonian helped build Tele-

Vue's sales.

Although Dobsonian telescopes can be entirely homemade, as indeed most of the San Francisco Bay area Sidewalk Telescopes were, many telescope makers preferred the look and ease of using commercially made focusers and mirror cells on their creations. As the Dobsonian revolution unfolded, the Kenneth Novak Company, a one-man operation based in Ladysmith, Wisconsin, became the supplier of choice. Since Novak designed and made all his own components, he could adapt quickly to the rapidly evolving market by providing the large spiders, diagonal cells, and low-profile focusers that Dobsonian builders wanted.

In three short years following its introduction, the Dobsonian swept America. As telescopes of 10, 12, 14, 16, and 17½-inch aperture became commonplace, a growing number of Dobs in the 20- to 24-inch range showed up at national star parties. Yet in the still larger apertures that observers dreamed about, it had become clear that the original Dobsonian design was simply not practical. The bulky solid tube, the heavy rocker box, and the massive ground board made the classical Dobsonian too big, too heavy, and too hard to build in large sizes. Dobson's 24-inch *Delphinium* required a crew of six or more helpers to set up. Although big operating crews were consistent with the aims and philosophy of the Sidewalk Astronomers, the new generation of serious deep-sky observers wanted a large-aperture telescope that one person, or at most two, could set up and use at remote dark-sky observing sites.

1.3.4 The Second Dobsonian Revolution

In the autumn of 1982, a two-page article entitled "[An] Extremely Portable 17½-inch Dobsonian" by Ivar Hamberg published in *Telescope Making #17* ignited the second Dobsonian revolution. Hamberg eliminated the cardboard tube, replacing it with an airy aluminum truss. The result was an extremely portable 17½-inch Dobsonian that could be hauled in the back of a compact car and assembled by one person at a dark observing site.

Hamberg constructed a lightweight ring to hold the secondary and focuser, and reduced the bottom of the tube to a short box. He replaced the solid tube with slender aluminum poles, and, because the balance point of a tube with a lightweight truss lies close to the bottom end, the side bearing could be attached to a shortened mirror box. This in turn reduced the height of the rocker, so everything was more compact.

To make the telescope portable, Hamberg broke down the tube and nested the mirror box into the already low rocker. The entire 17½-inch telescope collapsed into a 24 x 24 x 30-inch package and weighed 72 pounds plus 14 pounds for the mirror. Molded plastic knobs held the aluminum truss poles in place; the knock down and set up time was a few minutes.

Amateur telescope makers who saw Hamberg's article grasped the principles instantly, and realized that this design pointed the way to much larger apertures. Hamberg thus triggered a second Dobsonian revolution.

Fig. 1.20 *Ivar Hamberg's truss-tube telescope altered the course of telescope history. The open framework tube made of aluminum poles showed telescope makers that large Dobsonians could be as portable as much smaller telescopes with solid tubes. Photo by Ivar Hamberg.*

Over the next five years, telescope makers built and exhibited virtually every possible variation on the Dobsonian theme at Stellafane, the Midwest AstroFest, the Texas Star Party, the Riverside Telescope Makers Conference and the growing number of regional conferences and star parties around the United States. It was a time of experimentation, rapid evolution, and great excitement. Amateurs who had never dreamed of owning a telescope larger than 8 inches aperture suddenly found themselves building 17½-inch Dobsonians.

Despite the intense interest in building large telescopes, making large mirrors remained unpopular. Telescope makers preferred to focus their efforts on the mechanical structure and to purchase the optics. As Coulter Optical struggled to

Fig. 1.21 *Before long, there was a truss-tube Dobsonian version of every type of telescope. One of the most impressive telescopes in this time of innovation was Lee Cain's giant 17.5-inch binocular, a superb instrument for scanning the Milky Way.*

fill more orders for large mirrors than they had anticipated, the wait for mirrors grew. New optical companies appeared overnight, and in some cases, disappeared just as quickly. By hook or by crook, or by simply ordering a mirror and waiting a year for delivery, builders obtained big mirrors, completed telescopes, and began observing with them.

Amateurs accustomed to small-aperture mirrors with long focal ratios discovered how critical it was to collimate and mount their large, fast mirrors properly. In 1984, Eric Allen's article, "Updating Newtonian Collimation" (*TM#22*), described how to use a collimating tube and spots for precise alignment of a large mirror, emphasizing the importance of squaring on the focuser and aligning the secondary before collimating the primary. A year later David Chandler's article, "Flotation Cell Design" (*TM#26*), presented a computer program that worked out the precise dimensions of 9-point and 18-point flotation cells for any size mirror. With Chandler's software, a telescope maker could custom-design a high-performance mirror cell for any telescope in a matter of minutes.

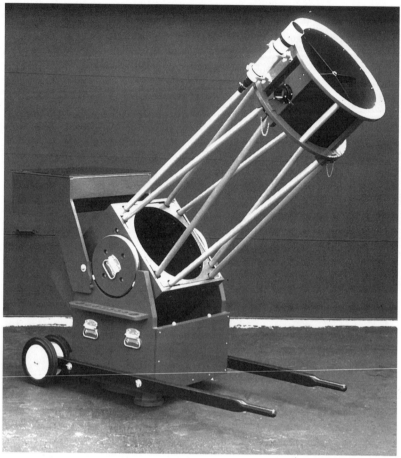

Fig. 1.22 *Many refinements in design set Ron Ravneberg's 17.5-inch Dobsonian apart from the crowd. Locking truss tubes, detachable wheelbarrow handles, and the distinctive "rooster tail" chart table (invented by Bill Burton) made set up easy and observing convenient. Photo by Ron Ravneberg.*

Although the Dobsonian opened the way to large telescopes, it also had broad application for more modest instruments. When Richard Berry's book, *Build Your Own Telescope*, was published in 1985, it gave novice telescope makers detailed plans to follow. Of the five telescopes described in the book, three were on Dobsonian mounts. Beginners could build a low-cost 6-inch *f*/8, a solid 10-inch *f*/6, or a classy 6-inch *f*/15 refractor solidly mounted on a tall, Dobsonian tripod. The book consolidated the small-aperture end of the Dobsonian revolution, and put plans for solid, functional, small-aperture Dobsonian telescopes into the hands of thousands of amateur astronomers.

The large-aperture end of the second Dobsonian revolution culminated at the 1987 Midwest AstroFest meeting in Kankakee, Illinois. At that meeting, Ron

Ravneberg displayed his home-built 17½-inch Dobsonian, a telescope that made believers out of large-aperture skeptics. Drawing on Ivar Hamberg's basic design, Ravneberg added a multitude of small refinements such as quick-acting clamps for drop-in truss tubes and detachable wheelbarrow handles for easy transport. Overall attention to detail made it one of the first large-aperture "portable" telescopes that was really easy to use.

That year David Kriege, a dentist by profession and coauthor of this book, attended AstroFest. As a teenager, he had been interested in astronomy and built several telescopes, but his interest lay dormant for over 20 years. At AstroFest, Kriege discovered that amateur telescope making had completely changed since his days in astronomy. Kriege realized he could now build the dream telescope of his youth, and set about making the largest he could afford, working non-stop, obsessively until it was done.

The resulting telescope, *Obsession 1*, had enormous side bearings, an ultra-light upper end, demountable truss pole fittings, an open, self-cooling mirror cell, an ultra-low rocker box, and a Formica laminate carefully selected for low friction and maximum smoothness. Kriege exhibited his telescope at the 1988 AstroFest, just one year after he reëntered amateur astronomy. Berry saw Kriege's telescope as a major breakthrough in Dobsonian design, and featured a portrait of it on the cover of *TM#35*, along with detailed articles by Ravneberg and Kriege.

1.4 The Dobsonian in Your Future

Contemporary developments point toward a long and exciting future for the Dobsonian. In 1990, an electronics engineer named Rick McWilliams developed the first commercially viable digital setting circles. Although his company, Tangent Instruments, did not sell direct to consumers, within a year virtually every major manufacturer and several of the larger distributors offered McWilliams-engineered units. All it took to use these units was attaching an angular encoder to the bottom and side bearings of the Dobsonian and stringing some wires to a hand-held computer box. On the box was a glowing display that showed the telescope's right ascension and declination.

Digital setting circles have greatly increased the speed and ease of finding sky objects, especially faint objects in unfamiliar parts of the sky. Many digital setting circles feature extensive internal databases or are linked to a computer database and a computerized star chart. Thus the novice can find thousands of faint galaxies, and the longtime observer can search for deep-sky objects that few humans have ever seen before.

In the optical arena, firms that offer amazingly large mirrors have entered the market. Some firms offer mirrors up to 1 meter (39.4 inches) in aperture. Designed for use in a modern Dobsonian and reasonably priced, these mirrors push amateur telescopes toward the ultimate limits of practicality. How much telescope can one person set up? 300 pounds? 400 pounds? 500 pounds? What is the point of diminishing returns in aperture? 36 inches? 40 inches? 48 inches? And when does a lad-

Fig. 1.23 *After a 20-year absence from amateur astronomy, Dave Kriege rediscovered astronomy and built the dream telescope of his youth in late 1987. The telescope that he called "Obsession 1" is the basis for the telescopes described in this book.*

der become too tall to climb in the dark? 12 feet? 15 feet? 18 feet? Whatever your answers to such "metaphysical" questions, enormous telescopes are possible. Dreaming of a 1-meter telescope to call your own is no longer dreaming the impossible dream. It can happen to you.

Finally, computer-driven Dobsonians have become a reality. The first firm to offer dual-axis "clock drives" for Dobsonians was Tech 2000 with their "Dob Driver," and more recently, amateur astronomer Mel Bartels built a pioneering 20-inch telescope with a computer-controlled alt-azimuth drive system comparable to those found on big professional telescopes. Without a doubt, high-tech Dobsonians will develop into a major class of amateur telescope in the coming decades.

Thus the future of the Dobsonian seems assured. Whether you opt for a simple, portable, hand-powered telescope or pursue the high-tech dream of big aperture coupled with fancy electronics, if you want a practical large-aperture telescope that you can build yourself, there's a Dobsonian in your future.

Chapter 2
Planning Your Telescope

When you build a telescope, no time is better spent than that invested in the planning stages that precede construction. Ideally, planning is something you take slowly—it's the time when your dreams and enthusiasm for your future telescope meld into a plan that you know will work.

During that time, you decide how big your telescope should be, how you will build it, and how you will use it when it's ready for the stars. You may be tempted to rush ahead—go ahead, be tempted!—but slow down and *really study this chapter*. After all, don't you want to know what you're getting yourself into?

Let's begin with a little scenario. It's the evening after a long day at work. Dinner is over and you're at home sitting on your comfortable living room sofa. The afternoon clouds are gone and the sky has turned sparkling clear. Now—do you have the energy to haul that dream telescope of yours to a dark site? Are you ready to set up in the dark? Do you feel enthusiastic about making fifty trips up and down a 12-foot stepladder? How do you anticipate feeling on the long drive back home?

Do you think you'll feel guilty when your wonderful big telescope collects dust instead of starlight? What does your spouse think of your nutty fascination with telescopes? You'd better think about that one carefully, because your spouse's attitude makes a big difference in the satisfaction you gain from astronomy.

If it sounds like we are trying to discourage you, rest assured that we are not. Rather, we want you to think about what building a telescope will mean to your life. We would rather see you happy with a small telescope than miserable with a big one.

So our advice to you is this: if you are the least bit apprehensive, stick with a telescope under 18 inches aperture. If you have serious doubts, then build an 8-inch or 10-inch telescope. Small scopes are easier to use, and if you find your interest building toward the obsession level, then you can always go bigger. The point is to build a scope that is easy for you to use, and then explore the universe with it.

So much for the Zen of telescope making. If you decide to build a telescope, **Section 2.3** of this chapter is very important. Why? We tell you in what sequence you should order, build, and assemble the components of your telescope. Other chapters give you the details of how to carry out the various steps of construction,

Table 2.1
Critical Reality Check for Dream Telescopes

Aperture (inches)	Total Weight (pounds)	Eyepiece Height (feet)	Cost of Optics (1997)
8	40	4	$200–400
12	75	6	$500–800
15	100	5.5	$900–1,800
18	120	6.5	$1,500–2,500
20	150	8	$1,700–3,000
25	250	10	$3,500–6,000
30	400	12	$4,500–8,000
36	575	15	$10,000–15,000
40	800	16.5	Your Call $

but here you learn what to build first, what next, and so on. Now we *know* that you can build a telescope any way you please and we can't do anything about it, but the reason we wrote this book was to save you a lot of time and trouble. Study it thoughtfully and we think you won't regret it.

2.1 Non-Telescopic Telescope Considerations

When you build a telescope, it becomes part of your life—and in the case of a large Dobsonian, a *large* part of your life. Somehow you must fit it in among your other activities, into your home, and into your automobile. This is not always easy, and what's practical should strongly influence what size telescope you build. Your family is a consideration, too. Can you really afford to build a big telescope, or might you and they be happier with a small telescope that would take you away from home less and cost less? These are important considerations that you should not ignore.

So here are some questions that you should ask yourself. There is nothing magic about this process and you certainly need not follow it step by step. If you're already an observer or telescope builder, you might have a lot of the answers already. Such planning encourages you to think through the reality of owning a big telescope.

- *What are your present observing habits?* Even if you've been an amateur astronomer for years, you may not want or need a big telescope. Carefully assess what you *actually* do when you go observing. If your deepest pleasure comes from visiting familiar celestial friends—Andromeda, the Dumbbell, the Orion Nebula—then a big telescope might complicate your hobby without adding to its enjoyment. But if you're restless and already push-

Fig 2.1 *Nothing, absolutely nothing, is so great for a serious observer as viewing the universe through one of "The Big Ones." However, for you the best may not be the biggest, so study this chapter carefully. You deserve the telescope that is best for you!*

ing hard against the limits of your present telescope as you search out fainter, deeper, less well-known objects, then you may need a bigger one.

- *Where you will observe?* If you live in an urban or suburban area, you'll want to take your telescope to dark skies at least some of the time. Sites that serve well are club observatories, the yard of a rural friend, a small plot leased from a farmer, parking areas in state parks, and roadsides in isolated areas. Wherever it is, you must feel safe and comfortable at your site.

- *How will you transport the telescope?* Next figure out how you will transport your telescope to your observing site. Even if you can squeeze a large telescope into a small automobile, is there space left over for other passengers? For a tent, stove, cook set, and sleeping bag? There's no lack of options beyond the family car; consider vans, pickups, and mini-trailers.

- *Where will you store the telescope?* Before you even start it, decide where you will keep your telescope. Garages, basements, attics, porches, sheds, and backyard observatories are the most common storage areas. Storage may set an upper limit on how big it can be. Work out how you will move the telescope from storage to the place you observe. Can you avoid steps? Carrying heavy equipment up and down stairs is enough to kill anyone's appetite for observing.

- *What happens "the morning after?"* The biggest enemy of amateur astronomy is sleep, or rather, lack of sleep. When you own a really large instrument, you can't observe for an hour—you're going to put in all-nighters. Just knowing that each time you stay up late, you'll have to wake up and get the kids off to school and then struggle through the day trying to stay awake can be a deterrent to observing. You will need to arrange your schedule to sleep undisturbed the morning after an observing session or you won't enjoy your large telescope.

- *What is the cost?* Although the mirror is the largest single expense, you will soon find yourself writing checks for goodies like hardwood plywood, aluminum tubing, varnish, eyepieces, and more than a few tools. Better that you plan ahead than run out of money halfway through the project. Be sure that you talk it over with your spouse before you're psychologically committed to a telescope plus four-wheel-drive truck—remember that almost any car can tow a small trailer, and that's all you'll ever need.

- *What is the **total** cost?* Don't act too surprised when your spouse decides to spend an amount equal to the cost of a telescope on his (or her) own hobbies. Wait until the kids decide to stake their claim to part of the family's hobby budget! The decision to build your dream telescope could add up to a figure considerably greater than the cost of the telescope alone.

Check out some big Dobsonians at a star party. Have you actually observed with a large telescope? If you have not, make sure you do. Every year there are dozens of star parties where you can see and look through a variety of telescopes including big Dobsonians: Riverside, Stellafane, The Winter Star Party, Astrofest, The Texas Star Party, Table Mountain Star Fest, . . . the list goes on and on. Check out upcoming dates and locations in *Sky and Telescope*. Even if you're not the type who normally likes these events, go to at least one. Spend the day looking at telescopes. Inspect them. Move them. Ask questions. Spend *all night* looking through telescopes. Look, learn, evaluate, and observe.

Going to a star party is a great reality check. Inform your dreams with hands-on experience about how it really is. Chilly, damp, bug-infested nights under a beautiful sky. Cold nights. Clouded-out nights. Nights when it rains and the tent leaks. Mornings when the car won't start. Does the excitement of observing outweigh the hassles?

If you go to star parties, you will see great telescopes. You will also see tele-

Fig. 2.2 *The open tailgate of the modern-style Dobsonian means there is lots of air circulation around the primary mirror, so it quickly cools after sundown.*

scopes that don't work *quite* right. You'll see telescopes that wiggle and jiggle. You'll try to look through telescopes that don't quite balance. You'll observe with telescopes whose images aren't quite sharp. Maybe not disasters—but scopes that fall a bit short.

You'll also learn how big "The Big Ones" really are. It's easy to draw up some sketches and tell yourself that a 10-foot tube isn't so long and 175 pounds isn't so heavy. We want you to imagine lugging each one into your back yard and assess the limits of your strength and patience. We don't want to scare you—we just want you to know what you're getting yourself into.

But don't let silly fears take over either. When you pick up a 25-inch Dobsonian on a pair of wheelbarrow handles, you're lifting less weight than someone hefting an 8-inch catadioptric. And if you do it right, it's actually easier to set up a big Dobsonian than it is to haul out a couple garbage cans to the curb on trash day. Check out the pictures in this book and you'll see how easy using a big scope can be.

Lastly, assess the rewards. Some people marvel when they look through a telescope and some people don't. If you are among those who would rather curl up beside the fire with a good book than shiver at the eyepiece, then stick with your books.

But should you find yourself craving more galaxies, fainter nebulae, and better views of the planets—the biggest telescope you can afford may be the right telescope for you.

2.2 Tough Telescopic Considerations

It is tempting to flip through a book like this one, get all excited, and place an order for a 36-inch mirror. *Don't do it.* Although you might succeed and build a fine telescope, the odds are against you. History says so. When the first big, affordable mirrors first came out—17½-inch mirrors—lots of telescope builders rushed to place an order. Time went by, and then one day the UPS truck pulled up and delivered a huge, heavy box. The proud new owners couldn't believe what they saw. The mirror was *far* bigger than they had ever dreamed!

Admittedly there was a lot less information available in those days. And because few of those eager-beaver builders had ever seen a 17½-inch telescope, they didn't have a clue how *BIG* it would be. Many of those mirrors languished in the back of a closet for *years*. A couple decades later, it is safe to say that most of them are now in telescopes, but we don't want to oversell you on big scopes. Instead, we want you to plan carefully. We want you to enter into this project with open eyes. Make no mistake: building a telescope is a big, time-consuming project. We want you to finish it and to enjoy using it for years in the future.

Let's talk about motivation: Why do you want a big telescope? If owning the biggest scope in three states is some kind of status thing or ego trip, have we got news for you! The average stargazer is drawn like a magnet to the largest telescope. You think you'll get your big scope built and be the center of attention at every star party forever. But some day somebody is going to set up a bigger telescope right next to you. Hah! Suddenly your telescope is not so big anymore. Where have all the stargazers gone? Next door to look through the "big" one.

Let's talk about using telescopes: How often do you go observing with your present one? Do you think you'll feel justified spending the same amount of time with your new 36-inch Dobsonian under the suburban skies in your back yard? If you can truthfully answer "yes," then a really big telescope may be for you. Otherwise, go for something smaller that you'll take out regularly. Even a good pair of binoculars is a lot better than a 36-inch telescope sitting in storage.

Let's talk about personal energy level. You have to be pretty hyper or pretty determined to take a big telescope out for a night of observing. If you have to load up, drive an hour, set up, observe, load up, drive an hour, and unload each time you want to observe, you'll start to feel tired and get a headache every time you think about your telescope. But aspirin doesn't cure telescope exhaustion.

Let's talk about health. This is a difficult topic, but if your joints ache on damp nights outdoors, and you are tempted to spend more time in the cold, the consequences are predictable. Allergies, arthritis, angina, and age all have the potential to reduce your observing pleasure. Take the state of your health into account as you plan for your telescope.

Fig. 2.3. *There's no doubt that a big telescope still attracts a crowd. If you take it to a star party, all day long people will gather to "ooh" and "ahh" at the mirror.*

Let's talk about size and strength. If you are 5-foot-2 and weigh 100 pounds soaking wet in your clothes, it's fair to say your 25-inch dream telescope is bound to be a lot larger and heavier than you are. The height of the eyepiece at the zenith is equal to the focal length of the primary mirror, so if you had a 12-inch *f*/5 Dobsonian, you could probably observe with your feet flat on the ground. Don't take this to mean you can't have a 25-inch telescope, but please understand, we want you to know that you're letting yourself in for a lot of late-night ladder climbing.

2.2.1 Telescope Aperture Considerations

How big is "big?" In the 1960s everybody's dream scope was an 8-inch *f*/8 Newtonian reflector on a machined equatorial mount. In the 1970s the dream telescope had escalated to 12½ inches in aperture. By the '80s, serious observers all lusted for a 17½-inch Dobsonian. As the '90s progress, everyone seems to want a humongous Dobsonian, and the bigger the better. Ten years hence, who knows what we'll want—probably something in the 1-meter class. But, despite the trend toward bigger and bigger, monster scopes are not for everyone. It may take some tough soul-searching to decide what size is best for *you*.

The important word here is *YOU*. Sure, it's prestigious to own a big telescope, but don't give in to perverse telescopic *machismo*. Aperture envy is a fatal disease. Build a telescope that's practical for you to use easily on clear nights. Don't feel guilty or insignificant because you don't own the biggest scope in your time zone. Any telescope makes poking around the sky a lot more fun.

How do scopes of different aperture stack up?

Well, an 8-inch is terrifically handy. Short and compact, it can sit unobtrusively on the porch or in the corner of the garage. Forty pounds to carry; set it up and be observing in five minutes. It is able to show you all the basic stuff in the sky. A 12-inch is a commitment. From porch to back yard, it takes two trips to carry the scope and a third trip for the eyepiece box. You need a minivan or station wagon to carry it. But it's a lifetime telescope for the serious observer, capable of heavy-duty planetary, lunar, and deep-sky observing. Hundreds of easy galaxies. Use it in your back yard or take it on field trips to dark skies. Make no mistake— you could be very happy with a 12-inch telescope.

Somewhere around 16 inches aperture, you break into a whole new class of observing—thousands of galaxies and globulars resolved to the core are yours for the viewing. But you now need a short stepladder. Observing becomes climbing. Set up in perhaps ten minutes if you're well organized, and no lugging a small person can't easily handle. A great scope for half-nighters and all-nighters under nearby rural skies.

Apertures in the mid-20s open another realm. Gaseous nebulae become more than faint smudges and smears. Wherever you point your telescope, you see galaxies. Your friends tell you that they've (gasp!) never seen the Whirlpool Galaxy so well before. On the down side, observing in your back yard is no longer worth the effort of setting up because the 12-inch will show you just as much, but under dark country skies, a 25-incher leaves a 12 eating dust. Technically, a 25-incher is a one-person set up, but it's a lot nicer to have a friend or two along to share the work—and the pleasure—of observing.

At 30 to 32 inches, you're into big-time ladder games. These are the telescopes we call "The Big Ones," in part because of the big ladder you need to reach the eyepiece. Before you become fully conscious of the tendency, you're choosing the next object near where the telescope is already pointing so you won't have to move the ladder again. Long ago you bottomed out even *Uranometria 2000.0*, and the most acute problem you face in your observing is figuring out what all those little galaxies are. Do you want to see a double quasar? Sure, go for it! You'll find it, too. And, although you need observing companions just to come up with enough new stuff to look at, you still begin winter observing sessions with a long, deep gaze at the Orion Nebula. The logistics of going to an observing site have become something just short of awesome, but you feel deeply rewarded by the views.

36 inches? 40 inches? You're starting to push Dobsonian technology to its limits. You fret about mirror flexure, cool down time, and you actually begin to fear tall ladders. Observing is ponderous—but the rewards are rich indeed. Star clusters look like they're made out of first-magnitude stars. When a faint supernova pops off in some distant galaxy, you look directly at it and estimate its color. Set up and take down are mandatory two-person exercises. At home you observe with a 12-inch scope. And when you go observing, you take the 12-inch for a few quick observing treats on the side.

Why should this be so? If you have a 40-incher, why even think about looking through a 12-inch? Due to the geography of the universe, the laws of physics,

Fig. 2.4 A modest-size telescope may be much more practical for you than a huge one. A telescope like this one easily fits in a small car.

and the design of the human eye and brain, the visual thrill you get from enormous apertures succumbs to the law of diminishing returns. Compared to the 0.2-inch aperture of the naked eye, the 2-inch aperture of a pair of 7x50 binoculars gives you a tenfold gain in aperture and a 100-fold growth in light—an astronomical jump into the universe. Another tenfold jump in aperture and 100-fold increase in light takes you to a 20-inch telescope, with so many celestial objects visible you won't see them all in a lifetime of observing. A third tenfold jump in size puts you on Palomar Mountain!

Well, those tenfold steps are really big steps. With more modest twofold steps, most observers agree that jumping from 10 inches to 20 inches aperture yields a fantastic gain in visual impact. In the 20-incher, you can see stuff you simply could not see with a 10-inch because the 10-incher was limited by its aperture. You might expect a similar gain in jumping from 20 to 40 inches, but the gain is much smaller than you expect. The reason is that you can already see things very well with a 20-incher and the atmosphere rather than the aperture limits what you can see. The 40-incher makes everything bigger and brighter, and you can certainly see more stuff, but the atmosphere sets the limit for both. You may feel that the rewards of observing with a 40-incher are not worth the added hassle and expense. But don't let us prejudice you. The best way to find out what thrills await you is to attend star parties and observe with the biggest of "The Big Ones" you can find.

Fig. 2.5 *At the Ultimate Star Party on Mount Locke, Texas, it's the American dream times three. For the aspiring telescope maker, star parties are the ideal way to explore using large-aperture telescopes, and perhaps even to discover a lifelong passion.*

Then you can decide on the basis of your own experience.

What's the bottom line? With this book, you can build any telescope in the whole size range. But we want you to be happy with what you build. Build a telescope you think you will feel comfortable with. Select a size you won't give a second thought to hauling out and setting up. Be practical. For many observers a 12-inch is better than a 25-inch. That's because the 12-incher doesn't need a ladder to reach the eyepiece, fits in the back seat of your car, and is all-around more "user friendly." On the other hand, the 25-inch takes you into the realm of the galaxies, and for you, that may be an absolute necessity.

Actually, it is best to own two telescopes. There's a big one for "all nighters" and a little one for quick observing sessions. Build a handy 8-inch *f*/6 for those nights when you need an astronomy fix but don't have the time or energy to use your 25-incher. A few views of the planets and a double star or two makes for relaxing and enjoyable astronomy. Best of all you can do it in your back yard under whatever kind of sky. Maybe it's not as good as big-aperture observing, but it's still a lot better than watching TV. And it partially assuages the guilt of owning a big telescope that, for a variety of perfectly valid reasons, you can take out only four nights a year.

Where super-big apertures really excel is on galaxy clusters, little-known exotic objects, and the owner's ego. If you have the money, don't mind the work, or find intellectual and emotional stimulation in hunting down never-before-seen stuff, then go for it. Build the monster telescope you have always dreamed of having and find Gomez's Hamburger!

We'll conclude this section with some hard-won advice.

- *Don't sell your old telescope.* If you already own an instrument, you may be tempted to sell it to raise cash for the new telescope. Don't do it! Over the years we have listened to dozens of serious observers wish they once again had their old scopes. Unless you really don't like your present telescope, hang on to it. Chances are that it's a smallish one that is easy for you to set up and easy to use. The only thing "wrong" with it is that it's smaller than the telescope of your dreams. Rest assured that if you keep it, in a year or so you'll find yourself using it again sometimes, delighted with the convenience of observing with it.

- *Read this book.* We're serious about this point or we would not say it again and again. This book has a lot of pictures in it that might tempt you to forego a careful reading. Do so at your own risk. Some of what we say may be old hat to you, but most of this book is based on practical experience. We understand that as a do-it-yourselfer you'll probably "wing it" on at least half the construction regardless of what we tell you. That's fine— but remember what you saw at the star party—telescopes that didn't quite make it. Good telescopes work well because they are designed to work well.

Before we close this discussion, we want to refresh your memory: the reason you build a telescope is to see the universe better. Choose a telescope that you'll look through often.

2.2.2 Are You Obsessed?

If you have read this far and still dream of big-aperture views, then let your mind spin free for a few moments. The sky is clear. Leaving a note for your spouse not to disturb you in the morning, you escape with your big telescope to your favorite dark site. You set it up and carefully tweak the primary into perfect collimation. As your eyes dark-adapt, you drink a cup of hot coffee and the mirror cools. The pink glow of sunset fades in the west and the glorious panorama of a perfect Milky Way unfolds overhead.

You pick an NGC object from *Sky Atlas 2000.0* and patiently star-hop over to it. You take that first long look and drink it in. Wow, you think, it feels good to be observing tonight. You open *Uranometria 2000.0* to the object in the eyepiece and star-hop to whatever galaxies happen to be nearby. After a couple more obscure galaxies, a PK planetary, and a faint globular, you realize it will take at least three more nights to finish off just one page. With more than 470 charts, you know you'll never run out.

Fig. 2.6 *Night falls and the stars swing silently overhead. Again and again you ascend the ladder to view one celestial wonder after another. If this is your dream, then go for it!*

The hours speed by and the sky swings to the west. New constellations rise in the east. You "discover" a few new deep-sky objects for your personal list and feel a sense of satisfaction knowing that only a handful of people have ever viewed them through an eyepiece.

You realize that a familiar Messier object has risen, but when you look, you're almost knocked off the ladder it's so bright. As the soft glow of dawn touches the eastern sky, you finally look at Jupiter. It totally blows away your night vision, but the seeing has gotten really good and the detailed image of the planet makes the whole night seem worthwhile. You pack up and drive home feeling great.

Does this describe you? If it does, then forget every caution we've raised. Go for the gold and build the telescope of your dreams.

2.3 How To Get Started

You have now reached one of the most important parts of this entire book, the crux of the whole telescope-building game. How should you go about building yours? Resist the urge to begin by building the mirror box and rocker. Although they are relatively easy to make and you can gain a great sense of accomplishment from seeing them sitting in your shop, their dimensions are nothing but a guess before you know how the telescope will balance. If you estimate wrong, you'll have to return to the lumber yard for material to rebuild them.

Instead, follow the plan listed below. There are good reasons you should undertake construction in this particular order. We have arrived at this sequence based on lots of practical experience making telescopes. Build yours according to this plan and you will save time, money, headaches, and a shop full of plywood scraps!

Step 1: Try out lots of different telescopes. If you are new to amateur astronomy, attend some star parties. If you are an old hand, test drive some big Dobsonians. Be sure you know what you're getting into.

Step 2: Determine the optimum size telescope for your needs. Take into consideration the size, weight, ease of reaching good observing locations, and impact on your checking account. Review this chapter.

Step 3: Consider the impact your proposed telescope could have on your life. Talk to your spouse, your kids, and your friends. Does it fit the way you live? If you have doubts, reduce the planned aperture by one third and repeat this step.

Step 4: Make the commitment: order the primary and secondary mirrors. Without taking this step, nothing else is going to happen. Ordering the optics at the outset gives the optician plenty of time to fill your request. See **Chapter 4** for our thoughts on selecting and ordering optics.

Step 5: Build the primary mirror cell. The dimensions of the mirror cell and

Fig. 2.7 *Kurt Vanderhorst with his Dob. Careful planning and attention to detail are the surest way to build a successful dream telescope. Build the components of your telescope in the order suggested in this chapter and it will go together quickly and efficiently.*

the tailgate define the size of the mirror box. You only need to know the diameter and thickness of the mirror—something the optician can tell you well in advance. See **Chapter 5** for details on constructing mirror cells.

Step 6: Build the secondary cage. Buy a couple sheets of high-quality plywood. After you have determined the inside diameter of the secondary cage, order the spider, secondary mirror holder, and the focuser. The secondary cage must be near completion before you can install the secondary mirror and locate the point of focus. The location of the focus point allows you to determine the distance between the mirror box and the secondary cage, and this distance and the weight of the secondary cage are required to determine the balance point of the tube assembly and thus the depth of the mirror box. See **Chapter 6** on this important assembly.

Step 7: Build the mirror box. The width of the mirror box depends on the width of the tailgate (which you have already constructed), and the depth (height) of the mirror box (see **Table 7.1**) depends on the weight and distance to the secondary cage (which you also know). Make the mirror box deeper than necessary so you can trim it to fine-tune the tube's balance. For now, forget about the side bearings. **Chapter 7** details the construction of the mirror box.

Hooray—the finished mirrors arrive on schedule! Now the pressure is on to perform the star test and check the quality of your mirror.

Step 8: Buy the truss pole tubing. Install the tailgate and the mirror cell into the unfinished mirror box. Install the primary mirror in the cell, and the secondary mirror in the secondary cage. Temporarily connect the secondary cage to the mirror box and determine where the focal plane falls. Adjust the distance between the two until the focal point lies at the top edge of the focuser draw tube. Measure the distance between the top of the mirror box and the bottom of the secondary cage. Determine if your tube assembly balances where you want it to. If it is top heavy, move the mirror cell tailgate upwards one inch into the mirror box and move the secondary cage an inch closer. Recheck the balance. After the tube assembly balances, trim the excess plywood off the bottom of the box with a saber saw or circular saw. (Be smart and give yourself a margin of safety by mounting the tailgate an inch lower than the "perfect" balance point. You can always add a little weight to the secondary cage if necessary.) While you are at it, try to star test your mirror. When this step is done, return the mirrors to safe storage. See **Chapter 8**, on the truss poles.

Step 9: Make the altitude bearings. It's time to finish the mirror box by adding the side bearings. Refer to **Chapters 7** and **9**.

Step 10: Construct the rocker. The inside dimensions of the rocker are determined by the outside dimensions of the mirror box. The height of the rocker sides is determined by the swing clearance. Cut the rocker bearing support surfaces to match the diameter of the altitude bearings. See **Chapter 10** on the rocker box.

Step 11: Build the ground board. The diameter of the ground board is determined by the width of the rocker. Consult **Chapter 10** for details.

Step 12: Make the mirror box dust cover and the wheelbarrow handles. See **Chapter 7** for the lowdown on components that should be attached to the mirror box.

Step 13: Make the truss pole connectors. For the details about these components which hold the truss tubes, see **Chapter 8**.

Step 14: Install the primary mirror. Take the beast outside and celebrate "first light." Before you invite all your friends over for a big star party, use it for a couple nights to work out the bugs. Make any needed adjustments and alterations, and then do the finish painting. See **Chapter 11**, on assembling and adjusting the telescope. Then throw a star party.

Step 15: Mentally change gears from builder to observer. This is a tough step: during the construction phase, you've come to feel comfortable in your shop, and now you must venture into the cool night. We wrote **Chapter 12** to help ease the transition.

The primary factors that control the order of construction are the size of the mirror cell and the weight of the upper tube assembly. These control how your telescope will balance. Little can be done before these two components have been completed and weighed. Although it seems counterintuitive, the order of construction that we recommend here is the most efficient way to build a telescope.

Chapter 3
Engineering the Dobsonian

Craftsmen, scientists, and engineers approach building telescopes in three very different ways. The traditional craftsman copies an existing design, building carefully and without question, sure that success lies in making a faithful copy. The scientist employs the pure principles of physics to make something entirely new, thereby risking failure, but sometimes winning spectacular success.

The third approach—that of the engineer—lies between the other two. It melds building from existing designs with working from first principles. We suggest that, while taking pains to understand the physical principles that underlie its performance, you base your telescope on a proven design.

Today, hundreds of telescopes demonstrate that the truss tube Dobsonian design works very well. Its success is based on hundreds of tiny details as well as solid principles of mechanical design. We know, however, that much of the pleasure in building a telescope comes from experiment and innovation. This chapter gives you the tools to experiment wisely, with one eye on the power of innovation and the other on the performance of a proven design.

This chapter is an overview of the engineering ideas behind the success of Dobsonian telescopes, an introduction to the principles that will govern their operation. We'll also look into the properties of the materials used in telescopes, and explore a few of the reasons why the materials of the Dobsonian revolution—plywood, aluminum, Formica, and Teflon—do their jobs so well.

3.1 Engineering for Performance

The most important message you can get from this chapter is this: good telescopes don't just happen. Those that work well do so because they have been designed to work well. It's not an accident. Simple principles of sound engineering underlie great telescopes. What makes a telescope great? After first-rate optics, nothing matters as much as a solid mounting. A mounting that bends under the weight of the telescope, shakes and shimmies, or moves when you push on it and bounces back when you let go, *simply isn't working right*. When you observe, the view through the eyepiece is rock solid. Each time you nudge the telescope, it moves obediently. When you let go, it stays. In short, a great telescope does exactly what you ask.

In principle, any type of mount can work equally well as long as it's properly engineered. The largest equatorial telescopes in the world, those built 20 to 50 years ago, are carried by enormous mountings designed above all else to reduce flexure. The alt-azimuth mountings of recently built giant telescopes are small by comparison, yet far more precise and far less prone to troublesome flexure than their equatorial predecessors. Nothing is wrong with equatorial mounts, but the simple fact is that it is a lot easier to build a really solid alt-azimuth mount. It's no accident that the largest telescopes in the world are alt-azimuth mounted.

The beauty of a properly made alt-azimuth is that the center of gravity of the scope is directly over the bearing surfaces and the base. On an alt-az, nothing overhangs. The key concepts—symmetry, a low center of gravity, compact supporting structures, weight over bearing surfaces, vibration damping materials, innovative bearing materials, and simple cheap construction—lifted John Dobson's simple alt-azimuth mount from obscurity to its present level of popularity. Amateur astronomers have come to realize that, pound for pound, no type of mount provides the smooth motion and stability of a Dobsonian.

Dobsonians excel in two areas: stiffness and movement. Build it right, and the Dob hardly sags or bends. It moves easily, yet stays where it's put. We'll examine each of these areas in turn, and show how engineering principles point the way to a solid telescope that moves well.

3.2 Statics: How to Make a Stiff Telescope

Problem telescopes seldom lack strength, but often lack stiffness. Most mountings can easily support hundreds of pounds, yet all too frequently they bounce, they wiggle, and they shake. Strength and stiffness are two different design goals. Strength depends on cross-sectional area and the material used; stiffness depends on distribution of material within the cross-section, length, and the material used. It is possible, therefore, to make a structure that is strong but not stiff or stiff but not strong. For example, a steel wire is strong but not stiff; a paper cylinder is stiff but not strong. To perform well, a telescope mounting must be fairly strong and very stiff.

3.2.1 Engineering Terms

Engineers use structural analysis to predict the strength and stiffness of telescopes. We're going to introduce you to some terms and ideas used in structural analysis—nothing fancy, just the basics. Even if you operate on a purely intuitive level, you can learn a lot. Let's get started by establishing a vocabulary that we can use to talk about telescope materials and structures.

- Mass
- Weight
- Force

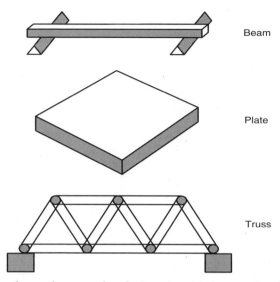

Fig. 3.1 *The beam, plate, and truss are three basic engineering structures found in telescopes. Thinking of a telescope as an assemblage of these basic forms makes designing it much easier.*

Mass is the amount of material in a body. Living as we do on the surface of a planet, in daily life we talk about the *weight* of an object and its mass as if they are the same thing. Strictly speaking they are not: mass is a fundamental property of matter whereas weight is the gravitational attraction between matter and the Earth. If you took your telescope to the surface of the Moon, its mass would remain the same but its weight would depend on the lunar gravity.

Intuitively, you already know what a *force* is—you push on something with a force and it starts to move, or more precisely, to accelerate. When you let go of an expensive eyepiece, it accelerates toward the ground. The weight of the eyepiece is the force between the eyepiece and the Earth. Because engineers build structures on the Earth, they treat mass and weight as one even though they know that technically mass and weight are different.

- Beam
- Plate
- Truss

Engineers use simplified structures to model the real world. Among the simplest structures is a *beam*. Place a broomstick, a wood 2x4, a steel I-beam, a strand of spaghetti between two supports and you have something an engineer would call a beam. The engineer reduces the complexity of the real object to a beam with certain mechanical properties and computes its response.

If you make a beam wide enough, it becomes a *plate*. In engineering terms, a sheet of plywood is a plate. A shade more complicated, if you bolt a bunch of beams together like an old-fashioned railway bridge, you have a *truss*, and once again, the

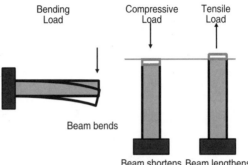

Fig. 3.2 *Loads on a structure can be tensile, compressive or bending. Bending loads generally cause a large deformation, while compressive and tensile loads cause very little. In engineering a telescope, therefore, it is best to make loads tensile or compressive.*

engineer's simplified truss predicts the behavior of such structures quite well. We'll come back to these "generic" structures later as we consider some important ways in which the parts of your telescope act like beams, plates, and trusses.

- Load

- Stress

- Cross-sectional area

- Deformation (or strain)

- Tension and Compression

Load is the total force that a structure must bear. The truss poles on your telescope carry a load equal to the weight of the upper tube assembly. The altitude bearings carry a load equal to the weight of the optical tube. Even the primary mirror carries a load: its own weight. When you move your telescope, the rocker box and side boards carry a changing, or dynamic, load. Engineers sometimes think of loads as bags full of birdshot hung on weightless structures, which is a useful method of analyzing how things work. In the U.S., engineers usually measure the load in pounds.

Stress is an engineering term that helps us calculate how a structure will respond to a load. Stress is the force per unit of area. You can think of the load as the product of the stress and *cross-sectional area* of a structure. Suppose you put a 500 pound weight on top of a marble column 5 inches wide and 10 inches deep. The cross-sectional area is 50 square inches, so the stress on the marble in the column is 10 pounds per square inch. The material in a big beam with a heavy load and a small beam with a light load may be subjected to the same amount of stress.

Deformation is the response of a structure to a load; that is, the amount by which the structure stretches, squeezes, or bends. Deformation is also called *strain*, which is confusing because in ordinary speech, stress and strain mean pretty much the same thing. Remember it this way: when you get stressed (from the outside), you feel a strain (internally). Stress is a load, and strain is a response. De-

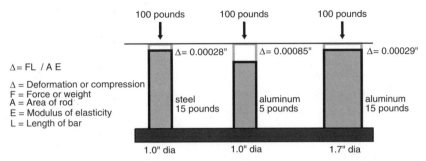

$\Delta = FL / A E$

Δ = Deformation or compression
F = Force or weight
A = Area of rod
E = Modulus of elasticity
L = Length of bar

Fig 3.3 *Pound for pound, aluminum is a better structural material for a telescope than is steel. In compression, a steel bar and an aluminum bar of the same weight compress an equal amount, but the aluminum bar is considerably stiffer because it distributes the weight over a greater area.*

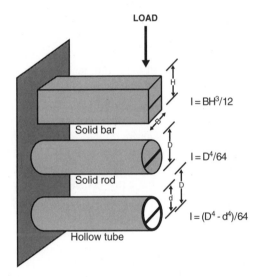

Fig. 3.4 *The greater its height or diameter, the greater the moment of inertia and the greater the resistance to bending in a cantilevered beam. Pound for pound, a hollow structure can be made stiffer than a bar or rod.*

formation is most often expressed in feet or inches.

Tension is a load that pulls a material apart; when materials are subjected to tension, they stretch. If this is confusing to remember, use this little mnemonic: "when you feel tense, consider whether you're getting stretched too thin."

Compression is the opposite, a load that presses a material together. It is useful to think of stress as a combination of tension and compression. When you bend a rod or beam, for example, the side bent out is under tension and the side bent in is compressed. A handy memory device for keeping the terminology clear in your mind is: "No wonder people respond to stress by getting bent out of shape!"

3.2.2 The Moment of Inertia

To build a stiff telescope, we need to minimize the deformation of the materials used in it. We want key parts—truss poles, telescope mirrors, rocker boxes—to carry their loads without deforming, or with as little deformation as possible.

How much a structure deforms under a load depends on three things:

- the *size and shape* of the structure,
- the *material(s)* it is made from,
- and, of course, the *load* it carries.

If you know these things, it is possible to compute the deformation of a structure. (We're not suggesting that you do this personally.) Running such computations is what engineers do, but understanding what works well is the key to building telescopes that perform well.

The *cross-sectional area* of a beam is the area exposed in a cut through it. Consider a beam 2 inches wide by 3 inches high and 65 feet long: the cross-sectional area is 6 square inches. Notice that the length doesn't matter. The cross-sectional area determines the strength of the beam, that is, how strongly you would need to load it in compression or tension to make it break.

The *moment of inertia* is a slightly old-fashioned engineering term for the resistance of a structure to bending and it is used by engineers differently than the term used in physics. More up-to-date engineers use the term "second moment of area." (You'll see both terms used in different text books.) It depends on the distribution of material in the structure. For beams, the moment of inertia depends on how the material is distributed in the beam. The basic rule is that the farther the material in the beam lies from the central axis of the beam, the larger the moment of inertia.

The cross-sectional area and the second moment of area are completely independent of the material in the beam. A rod 1 inch in diameter made from balsa wood has the same second moment of area and cross-sectional area as a 1-inch rod made of steel. We expect the steel rod to be stronger because of composition, but the cross-sectional area and moment of inertia of the steel beam and the balsa wood beam are the same.

3.2.3 Elastic Deformation

All materials deform. Primary mirrors bend. Rocker boxes flex. Truss tubes sag. When you subject a body—a mirror, a truss tube, a rocker box, or anything—to a load, it always deforms. The deformation may be large or it may be small, so engineers devised ways to measure how much various materials resist deformation.

The *modulus of elasticity* is the ratio of stress to strain. (Recall that stress is the load per unit of area and the strain is the resulting deformation.) To see this more clearly, consider a block of foam rubber sitting on a table. Atop the foam rubber block is a weight. The stress on the block is the weight of the load divided by the cross-sectional area of the block. Given a block 4 by 4 inches in cross section and 6

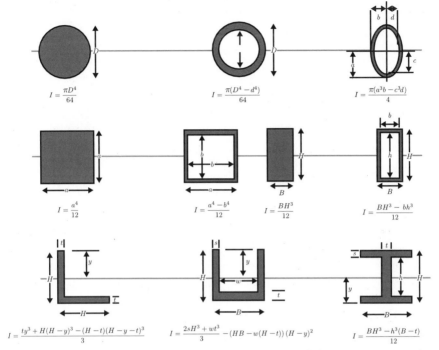

$$I = \frac{\pi D^4}{64}$$

$$I = \frac{\pi (D^4 - d^4)}{64}$$

$$I = \frac{\pi (a^3 b - c^3 d)}{4}$$

$$I = \frac{a^4}{12}$$

$$I = \frac{a^4 - b^4}{12}$$

$$I = \frac{BH^3}{12}$$

$$I = \frac{BH^3 - bh^3}{12}$$

$$I = \frac{ty^3 + H(H-y)^3 - (H-t)(H-y-t)^3}{3}$$

$$I = \frac{2sH^3 + wt^3}{3} - (HB - w(H-t))(H-y)^2$$

$$I = \frac{BH^3 - h^3(B-t)}{12}$$

Fig. 3.5 *You can compute second moments of area for many different cross sections for your telescope's truss tube using these formulas. When you take weight and availability into account, thin-walled tubes are almost always the best choice.*

inches high, loaded with a weight of 48 pounds, the stress is 3 pounds per square inch.

The strain is how much the block of foam rubber has been compressed under the load. Since the block was 6 inches tall before the weight was applied and 5.4 inches tall afterward, the block has been compressed by 0.6 inches. The modulus of elasticity is the ratio of stress to strain, so 3 pounds per square inch stress divided by 0.6 inches deformation gives a modulus of elasticity of 5 pounds per square inch. By comparison the modulus of elasticity for steel is 30,000,000 p.s.i.

The modulus of elasticity is a property of the material, and only the material. Stiffness is a measure of *elastic* resistance to load. The key word is elastic; the modulus of elasticity measures how much a structure can deform under load and still return to its original dimension after the load is removed.

3.2.4 The Strength of Materials

Now that you have the terminology, let's consider a simple case: a wooden beam sticking straight out from a brick wall. To the engineer, this is the classic case of the rigidly supported horizontal beam. We hang a load on the end of the beam. Question: what happens?

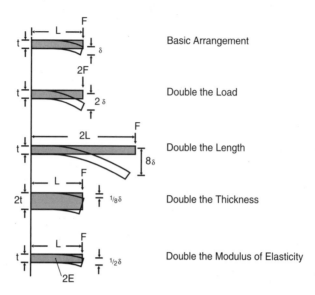

Fig. 3.6 *These beam-bending thought experiments reveal much about building rigid structures. For a cantilevered beam of length l, thickness t, carrying a load F, deformation is represented by* δ. *Note how the deformation changes when you double the length, thickness, and modulus of elasticity. Short, stubby beams are the stiffest by far.*

Well, from your own daily experience, you know that the end of the beam will bend downward under the load. The engineer knows that the end of the beam will deflect by an amount:

$$\text{deformation} = \frac{\text{force} \times \text{length}^3}{3 \times E \times I}.$$

In this formula, *length* is the distance between the mirror box and the secondary cage, *force* is weight of the cage with all accessories, *I* is the second moment of area of the truss poles, and *E* is the modulus of elasticity. The most important thing to notice is that the deformation depends on the cube of the length of the beam. Double the length of the beam, and you increase the deflection eight times. There is no need to take our word for this: Hang a ruler over the edge of a desk and apply some test loads. You will soon realize how critical it is to keep beams in a telescope as short and stubby as possible. If you learn only one thing from this chapter, let that be it.

3.2.5 Stiffness of a Truss Pole

Let's examine truss tubes more carefully. Exactly how does this moment of inertia stuff work? Consider an aluminum tube that is supported horizontally at one end with a load hanging on the other.

What happens as you change the second moment of area? The expression for the second moment of area of a hollow tube:

$$I = \frac{OD^4 - ID^4}{64}$$

where *OD* is the outside diameter of the tube and *ID* its inside diameter. The deflection formula thus becomes:

$$\text{deformation} = \frac{\text{force} \times \text{length}^3}{(3/64) \times E \times (OD^4 - ID^4)}.$$

If you want to do the math, go ahead and enjoy yourself comparing different geometries and tube materials. You will soon see that the bigger the tube diameter, the lower the deflection. For making comparisons, the following formula is quite useful for round tubes:

$$\text{deformation is proportional to} \quad \frac{\text{force} \times \text{length}^3}{\text{diameter}^4}.$$

Note that deformation is proportional to, not equal.

What we see is that *the bending is directly proportional to the cube of the length* of the truss. If you compare a big Dob with ten-foot poles with a small Dob having five-foot poles, you find the same diameter poles would flex eight times as much. (Twice the length cubed is a factor of eight.) When you consider that the end components are also heavier, you see that the poles that work well in a five-foot length won't work well at all in a telescope where they must be ten feet long.

For tubing, *the stiffness of the truss is proportional to the fourth power of the diameter of the poles.* Bigger diameter, stiffer truss. Not only is it stiffer, it is stiffer to the fourth power. The fourth power of anything greater than one or zero is considerable.

Imagine two hollow aluminum tubes with the same wall thickness and the same length. One tube is an inch in diameter and other tube is 1½ inches in diameter. The larger tube is 1.5^4, or 5.06, times stiffer! By increasing the diameter 1.5 times, we beef up the truss by a factor of five. This extra stiffness makes a big difference at the eyepiece. What good are diffraction-limited optics if the image is dancing around on wobbly truss poles?

3.2.5.1 Flexure in the Rocker Box

The truss tube Dobsonian pushes the design of the rocker to the limit. Because the mirror box has huge side bearings, the rocker box can have a very low profile. These ultra-short sides produce an extremely rigid rocker. Let's examine why.

As we have seen, flexure is proportional to the cube of the unsupported length of a beam or plate. Because of this, a telescope with 20-inch-high rocker sides (and 6-inch-diameter side bearings) suffers eight times the rocker-box flexure of a telescope with 10-inch-high rocker sides (and 26-inch-diameter side bearings). For all practical purposes, all you need to remember is that when the height of the rocker sides is doubled, the flexure increases greatly. Keep the rocker sides low for a stiff mounting.

Keeping the height of the rocker sides minimal is half the story. The other half is the thickness. In a flat plate, the stiffness is proportional to the cube of the thickness. Doubling the width of the rocker sides—simply using two layers of ¾-inch plywood bonded together—stiffens the rocker sides by a factor of eight. Even bonding a layer of ¼-inch plywood to ¾-inch plywood produces a rocker that is considerably stiffer than one made with ¾-inch plywood alone.

3.2.5.2 Flexure of a Telescope Mirror

The laws of physics that apply to the stiffness of the rocker also affect the primary mirror. As you will see, flexure grows very rapidly as mirrors become larger. Flexure in a flat plate supported at the edges is proportional to the cube of the diameter divided by the cube of the thickness. Consider two mirrors made from 2-inch solid Pyrex, one of 20 inches aperture and the other of 40 inches aperture. If they could somehow be the same weight, the larger mirror would flex eight times more than the smaller mirror, simply because of its larger diameter. Of course, they cannot be the same weight; there is simply a lot more Pyrex in the larger mirror. When you include the difference in weight and diameter, the larger mirror flexes 16 times more than the smaller one. This is why big mirrors need fancy mirror cells for support.

3.2.5.3 Why Simple Scaling Doesn't Work

Building small telescopes is easy. Building large aperture telescopes is not. As components get bigger, you rapidly encounter lots of severe engineering problems. To go from a 10-inch scope to a 20-inch one, you can't just double the dimensions. In the jargon of engineering, telescope designs don't "scale" in a simple way.

Bigger rocker components suffer more overall flexure not only because they are bigger but also because they must carry a heavier load. Suppose that you double all the dimensions of a 10-inch rocker box. Doubling the distance across the base and sides without increasing the thickness, you increase their flexure eight times if the 20-inch telescope weighed what a 10-incher does. Beefing up the thickness restores the lost factor of eight—but the weight of the rocker has risen by a factor of eight and so has the weight of the telescope. Simple scaling has given you an incredibly heavy 20-inch Dobsonian.

For all you folks who slept through high school physics, here is a refresher. The area of an object increases by the square while its weight increases by the cube. Even when you work hard to control it, the weight of a telescope tends to increase a lot faster than its aperture.

From building a lot of telescopes, we know that the optical tube assembly of a carefully-designed 40-inch Dobsonian weighs about four times as much as the tube of a 20-incher, so the flexure of the rocker increases by a further factor of four. To make matters still worse, for a given eyepiece the magnification of the telescope doubles, so that flexure in the mount becomes grossly apparent. The bottom line for experienced telescope builders: the 20-inch Dob that you built last year didn't present anything like the structural demands of a 40-incher you're dreaming about. By comparison with a 40-incher, it is *easy* to build a stiff 20-incher.

Fortunately, you aren't stuck with simple scaling. You can increase the thicknesses of the rocker sides and bottom to regain some stiffness. You can make the bearings even bigger and reduce the rocker sides a bit more. That's really what this book is about: beating scaling by using the laws of physics. If you make the azimuth and altitude bearings about the same size as the rocker bottom, you score a touchdown. If you beef up the rocker walls and bottom you have won the game.

3.2.6 Dobsonian Dynamics: How Dobsonians Move

Dobsonian telescopes depend on balance and friction. Because it is balanced, the Dobsonian remains pointed where you want it to point. A good Dobsonian should have just enough friction to hold it in place against a little imbalance and the pressure of the wind, but little enough that a light touch on the tube near the eyepiece is sufficient to move the telescope to follow a star.

The goal for the dynamic Dobsonian is thus twofold. The first goal is to design the telescope so it balances and stays where you put it, and the second, to design it to have the right amount of frictional resistance to movement.

3.2.6.1 Balancing a Dobsonian

From an engineer's point of view, the tube of a Dobsonian is like a child's teeter-totter—a bar with weights on both ends. From your personal experience on the grade school playground, you have an intuitive understanding that the big kid has to sit close in and the little kid has to sit far out to achieve balance. The engineer says that it takes equal torques to balance the seesaw.

Torque is the product of the weight of the kids and their distance from the center of the seesaw, or in engineering terms, the sum of the products of the acting forces and their moment arms. If the total of all the torques acting on a body is not zero, then the resulting torque causes the body to rotate. (Translation: When you put a big Nagler eyepiece in the focuser, the front of the telescope sinks.) Your goal is to make the sum of all the torques equal to zero.

Over the years, a lot of folks have built telescopes using the "by-guess-and-by-gosh" method, balancing the tube either by moving the altitude bearings or by adding weights. Not only does 32 pounds of lead on the back end of your telescope brand you as a novice, but it also means that you're going to lug that 32 pounds of lead out every time you use it. We're not going to let you do that.

Here's how balancing by computing torques works:

Consider the telescope as a set of weights at fixed distances from the axis of rotation. In the modern-style Dobsonian, the axis of rotation is at the top of the mirror box. Thus the moment arm for each component is its distance from the axis. Life is made simpler because even though the weight of a component is spread over its length, we can treat the weight as if it were concentrated at the center of gravity of the component. For example, each truss tube acts as if its entire weight was at the center of the tube, and indeed, the whole system of eight truss tubes acts as if its entire weight was located at the center of the truss.

A balanced telescope satisfies the following condition:

$$\sum_{i=1}^{n} W_i d_i = 0.$$

This equation says: for a telescope with n components to balance, the total of the torques of the individual components must equal zero. To obtain the torque of each component, multiply its weight times its distance from the axis of rotation.

One more wrinkle and we're through. Remember René Descartes, the great French mathematician who invented Cartesian coordinates? Using Cartesian coordinates, the distance from the axis of rotation is positive if you go toward the front of the telescope from the balance point and negative if you go toward the back from the balance point. However, you can make life easier for yourself by keeping all of the distances positive by setting the front components on one side of the equal sign and the back components on the other side of the equation.

You will be ready to put all this good knowledge to work in **Chapter 7**, when you are building the mirror box. At that time, you should know:

- the weight of the mirror, W_m
- the weight of the mirror cell, W_{mc}
- the weight of the mirror box, W_{mb}
- the weight of the truss tube, W_{tt}
- the weight of the secondary cage, W_{sc}

and you should have a pretty good idea of the distance each of these components will be from the rotational axis located at top of the mirror box. You've got balance when:

$$W_m d_m + W_{mc} d_{mc} + W_{mb} d_{mb} = W_{tt} d_{tt} + W_{sc} d_{sc}.$$

If you want to avoid doing the algebra, you just fiddle with the height of the mirror box until the torques from the bottom end balance the torques on the top end. When you make the mirror box one inch deeper, the distance to the mirror and mirror cell, d_m and d_{mc}, increase by an inch, but the distance to the center of the mirror box, d_{mb}, increases by only half an inch. The distance to the secondary cage, d_{sc}, decreases by one inch, but the distance to the center of the truss tubes, d_{tt}, decreases by only one half inch. On the third or fourth try, you have balanced your teeter-totter to the stars!

3.2.7 Friction in a Dobsonian

The force necessary to move your Dob depends on the weight of the telescope, the

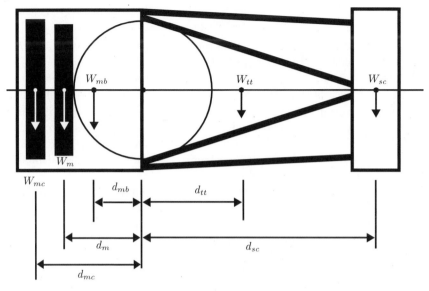

$$W_m d_m + W_{mc} d_{mc} + W_{mb} d_{mb} = W_{tt}\, d_{tt} + W_{sc} d_{sc}$$

Fig. 3.7 *To balance a Dobsonian, the torques on opposite sides of the balance point must be equal. Each of the five major components of the telescope—the mirror cell, mirror, mirror box, truss tube, and secondary cage—has a torque equal to the product of its weight and the distance from the balance point to the component's center of gravity.*

bearing materials, and the geometry of bearing surfaces. Moving a telescope that weighs several hundred pounds should take a force of just a couple pounds. Building such a telescope is not a hit-or-miss proposition: you have to engineer the frictional resistance.

To understand friction, suppose that you have a block of one material resting on the surface of another material. To move the block on top, you must pull (or push) the block sideways with a certain amount of force. (Imagine using a spring scale to measure how hard you must pull on the block. The spring scale measures the *frictional force*.) The frictional force is the force necessary to slide the block across the surface at a constant speed.

3.2.7.1 Friction Defined

For a given pair of materials, the frictional force is always the same fraction of the force that the block, because of its weight, exerts on the surface. (Imagine using the spring scale to lift the block straight off the surface. If the surface is level, the scale now measures the *normal force* between the surfaces. In the everyday world, we call the normal force on an object its *weight*.) The ratio between the frictional force and the normal force is the *coefficient of friction*.

Here's the corresponding equation:

$$f = \frac{F}{N}.$$

In this formulation, f is the coefficient of friction, F is the frictional force, and N is the normal force between the materials. On a flat surface, the normal force is the weight. You can measure the coefficient of friction between two materials using a spring scale to measure the forces; the ratio of the forces is the coefficient of friction.

For most combinations of materials, the force needed to start the motion is greater than the force needed to maintain sliding. The friction between non-sliding surfaces is called *static friction*; between sliding surfaces it is *dynamic friction*. The corresponding coefficients of friction are called the coefficient of static friction and the coefficient of dynamic friction.

Let's see this in action: You have a spring scale, a block of wood, and a table with a Formica top. You weigh the wooden block and the scale reads 10 pounds. Placing the block on the table, you pull on it gently with the spring scale. Nothing happens; the block just sits there. You increase the pull until at 3.7 pounds pull, the block suddenly starts to slide. After a bit of experimentation, you learn that you can keep the block moving at a constant speed by pulling with a 2-pound force. This means that the static coefficient of friction is 0.37 and the dynamic coefficient of friction is 0.20. Try this with a variety of materials and you'll find that dynamic friction is always less than static friction.

For Teflon sliding on smooth Formica, the static coefficient of friction is 0.10 and the dynamic coefficient of friction is 0.08. For comparison, Nylon on Formica has a static coefficient of 0.20 and a dynamic coefficient of 0.19—roughly twice that of Teflon. Only a few materials —such as ski wax sliding on snow— have lower coefficients of friction than Teflon on Formica.

So what do sliding blocks have to do with a telescope? Well, the bearings in a Dobsonian telescope are nothing but blocks sliding on each other. The bottom bearing is a chunk of tabletop laminate sliding on three blocks of Teflon plastic. Same for the side bearings: they are laminate sliding on Teflon. Simplicity: it's what makes a Dob so great. With seven simple little chunks of plastic and ten bucks worth of Formica you replace a machine-shop job that would have cost a thousand dollars and wouldn't move so nicely.

3.2.7.2 Dobsonian Friction

You can already see what's going to happen when you put these laminate-and-Teflon bearings into a big Dob. The telescope will act like a lever that passes the force you exert on the tube down to the bearings. When the force exceeds the static friction, the telescope starts to move. As long as you continue to press on the telescope with a force that is enough to overcome dynamic friction, the telescope moves at a constant speed.

Because the telescope moves independently in two axes, altitude and azimuth, you could calculate the force required to move the telescope in each axis. When these two forces are close to the same value, observers cannot distinguish

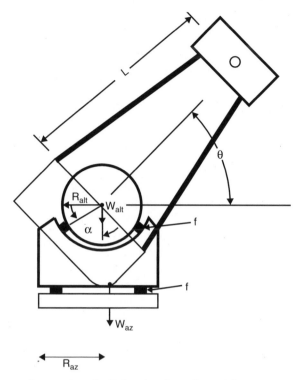

Fig 3.8 *The ease of movement of a Dobsonian depends on eight parameters: the length of the tube L, the weight of the tube W_{alt}, the radius of the side bearing R_{alt}, the angular position of the altitude pads α, the elevation angle of the telescope θ, the total weight of the telescope W_{az}, the radius of the azimuth bearing R_{az}, and the coefficients of friction f, of the bearing materials.*

the individual axes: they simply feel a telescope that moves easily in any direction.

Let's define a few symbols and get started. Call the tube length from the altitude bearing to the eyepiece L, the radius of the altitude bearing R_{alt}, and the distance from the pivot point to the Teflon pads in the azimuth bearing R_{az}. The altitude angle of the telescope we'll call θ, and half the angle between the Teflon pads in the altitude bearing as α. (This makes for greater convenience later.) The weight on the altitude bearing, W_{alt}, is the weight of the telescope tube. The weight on the azimuth bearing, W_{az}, is the sum of the weights of the tube and the rocker box.

Now for the azimuth axis. What is the force F_{az} that you will feel when you press on the tube near the eyepiece? The friction at the pads on the bearing is simply W_{az}. (In case you're wondering, the number of bearing pads cancels out.) The telescope and bearing act as a lever to reduce this force. The short end of the lever is the radius of the bearing and the long end is the length of the tube times the cosine of θ, the altitude angle. The result tells you how much force you must apply to move the telescope in azimuth:

$$F_{az} = \frac{W_{az} \; f R_{az}}{L \cos \theta}.$$

Altitude works much the same way. The frictional force is Nf, but because the pad is tilted at an angle α from the horizontal, the normal force is increased to $W_{alt} \sec \alpha$. The tube and bearing again act as a lever to reduce the force required at the eyepiece. The force required to move the telescope in altitude is:

$$F_{alt} = \frac{W_{alt} \; f \sec \alpha R_{az}}{L}.$$

So far we've been talking physics, but now there's room for some engineering creativity. We've made up a table showing what you need to do to change the ease of motion in each axis. If your telescope doesn't move as easily as you would like it to, or it moves too easily, just apply the rules in **Table 3.1**.

For example, suppose a telescope moves just fine in altitude but is a little too stiff in azimuth. For a finished instrument, you cannot change the length of the tube or do very much about its weight, and besides, changing those would mess up the altitude motion. So you either move the pads on the bottom bearing closer to the center or change the bottom bearing to material with a lower coefficient of friction.

3.2.7.3 Curing Stiction

If you were paying close attention earlier, you might have spotted a potential problem. As you start to push a stationary telescope, what happens? The force builds up against static friction until suddenly the scope "pops" free and starts to move. The force to move the scope suddenly drops—now you have dynamic friction—so the scope accelerates, or zooms forward. You relax the pressure of your hand and the scope suddenly stops as static friction kicks in. Stiction—bad motion from sticky bearings—can be a significant problem unless you use the right bearing materials.

The reason Dob builders like virgin Teflon and Stardust, Ebony Star, or glassboard laminates so much is that for these particular materials, the coefficients of static and dynamic friction are nearly identical when the surfaces are dry, and coating the bearing surface with silicone car wax reduces the difference to zero. There is no "pop" when the telescope starts to move, only the "buttery" slide of a true Dobsonian.

In addition to starting smoothly, the best Teflon and laminate combinations have another beneficial property: the faster the bearing surfaces slide, the higher the coefficient of friction. The bearings become harder to push when you push harder. As a result, the telescope resists motion that is too fast. Speed regulation is another key component in the motion of a true Dob.

Other bearing materials such as nylon, polycarbonate plastic, and smooth Formica may be cheaper, or maybe you've got a friend with warehouse full of them, but they just aren't the same. Use them and your telescope will pop and stick.

Table 3.1
Controlling Friction in Altitude and Azimuth

Factor in Telescope Design	To Increase Ease of Movement	To Decrease Ease of Movement
Telescope's Total Weight	Decrease weight	Increase weight
Length of Tube	Increase length	Decrease length
Coefficient of Friction	Decrease friction	Increase friction
Size of Bottom Bearing (azimuth only)	Decrease size	Increase size
Size of Side Bearing (altitude only)	Decrease size	Increase size
Angle Between Side Bearing Pads (altitude only)	Decrease angle	Increase angle

The whole scope will chatter when it starts, vibrate as it moves, and screech to a stop. You may be tempted to use alternative materials in it, but don't do it. You'll end up with a telescope that looks like a Dob on the outside, but where it counts—performance—it won't act like a true Dob. Dobsonian telescopes depend on having the right amount of friction in their bearings to function properly. It pays to get the friction right.

3.3 Materials for Telescopes

Plywood, Formica, Teflon, and aluminum are the basic materials of modern homebuilt telescopes. We have not selected these materials just because they happened to be lying around handy. Each of them offers specific advantages to the telescope maker. In this section we focus on each of these materials in turn. If you take the time to select and use the right materials, you'll be happier with your finished telescope.

3.3.1 Wood

The basic construction material you will use to build your telescope is plywood. No other material works as well. Furthermore, it is cheap and easy to work. Wood, in the form of plywood, is the material of choice for most telescope makers.

Why wood? Compare it to metal and you'll see why. Wood requires no expensive tooling: you need only a power drill, a saw, and a router. Wood is forgiving: tolerances are measured in sixteenths of an inch, rather than in thousandths of an inch that metal sometimes requires. Wood is so important that we have includ-

Table 3.2
Mechanical Properties of Telescope Materials

Material	Modulus of Elasticity (millions of PSI)	Density (pounds per cubic inch)
Aluminum	10	0.097
Baltic birch ply*	1	0.026
Brass	14	0.310
Copper	17	0.321
Steel	30	0.283
Titanium	16	0.162
Fiberglass	2 to 5	0.066 to 0.071
Pyrex	9	0.081
Zerodur	13	0.091
White oak	1.7	0.029
Ponderosa Pine	1.2	0.017
*3/4-inch 13-ply panel, load parallel to face veneer.		

ed **Appendix A** which is devoted to its properties.

Wood components can be assembled fast and easily with nails, screws, bolts and adhesives. Metal components need expensive boring for every fastener or they must be welded together. Wood is also self-damping. Vibrations from wind and hand guiding die out in a second on even the largest well-made plywood Dobs. Some of the massive metal monsters we've observed with oscillate for minutes. By the time the telescope stops dancing, the object has drifted out of the field of view.

Wood is also light. Pound for pound, no other common construction material can match it for strength and stiffness. Only fiberglass and foam-core composites, which are expensive and difficult to work, beat wood.

And last but hardly least, wood is beautiful. A telescope with a lustrous wood-grain finish is as almost as much a pleasure to look at as it is to look through.

3.3.1.1 Wood Is a Natural, Organic Composite

"Wood" is a very general term that covers a multitude of variations. Sawn wood, or lumber, was once part of a tree, so there as many types of lumber as there are types of trees. Not only are there thousands of species of trees, but variations in age, moisture, soil conditions, and growing space effect the properties of lumber. When you think about wood, remember that it is a highly variable material.

But all types of wood share key properties. Under a magnifier, you can see long, hollow fibers of cellulose, a natural plastic that formed the walls of the individual cells in the once-living wood, arranged in a complex three-dimensional structure. Layers of dense, hard material (xylem cells) alternate with softer, spong-ier material (phloem cells). As it grows, the tree lays down material just under the bark; in the winter the material is hard and in the summer it is soft. These layers are called growth rings. The resulting composite is light, strong, and rigid.

3.3.1.2 Wood Is Orthotropic

Wood has unique and independent mechanical properties in three mutually per-pendicular axes (that's what orthotropic means). The axis that runs parallel to the trunk of the tree, along the long dimension of the cell structure or the grain of the wood, is called the longitudinal axis. The radial axis of the wood extends outward from the center of the tree; and the tangential axis lies perpendicular to the grain and tangent to the growth rings.

Wood is stronger in the longitudinal axis than in the other two axes. For con-struction, the longitudinal and tangential axes concern us the most. Think of a pine board; to build anything useful with it, you must orient it to exploit its strength along the grain without exceeding its strength and thereby cracking or splitting it across the grain.

Plywood overcomes this limitation of sawn lumber. By laminating thin plies (layers) of wood oriented in different directions, we obtain a uniform panel that has the longitudinal strength of wood in all directions. Plywood is thus a compos-ite structure built up from composite materials. It is important to understand that a plywood panel is only about half as strong as a solid board of the same species and thickness. The grain of a board is homogenous and runs the same direction throughout, while plywood has half the plies running parallel to the face veneers and half perpendicular. For a simple beam a common board is superior. For panels like those used in the mirror box and rocker, plywood is the material of choice be-cause it is stiff in all directions.

It would take dozens of fancy parameters to fully describe the mechanical properties of wood. We need only concern ourselves with three: stiffness, hard-ness, and specific gravity.

Stiffness: Resistance to bending is expressed as the modulus of elasticity, E, the ratio of unit stress to unit strain. In other words, the modulus of elasticity de-scribes the amount of bending (elastic deformation) in a material under a given amount of applied force.

Because wood is orthotropic, it has three distinct moduli of elasticity: along longitudinal, radial, and tangential axes. They vary within and between species, and with the moisture content and specific gravity of the wood. For the plywood in our telescopes, only the longitudinal and tangential moduli of elasticity matter, because the plies are arranged to alternate between these orientations. The larger the numerical value of the modulus of elasticity, the stiffer the wood and the less it will bend under a given load.

Hardness: Resistance of wood to denting, wear, and marring is called its hardness. Officially, hardness is determined by the force required to embed a

0.444-inch diameter steel ball to one-half its diameter in the wood. The values listed in **Appendix A** are the average of radial and tangential hardness. The larger the hardness number, the more resistant a given type of wood is to dents, scratches, and scuffs.

Specific Gravity: This is the weight per unit volume compared to that of water. Since the mass of water is 1 kilogram per liter, the numerical values of specific gravity and density are the same. Regardless of the species, wood is composed largely of cellulose with a specific gravity of 1.5. But dry wood floats on water because wood is full of air-filled cavities and pores.

Some species have more cellulose per unit volume than others, and therefore have higher specific gravity. So long as a piece of wood is clear of knots, straight grained, and free from defects, the specific gravity is an excellent index of the amount of substance it contains; it is a good index of mechanical properties. In general, the higher the specific gravity, the stronger the wood.

3.3.1.3 Selecting Wood

The reason for knowing the modulus of elasticity, the hardness, and the specific gravity is to aid in selecting lumber or plywood that has the best properties for a given task. Here are a few examples.

For the truss pole split blocks, you need a wood that is very strong. After all, you'd be embarrassed if the blocks fractured as you proudly set up your telescope at the big star party you drove to 500 miles to attend. Pick wood with a high specific gravity: shagbark hickory would be great. Many other hardwoods will perform well, and almost any hardwood is better than a softwood. They will also provide a beautiful contrast to the wood of the mirror box if you mount the blocks on the outside.

How about the three feet that the entire telescope sits on? Since they sit down in the gravel and dirt and take quite a compressive beating, pick a wood with a high surface hardness. Sweet birch and sugar maple are good choices for the feet.

The plywood rings that form the backbone of the secondary cage need to be stiff, strong, and remain very flat. To achieve this goal, choose plywood because it lies flat. But don't use any old plywood; choose one made from a species with a high modulus of elasticity, such as sweet birch or white oak.

On the other hand, the rocker bottom does not need to be particularly strong, but it is nice to keep the weight of the rocker down so you don't get a hernia when you lift it. Select a softwood ply made from eastern or western white pine. Although both species have a relatively low modulus of elasticity, they also have a low specific gravity. One of the firs would also do nicely. Select a less expensive grade of plywood like B-C and bond the two C faces together.

Although you can make the rocker *bottom* from a softwood, the rocker *sides* are a different matter. The sides get nicked by shoes, bumped with your ladder legs, and banged against the sides of your vehicle. Why not use a plywood that is made from species with a high surface hardness? Select plywood that has surface

Table 3.3
Softwood and Hardwood Plywoods

Thickness in millimeters	Thickness in Inches	Hardwood Veneer Hardwood Core Plies (HVHC)	Hardwood Veneer Softwood Core or Softwood Veneer Softwood Core Plies (HVSC or SVSC)
3	⅛	3	na
4	⁵⁄₃₂	3 or 4	na
5	³⁄₁₆	4 or 5	3
6	¼	5	3
9	⅜	7	3, 4, or 5
12	½	9	3, 4, or 5
15	⅝	11	5 or 7
18	¾	13	5 or 7

veneers of sweet birch, sugar maple, or one of the red oaks, and for the very best results, select plywood with a hardwood core.

To help you purchase wood for your telescope, **Table A.2** lists the mechanical properties of different species. It reveals the common-sense fact that hardwoods are harder, stiffer, and denser than softwoods. Flip back to **Appendix A** and compare different species within a each group.

Sugar maple is stiffer, harder, and denser than any other type of maple. The mechanical properties of the maples are considerably superior to aspens, another species often used in hardwood plywood. Want to know why bowling alleys prefer maple for the lanes? Compared to softwoods, a maple board is a lot stiffer, nearly twice as dense, and three times harder.

Buy wood that has been properly dried. Wood with a 15% moisture content should assemble easily and remain dimensionally stable. A telescope made from wood with a moisture content that is too high will buckle and warp. Know what you are buying, and buy dry.

When you purchase lumber or plywood, ask the dealer what type of wood you are getting and if it has been adequately dried. A sheet of "pine," "oak," or "birch" plywood can be made from many different subspecies with varying degrees of strength, hardness and density. The filler layers in plywood are often made from a weaker species to save costs.

The bottom line: find out what you are buying. At a good lumber yard you can get whatever you want. Don't settle for whatever you get.

Table 3.4
Comparative Stiffness and Density of Materials

Material	Stiffness*	Density*
Aluminum	1.0	1.0
Steel	3.0	2.9
Brass	1.5	3.2
White Oak	0.2	0.3
Fiberglass	0.4	0.7
Pyrex	0.9	0.8
Plywood	0.1	0.2
* relative to aluminum = 1.0		

3.3.2 Aluminum

Aluminum is readily available in numerous alloys. These are made for various industrial applications. Whereas they differ greatly in yield strength or resistance to permanent deformation, they all have nearly the same modulus of elasticity, and so will make an equally stiff truss. For the telescope maker, ordinary cheap 6061-T6 aluminum is a dream material. You can buy the stuff in pretty, ready-to-use lengths from most hardware stores, and it's a good value.

Why aluminum? Keeping the weight of the truss low is extremely important in portable telescopes. Because the entire truss assembly is above the fulcrum, every pound saved in the truss translates into a shallower and lighter mirror box. The choice of metal for the truss poles is therefore easy: you can't beat aluminum. Just make sure that the moment of inertia of the truss pole is large enough to give the stiffness you need. Strength? Well, the poles don't need to be strong. Aluminum is plenty strong enough and it is considerably lighter than steel, copper, or practically anything except titanium, magnesium, and exotic carbon-fiber composite aircraft materials.

Other materials are *stronger* than aluminum, but don't confuse strength with stiffness. Yield strength means resistance to structural failure, collapse, catastrophe. Modulus of elasticity means elastic resistance to deformation. Our needs are the same as manufacturers of boats, planes, and beer cans: we choose aluminum because we need stiffness but not weight. And we all like aluminum because it's waterproof, easy to cut, relatively cheap, and it needs no protective paint. Steel tube is just the opposite.

Copper pipe has many of the advantages of aluminum—availability, cost, resistance to corrosion and workability—but it is three times heavier than aluminum and only 1.7 times stiffer.

Wood isn't a bad alternative to aluminum. Pound for pound, pine compares favorably to it, but hollow wooden dowels are not feasible, so wooden truss tubes

Table 3.5
Laminate Selection Guidelines

Aperture of telescope	Laminate	Recommended Pad Loading
up to 18 inches	#1782 Stardust Formica or Wilsonart Ebony Star on both axes	15 PSI on the pads
20 to 24 inches,	Ebony Star or glass-board on altitude ax-is, glassboard on azimuth axis	15 PSI for Ebony Star, 12 PSI for glassboard
25 inches and over	Glassboard on both axes, central pivot pad mandatory	12 PSI on the pads

come out heavy compared to aluminum truss poles. Nonetheless, lumber and ply-wood "poles" have been used, as have pine dowels, broom sticks, and domestic clothesline poles. Use them if you can't afford aluminum.

You'll find more about selecting aluminum tubing for truss poles in **Chapter 8**.

3.3.3 Teflon

Teflon is Dupont's brand name for polytetrafluoroethylene plastic, called TFE or PTFE by the rest of the world. Two grades are available: virgin and mechanical. Virgin Teflon—as the name implies—is pure and has never been used. Mechani-cal Teflon is a mixture of recycled Teflon with virgin Teflon. When they are blended, the recycled Teflon never fully melts to an homogenous mixture. As a result, particles of recycled Teflon stick out of the mix, and it feels rougher. The-oretically both have the same coefficient of friction, but virgin Teflon beats the mechanical Teflons we've tested.

Because hardly anything sticks to Teflon, it is hard to hold in place. Most telescope makers nail thick pieces of Teflon to their bearings, but etched virgin Teflon is easier to use because you can glue it in place. The material is acid-etched on one side and a carbonized coating is deposited on the etched surface. The coat-ing allows the Teflon to accept adhesives. You can use contact cement to attach the pads of etched virgin Teflon directly to the plywood, thereby eliminating the possibility that nail heads will someday work loose and scratch the laminate.

Teflon is expensive. Why pay for thick Teflon when only the top few layers of molecules do all the work anyway? See **Appendix E** for suppliers. We recom-mend using $\frac{3}{32}$-inch-thick etched virgin Teflon for the side and bottom bearings.

3.3.4 Plastic Laminates

When Dave built his first Dobsonian, very few people had telescopes larger than 17½ inches aperture. At the time, it was simply taken for granted that the bigger a Dob was, the stiffer it moved. Dave was among the first to seek a way to reduce the friction and to enjoy the hallmark smoothness of a smaller Dobsonian by experimenting with different bearing materials.

In *Telescope Making #8*, Richard Berry had described how friction effects the way Dobsonian telescopes move. (The essence of his article appears in **Section 3.2.7.**) Since Teflon was already close to the ideal material, Dave knew he could only find a lower coefficient of friction in a new laminate. What he did was carry a small piece of Teflon around in his pocket. He placed the Teflon against the best existing material and then rubbed the combination on a new material. Whichever laminate slid on the Teflon was the one with a lower coefficient of friction.

Dave ran around for a year testing every laminate he thought might work—in restaurants, bars, kitchens, and bathrooms. In addition, he earned a lot of funny looks from clerks at home improvement centers rubbing his test Teflon against wall coverings, countertops, and every countertop laminate chip he could find. He discovered that a laminate called #7181 Beige Tweed Formica had a lower coefficient of friction than any other surface he could find. So fast did the word spread that within months, every new Dobsonian built in the United States was made using Beige Tweed Formica. Later, Dave found even better laminates. Ebony Star, Stardust and glassboard.

3.3.4.1 "Ebony Star" and "Stardust" Laminates

It is a pleasing coincidence that the names of the two best all-around laminates for telescope bearings contain the word star. They are *Ebony Star* and *Stardust*. Wilsonart Inc. manufactures Ebony Star #4552 and Formica Inc. makes Stardust #1782 for countertops. They are significantly slicker than #7181 Beige Tweed Formica. They have a hard stippled surface like an orange and are speckled black in color.

Ebony Star and Stardust are made, like most countertop laminates, from a top layer of decorative paper impregnated with melamine resins, pressed and bonded to kraft-paper core sheets impregnated with phenolic resin. The sheets are bonded at pressures greater than 1,000 pounds per square inch at temperatures approaching 300 degrees Fahrenheit. The back is sanded to facilitate bonding it to plywood kitchen countertops.

One of the remarkable discoveries that came from Dave's research is that textured laminates generally have less friction than smooth laminates. Ebony Star's and Stardust's texture undoubtedly accounts for an important part of their performance. Wilsonart Inc. labels the surface texture "Touchstone." The Touchstone series of laminates includes #4552 Ebony Star, #4406 Diamondhead and #4408 Surfside. Formica Inc. labels their surface texture "Quarry Finish." The

Quarry Finish series of laminates includes #1782 Stardust, #1816 Fogdust, #1783 Firedust, and #680 Granite. If your local building supplier does not carry these laminates try the manufacturers' customer service numbers listed in **Appendix E** to find a source near you.

3.3.4.2 Glassboard

Another laminate we recommend highly is "FRP." Called fiberglass board, bead-board, glassboard, but more properly Fiberglass Reinforced Panel, it is sold as a wall covering for the food processing industry. The hard, nonporous surface is impervious to most solvents so it can be cleaned efficiently. You have probably already seen FRP on the walls of public rest rooms. Its texture and slipperiness makes FRP virtually graffiti-proof.

Glassboard is a fiberglass product made by Sequentia in Ohio. They call it "Structoglas." It is a laminate consisting of a uniform mat of high-strength glass fibers imbedded in an organic resin. The composite structure is cured under heat and pressure to a solid material which is lightweight, strong and shatter resistant, with excellent physical properties. It has a smooth mounting surface. The business side has an extremely hard, pebbled surface.

The bumps on FRP are so large that, unless the Teflon pads under your telescope are also fairly large, you can feel them at the eyepiece when you are guiding the scope. On the bigger Dobs this is no problem because the pads are large enough. FRP is sold at many lumber yards in 4 by 8-foot sheets 0.090 inches thick in white, gray, almond or beige for about $30 a sheet. Even though the texture of all four colors is the same, the gray glassboard seems to work best.

FRP works well on the side and bottom bearings for scopes in the 18-inch to 40-inch range, that is, FRP is best for large and heavy telescopes. In small instruments, FRP often has too little friction. Breezes over 5 miles per hour spin the telescope like a giant weather vane. If you crave a telescope that moves very easily, then use FRP.

As a general rule, a bearing loading of 15 pounds per square inch on a Teflon and Formica bearing results in the lowest friction. Deviating from 15 psi in either direction makes the telescope stiffer and more lethargic. However, with Teflon and glassboard bearings a load of 12 pounds per square inch works better, so use larger Teflon pads with glassboard.

You can apply any of these laminates with contact cement. For glassboard, clean the bonding surface with acetone and it will accept the adhesive better. Panel adhesives also work with glassboard if they are applied as a very thin layer.

3.3.4.3 Put Car Wax on Laminates

Car wax is the secret of great scope bearings. Sometimes at star parties, Dave tells people that the "buttery" feel of a good Dobsonian *is* butter (after all, Wisconsin *is* "the dairy state"), but the real secret is silicone car wax.

Silicone grease eliminates the difference between static and dynamic friction, but it has two severe disadvantages: it becomes stiff in the winter, and it holds dirt. The silicone sprays intended for use as dry lubricants don't work either. Something in them gums up after a couple of days, making the bearings become stiff again. What you need is silicone without a carrier: an automotive wax. Turtle Wax works very well, but almost any brand of car wax seems to work as long as the main ingredient is silicone. Spread it on with a rag, let dry to a haze, and then buff. Silicone car wax will noticeably decrease the friction on any Dobsonian, large or small. When you wax your car in the spring and fall, wax your Dobsonian's bearings too.

To make the telescope move even easier yet, apply a little Armor All on the bearing surfaces *after* you put on the car wax. Armor All is available at most hardware/discount stores in the automotive department. Always put some silicone car wax on before applying the Armor All.

3.4 Protective Coatings

When it comes to appearance, we prefer the natural look of wood. It's a real winner. Wood is a building material with a grain, texture, and character all its own. Seeing wood sanded smooth, stained, and varnished properly sets one's inner emotions afire. People gravitate to wood. They like to run their hand over it and feel it.

When was the last time you saw someone fondle a fiberglass or metal telescope at a star party? Artificial materials are cold and uninviting. Mass produced from machines or molds, they lack the character and warmth of wood. If you put ten wooden telescopes built from the same plans side by side, each one looks unique. Its grain, the way it takes on a stain, and its growth variations make wood interesting to look at. You don't hear folks say, "What a nice metal tube." Instead they say, "I like your woodwork" or "That sure is beautiful wood." Why do you think cheap furniture, automobile dashboard trim, and microwave ovens have simulated wood-grain finish? Humans like wood.

Use paint to hide and cover defects like wood putty, pencil marks, tool dings, or the grain that runs the wrong way. Paint inside surfaces black to minimize the reflection of light. But for exposed wood, use stain and varnish to show off your craftsmanship and your care for the finished product. Varnish hides nothing and tells all.

For telescopes that are intended to be portable, polyurethane varnish is far superior to lacquer or paint. Bumps and scrapes show on a painted telescope much faster than they do on one that is varnished. Paint tends to chip, leaving conspicuous defects. Varnish is forgiving. A gouge or scrape in varnish exposes the stained wood beneath, so the defect is hardly visible. If the trauma is severe enough to remove wood, all you have to do is rub the area with a Q-tip and some matching stain, and the defect becomes almost invisible. It doesn't take long before a painted telescope needs a new paint job, while a varnished telescope still looks pretty good.

Of the hundreds of types of varnish on the market, we recommend only poly-urethane. It is relatively waterproof and doesn't deteriorate from the acid dew in-dustrialized countries generate. Polyurethanes are mechanically tough and stand up to constant handling. Select one of the more expensive grades of exterior poly-urethane varnish and you'll get what you pay for.

Some people prefer a deep glossy finish, the type that looks like your tele-scope got caught in an ice storm. To achieve this it is necessary to build up the fin-ish with three or four coats. Sand each layer after it has thoroughly dried with wet/dry sandpaper. Start with a #300 or #400 grit and then fine sand with a #600 grit. Rinse, wipe dry, and apply the next coat. If you are using a brush to apply the var-nish, brush it on thoroughly with vertical strokes. Horizontal strokes tend to sag before the varnish sets.

Covering all exterior surfaces in Formica is another option. Formica doesn't look quite as pretty as the wood underneath, but it is the ultimate finish for people who must observe in the rain. Because Formica is waterproof, heavy dew or frost won't affect it. However, there are some drawbacks you should be aware of. Al-though it is not hard to work with laminates if you know how, a telescope is a poor place for learning how. It takes practice to apply and trim properly. If you don't know how, don't do it. Furthermore, nothing covered in Formica should ever be left sitting in the hot sun. You take your telescope to a star party and put it on show—only to find your beautiful Formica falling off under the hot sun.

Paint the inside of the mirror box and rocker base black. Ask for any solid color latex exterior stain and tell the paint man at the hardware store to add as much black pigment as possible. It should dry very black. Be sure to apply the black stain *after* you have varnished the outside of each component. If you do the inside first and get some black on the unfinished outside surface, you cannot fully remove it. But if you spill a drop of black on the already varnished surface, it wipes clean with a damp rag.

Chapter 4
Optics for Dobsonians

Telescope makers often have unreasonable expectations about mirror quality, delivery time, and price. They want a great optic, they want it right away, and they want it to be low cost. No one that we've ever met has complained that a telescope mirror was too good or was delivered too promptly. Quality costs more and you sometimes have to wait longer to get it. But you only need one mirror. They don't wear out. And you will never regret having a good mirror.

Because the choice of the primary mirror governs almost all of your subsequent decisions, read this chapter carefully. Don't just sit at home—get out and learn about *big* telescopes. Attend star parties, mooch off your friends, and test-drive every big telescope you see. Study their construction. Shake their mountings. Star test their optics. Ask questions of people who seem to know something, but never take their answers as gospel. Think for yourself. Learn enough to trust your own judgement.

Before you place your order, talk to several different optical shops. Describe what you want, listen to what they say, and make sure you feel comfortable with the people and their answers. Be polite and businesslike in your contacts with them, and insist on polite and businesslike treatment from them.

When you order the mirror, put what you have discussed in writing. Specify the aperture, focal length, the date of delivery, and the quality you expect based on your discussions. In accepting your order, make sure the people at the shop understand that you consider your order a contract. Ask them to notify you promptly if for any reason they will not be able to deliver the mirror you have ordered at the time that they have promised.

For your part, when the mirror arrives, be ready for it. Your telescope should be close enough to ready that you can star test the mirror within a few days (or at most, a few weeks) of receipt. Try to be objective, but remember that this is the honeymoon. Sure, the images are big and bright, but that comes with the territory. All big mirrors make big, bright images. Quality takes both effort and objectivity to assess. Have knowledgeable friends check the mirror. Ask them to be critical. If the mirror has a problem, discuss it objectively with the optical shop. Be sure that the vendor understands that you require a few weeks to assess the mirror's performance under a variety of conditions.

If the mirror is great, then rejoice. Notify the optical shop in writing that you are satisfied with the mirror. Then put the finishing touches on your telescope and start observing.

4.1 Basic Newtonian Optics

Before selecting a mirror, you need to know something about the optics, specifically about Newtonian optics. If you have already built one or more telescopes, you probably already understand the basics and might appreciate a chance to brush up. If you have never built a telescope before, then this section is mandatory reading.

The optics in a Newtonian telescope are very simple. The *parabolic primary mirror* sits at the bottom of the tube; the *flat secondary mirror* is held at a 45-degree angle near the top of the tube. Light from a distant planet, star, or galaxy crosses space, passes through the earth's atmosphere, and finally travels down the tube to the mirror. It reflects off the paraboloidal surface and heads back toward the front of the tube, now converging toward a focus. The secondary mirror intercepts the converging beam and reflects it to the side of the tube. It comes to focus and forms a *real image* just outside the tube.

If you hold a piece of paper at the focus, you will see a picture on the paper of the distant trees, or the moon, or a planet, or whatever happens to be in front of the telescope. Remove the paper to view the image with an *eyepiece*, and you'll see an enlarged view through the eyelens. The eyepiece is just a fancy magnifying glass that enables you to study the real image. The whole purpose of the telescope is to support two feather-light wisps of aluminum with the correct spacing and orientation to show you a picture of the universe.

The primary mirror of a Newtonian telescope is usually ground and polished from a disk of glass that has a low coefficient of thermal expansion, such as Corning's Pyrex glass. The polished glass surface is coated with a very thin but highly reflective layer of aluminum. Because water vapor and oxygen react with bare aluminum, it should be overcoated with a barrier of inert material such as silicon monoxide or magnesium fluoride that protects the metal and also enhances the reflectivity of the metallic surface. The secondary mirror is an elliptical slab similarly coated and protected.

A fresh coating of bare aluminum reflects about 88% of the incident light at the peak sensitivity of the human eye (around 550 nanometers wavelength). It will last about 10 years. A properly designed enhanced aluminum coating can increase reflectivity to 96% but will have a lifetime of about 5 years.

The job of the primary mirror is to bring the impinging starlight to a focus. To form a sharp image at the focus, an optical system must be constructed so that the length of the path that each ray travels from the star to the focus is the same. Some telescopes have multiple mirrors and lenses to achieve this goal. The strength of the Newtonian is that it employs only two optical elements: one to focus light and a second to reflect it to the eyepiece.

If the primary mirror has an accurate surface, then (apart from the atmo-

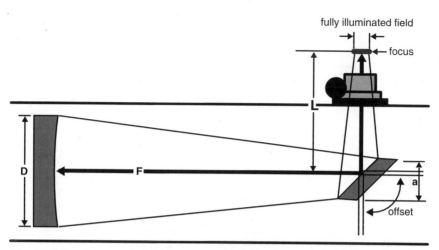

Fig. 4.1 *Dobsonian optics are simple: they consist of a primary mirror with a diameter D and a focal length F, and a flat secondary mirror with a minor axis a. The image is formed at the focus at a distance L from the optical axis.*

sphere) light itself is the limiting factor in the quality of the images your telescope can give. The reason is that light acts both like particles (photons) and waves. As a particle, we say that light consists of photons, and that each photon has a certain energy. As a wave, yellow-green light has a *wavelength*, or crest-to-trough-to-crest distance of 550 nanometers, or 22 millionths of an inch. The fact that every ray of light must travel the same distance to the focus ties in perfectly with the wave nature of light: it means that waves of light that leave the star in step and strike a paraboloid will arrive at the focus of the mirror in step. When they arrive in step, the waves add constructively to form a bright image of the star.

But nothing is simple with light waves. As you may have noticed at the beach, waves spread when they travel and bend around obstacles. Light does this too; the phenomenon is called *diffraction*. When the waves of light arrive perfectly in step at the focus, they add constructively and there is a spot of light. But immediately beside the focus, in the not-quite-localized way of waves, waves from the mirror arrive nearly in step, so beside the exact point of focus there is still some light. A tiny bit further away, fewer of the waves reinforce and the intensity of the light is lower. Beyond that, there's a point where the waves cancel and the light intensity drops to zero.

Thus, because of the wave nature of light, an image of a point source such as a star is not a perfect point, but a tiny circle called the *Airy disk* after Sir George Airy, the English Astronomer Royal who was the first to describe it mathematically. The Airy disk is surrounded by a system of faint *diffraction rings* that decline rapidly in brightness as you look further from the point of focus.

The effects of diffraction are small: in a typical Dobsonian, the Airy disk is only 0.0003 inches across. Details smaller than the Airy disk cannot be represent-

Table 4.1
Properties of Typical Dobsonian Telescopes

Optics	Lowest Magnification	Limiting Magnitude	Res	Galaxies	Weight	Height	Cost of Optics ($)
8 *f*/6	29	14.7	0.68	1,000	40	4	300
12.5 *f*/5	45	15.5	0.43	5,000	75	6	700
15 *f*/4.5	54	15.8	0.36	8,000	100	5.5	1,400
18 *f*/4.5	65	16.2	0.31	12,000	120	6.5	2,000
20 *f*/5	73	16.4	0.28	15,000	150	8	2,600
25 *f*/5	91	16.8	0.22	20,000	250	10	4,500
30 *f*/5	109	17.2	0.18	many	400	12	7,000
36 *f*/5	131	17.5	0.15	very many	575	15	12,000
40 *f*/5	145	17.7	0.14	way too many	800	16.5	16,000

Key To Headings: Optics: aperture in inches. **Lowest Mag:** lowest useful magnification (7 mm exit pupil). **Lim Mag:** limiting magnitude; magnitude of faintest star visible. **Res:** Resolving power in arcseconds under ideal conditions; the Rayleigh criterion. **Galaxies:** estimated number of deep-sky objects visible. **Weight:** total weight in pounds in a well-built Dobsonian. **Height:** eyepiece height in feet. **Cost:** for a good primary and secondary with standard aluminum coatings.

ed realistically in a telescope regardless of the magnification. Diffraction is thus a fundamental limitation in the performance of a telescope. It is a compliment to say that a mirror is *diffraction limited* because it means that the optics are so well made that only diffraction limits its performance.

We have seen what happens when waves from a perfect mirror add constructively to form an Airy disk. The result is a clean, hard, sharp Airy disk at the focus. But suppose a mirror has severe low and high areas so that half of the waves arrived at focus out of step with the other half, that is, so that some waves would arrive as crests and others would arrive as troughs. The waves would then cancel out and there would be no light at the focus of the mirror. The star will not disappear from view because the light will add constructively somewhere nearby, but the defective mirror would form a mushed-out smear of light instead of a clean Airy disk. Niceties such as the diffraction limit matter less with a mirror this terrible; all you can see are *figure defects*.

Mirrors seldom depart so dramatically from the ideal, but mirrors are never perfect—and they don't need to be perfect. The great English physicist John William Strutt, Lord Rayleigh, stated that to give "sensibly perfect" performance, the greatest departure from equal path length should not exceed one-fourth the wavelength of light, or in modern jargon, "no more than a ¼-wave peak-to-valley error on the wavefront." Judging a mirror by the peak-to-valley error can be misleading, however, because the peaks and valleys may occupy a tiny fraction or most of the mirror's area—the peak-to-valley criterion does not specify.

Today astronomers prefer to specify the area-weighted average departure from perfection, or the root-mean-square error, known as the r.m.s. error, to describe the quality of the images formed. When the wave front reflected from a mirror has a $1/14$-wave r.m.s. departure from perfection, the mirror is considered diffraction limited, or capable of giving "sensibly perfect" performance limited not by errors in its surface but by the wave nature of light itself.

Unfortunately, performing tests in the optical shop to determine the r.m.s. figure error of a mirror may cost several hundred dollars. Furthermore, opticians have very mixed feelings about such testing because after they have striven to make a mirror as perfect as they can, the test always reveals some level of residual error. Nonetheless, as the purchaser of the mirror, if you *must* know how good a mirror really is, *pay* for testing.

Later in this chapter we discuss how you can tell whether a mirror is diffraction limited, and if it is not, how you can diagnose—and even correct—some types of error. It is important that you understand that it is not easy to measure the quality of a mirror. Atmospheric seeing, air currents, collimation, and flexure all mimic imperfections in the optics. Only when you have tuned up the mechanics of your telescope and observed on nights with steady air can you truly judge the quality of the mirror itself.

4.2 Choosing the Mirror

There is no truly rational way to decide what optics you want for your telescope, although it is possible to describe factors that make up the rational part of the decision. The aperture, the focal ratio, the type and thickness of the glass, and the assurances of quality offered by the optical shop all play leading roles. But other factors—greed, envy, lust, and the size of your savings account—almost always play bit parts. Accept the fact that your decision will not be entirely rational—and do the best you can.

4.2.1 Choice 1: What Aperture?

Observationally, there's no place to hang your hat when it comes to choosing an aperture. The increase in observing capacity is seamless and smooth. Whatever aperture you chose, a little bigger aperture will do more. In the standard progression of sizes—3, 4¼, 6, 8, 10, 12½, 14¼, 16, 17½, 20, 25, 30, 32, 36, 40, . . .—there is no place where you suddenly reach nirvana with a mirror. When it comes to light gathering power and resolution, bigger is *always* better.

Therefore, your decision inevitably must be based on factors other than pure observing capacity. Size, weight, and price usually turn out to be the deciding factors. With a 16-inch telescope you can observe with your feet on the ground. With a 25-incher you take three quick steps up a short ladder. With a 40-inch one you must ascend a very tall ladder.

Table 4.2
Newtonian Optics Workshop

Parameter	What It Is	How To Find It
D	The aperture, or diameter of the primary mirror.	Measure directly.
F	The focal length of the primary mirror.	Look for the focal length etched or written on the back or side of the mirror, or measure optically.
f-number	The focal ratio of the primary mirror.	$f\text{-number} = \dfrac{F}{D}$
H	Height of eyepiece holder.	Measure from the inside of the focuser board to the top of the focuser, with the focus tube slightly racked out.
T	The inside diameter of the secondary cage (described in **Section 6.1.2**).	$T = D + \dfrac{F}{100}$
L	Folding distance of Newtonian—the distance from the center of the tube to the diverted focus.	$L = H + \dfrac{T}{2}$
d	Diameter of the fully illuminated field of view.	For 1.25-inch eyepieces, d = 12 mm. For 2-inch eyepieces, d = 20 mm.
a	Minor axis of the secondary mirror.	$a = d + \dfrac{(D-d)L}{F}$
O	Secondary mirror offset.	$O \approx \left[\dfrac{a(D-d)}{4F}\right]$
O'	Center dot offset.	$O' = \sqrt{2}\,O$

How to use this table: start with a mirror of known diameter and focal length, and then calculate each of the parameters in turn.

The weight of a telescope generally rises with the 2.5 power of the aperture, while cost rises with the cube of the aperture. Although the observational capacity of a telescope yields a fixed answer, weight and price do not. Your muscles are only so strong, and your bank account only so fat. Historical facts: George Ellery Hale dreamed of placing a 300-inch telescope atop Palomar Mountain. What he got was a 200-incher. NASA planned for a 120-inch Large Space Telescope. They ended up flying the 94-inch Hubble Space Telescope. It's okay to dream big, but think practical. In astronomy there is no substitute for aperture. So pick an aperture you can afford and will be comfortable using.

To help select the most appropriate scope for you, here are some scenarios built around telescopes of different sizes.

Apertures 8 to 10 inches: Just two short decades ago, telescopes in this size range this were considered large. Nothing has changed about the scopes, so what happened? Scopes in the 8- to 10-inch range can give you a lifetime of observing and not break the bank. In fact, one of the secrets among big Dob owners is an undeniable yearning to go back to the good old days when they owned nothing but a modest backyard telescope.

Take the planets. On nights when the seeing is poor, you still get a good view of Jupiter with an 8- or 10-inch telescope. If Saturn has a new white spot on the North Equatorial Belt, you'll see it, and during an apparition of Mars you can track the dust storms on its surface. Best of all, observing is hassle-free. You hear about something and pretty soon you're at the eyepiece.

A year ago last fall, just before school started, the 8-inch telescope accompanied you, your spouse, and three small children for five nights camping in the high Sierras. It was wonderful. Mercifully, the kids ran around all day like maniacs and conked out at dusk, leaving you a few hours for observing each evening. For two lovely nights you swept the Milky Way from Sagittarius to Cygnus and never ran out of things to see. When you felt tired, you popped a 28-gallon trashbag over the telescope and crawled into the sack. The last morning you woke before dawn to renew your acquaintance with the Orion Nebula. You found Messier 78 and glimpsed the Horsehead, too. All with an 8-inch telescope.

Apertures 12 to 16 inches: Everything about it says, "This is a serious telescope." You like the solidity of the mounting and the ease with which you can point to one object after another. It's the perfect size for quick, precise observing. Three years of Messier Marathons and you got it right—with the help of a crystal-clear sky, you bagged M-30 and got every object. You have confidence in this telescope—with your red flashlight and charts, you can find anything plotted in any star atlas. Sure, stuff looks bigger and brighter in those monster scopes, but you don't have to host a public star party whenever you show up at the astronomy club's observing site. Instead, you quietly knock off another couple dozen deep-sky objects and go home at midnight with no fuss and no muss.

Apertures 17½ to 22 inches: This is the life! About 15,000 galaxies to call your own and a telescope that fits in the back of a compact car. With enough space left over for a ham sandwich. Almost anyone can reach the eyepiece of the ubiquitous 17½-inch *f*/4.5 without straining, and the biggest telescopes in this class— the 22-inch *f*/5s—need only a short ladder. You can literally see everything in the New General Catalog. No place in the spring sky is free of pesky background galaxies.

What can't you see? The Messiers are too easy, and everything plotted in *Uranometria 2000.0* is visible, though some things look pretty faint. Nebulae? No problem, especially with the OIII filter that has proven to be such a fine investment; it helps every nebula even under the black skies of western Texas where you trailered for the Texas Star Party last spring. One of your observing buddies specializes in finding dark nebulae; the year before last at Riverside Telescope Make's Convention he turned up over 30 objects like drops of black ink against

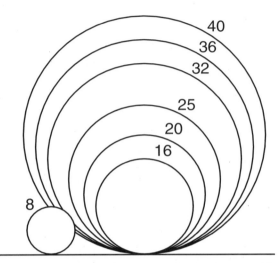

Fig. 4.2 *Think big—your quest for more light will lead you to larger and larger mirrors. Here, beside a "standard" 8-inch amateur aperture are the mirrors of Dobsonians up to 40 inches aperture, promising unparalleled views of faint deep-sky objects.*

the starclouds in Sagittarius and Scorpius.

The planets are no slackers either. You lucked out with a truly fine mirror that shows Cassini's Division crisp and clear whenever the air is steady. At a meeting in the midwest last summer someone disdainfully called your telescope a light bucket. "Sure it's a light bucket," you said, "but it's also capable of giving fine images." And you challenged him to a observing duel on the planets. Your telescope outresolved his hands down, and showed more satellites around Saturn, too. In terms of sheer return in observing pleasure for the dollar spent, you can't beat a scope in this size range.

Apertures 25 to 32 inches: Big but not crazy big. With your 25-inch scope, you can arrive at a site and be set up in 15 minutes. By yourself. Twenty minutes if people try to help you. Allow an hour for the scope to reach full thermal equilibrium and you're "go" for the night. You prefer observing far from civilization, by yourself, with your music. Among the stars you have many old friends, and you visit them during the dark of the moon.

There's method in your madness. Without quite telling anybody, you're searching for supernovae. You have a list of galaxies, old friends whose star patterns you have memorized, and each month as soon as the moon leaves the sky you check on as many as you can. If a supernova brighter than 16.5 magnitude occurs, you plan to be the first to see it.

Others seem to crave newness, always more new objects to observe, but you like the tranquillity of watching and waiting for an event that may happen or may never happen. It is a pleasure to swing the telescope, ascend the ladder, gaze through the eyepiece at each distant swirl of stars. You like the telescope. It is capable of showing you things that few other eyes have ever seen.

Apertures 36 to 40 inches: The telescope is a guaranteed hit at any star party. You roll into a dark-sky site towing a 12-foot trailer with the scope inside. You drop the back gate, attach the handle, release the hold-downs, set the ramp in place, and roll out the rocker. (Watch out, though, because at 600 pounds it won't roll on soft ground!)

You crank up the wheels, slide in 12-foot truss tubes, climb the ladder to attach the upper ring. With the truss assembled, you lift the secondary cage (35 pounds) and clamp it in place. Uncover the mirror, check the alignment, attach the finder, and you're in business. You can do everything yourself, but it's nice to have some helpers tonight. Twenty minutes is a great set up time for utterly fantastic observing—and some of the folks with SCTs are still polar-aligning long after you've started observing. Heck, your *finder* telescope is an 8-inch.

Your constant companion is a lightweight fiberglass ladder 12 feet tall. The sheer logistics of moving the ladder and telescope incline you to "work" a dozen objects from one *Uranometria 2000.0* chart before slewing to another part of the sky. The scope is great for three observers—you get lots of big-scope time and three brains is a lot better than one for locating new objects. You count a good four-hour observing session as more than adequate reward for the effort of setting up. You spent over $12,000 on the telescope and have another $2,500 in the trailer. Big deal. Money was made to be spent! Life was made to be lived!

4.2.2 Choice 2: What Focal Ratio?

You have made that big decision and decided what the aperture of the primary mirror will be. The next factor to consider is the focal ratio. Since most large-aperture commercial mirrors have focal ratios between *f*/4 and *f*/5.5, the decision is not driven by strong technical considerations. The decision depends on your personal preference for telescope size *versus* image quality. It depends on the type of observer you are, what you like to look at, and the skill of the optical shop.

Fast—f/4.5 and Below: Short-focus mirrors make for short, friendly telescopes. You might not need a ladder, and the wide field of view and lower magnification make objects easier to find. In really large apertures—30 inches and up—*f*/4.5 is the only realistic focal ratio. Longer focal lengths are too tall.

But watch out—coma can be a problem. An *f*/4 mirror has twice the coma of an *f*/5 mirror because the coma-free field is proportional to the cube of the focal ratio. For sharp images to the edge of the field, you'll need a coma corrector lens. Coma correctors generally decrease the field of view and put more glass between you and the universe—but they do get rid of the coma. Given the easy-to-reach eyepiece and the bright, wide field of view, we feel that *f*/4.5 and faster mirrors are ideal for large-aperture telescopes that will often be used at public star parties or for ladder phobics.

Normal—f/4.5 to f/5.5 Mirrors: If such a thing as a standard Dobsonian existed, it would have an *f*/5 mirror. Compared to faster mirrors, you get less coma and higher magnification. Collimation is easier, the central obstruction smaller, and you get better lunar and planetary viewing because the slower *f*/5 optic is easier for the optician to make than a faster one of the same aperture.

Against the pluses, moving from *f*/4 to *f*/5 could mean you'll need a ladder, or that it will take a few more steps up the ladder to reach the eyepiece. The field of view, though sharper, is also narrower, and the telescope is overall less suitable for public star parties. After weighing the pluses and the minuses, we feel that an *f*/5 mirror rates as the best focal ratio for a large-aperture personal telescope.

Slow—f/5.5 and Above: Somewhere around *f*/6, Newtonians become virtually free of coma, and opticians start to make serious promises about "better than ¼-wave" optics. Through a modern multi-element wide-field eyepiece, the "porthole effect" can be awesome. If you are a serious lunar or planetary observer, your chances of getting an absolutely top-notch diffraction-limited optic rise dramatically at *f*/6 and above.

The question to consider is how tall you want your telescope to be. Some people do not like ladders, so if you're one of those types, you'd better not be a planetary observer in quest of a large-aperture optic with a long focal ratio. Just to be sure, test yourself. Before you spend any money on a mirror, borrow a ladder like the one you will need for observing with the finished telescope. Set up the ladder outdoors at night and climb to the top. Climb down and the climb up again ten times. If it's too high for comfort, specify a faster optic.

Suppose that you climb that tall ladder and still get excited about owning the finest 1-meter *f*/6 telescope in your state—what then? Ask yourself, "Who else will use this telescope?" If you don't pay attention to this facet of observing you might find yourself offering fantastic high-resolution coma-free views of the universe, but no one dares to climb to an eyepiece 18 feet above the ground.

4.2.3 Choice 3: Make or Buy?

Mirror making has long been an integral part of amateur telescope making. In the beginning, amateur telescope making grew through the efforts of men like Russell Porter, a tool designer, and Albert "Unk" Ingalls, an editor at *Scientific American* during the 1930s, when money was tight. By the 50s, *Sky and Telescope* had taken over the role of promoting telescope making in America. The "Gleanings for ATM's" column, conducted by Earl B. Brown and later by Robert E. Cox, placed considerable emphasis on optics and mirror making.

During those decades, amateur telescope makers ground thousands of mirrors. There are some interesting reasons why so many amateurs learned this specialized skill. There were fewer optical shops back then, and before World War II, money was tight. Folks who wanted a telescope generally *had* to make their own mirror because there was no other option. Since it is fairly easy to grind and figure a small mirror with a long focal ratio, even the novice could grind, polish, and figure a 6-inch *f*/8 mirror on the first try.

Although generally not as good as a top-notch refractor, these home-built telescopes showed the moon and planets very well indeed, and they got people

Fig. 4.3 *They're not as light as they look! Dave Kriege holds two large mirror "dummies" of foamed styrene plastic. Anyone contemplating making a large Dobsonian needs to come to grips with the bulk and weight of a large telescope.*

started in astronomy. For the amateur optician, telescope making was the essence of astronomy. Shaping the precise optical surface of the primary mirror—with his or her own hands—was the intoxicating experience that made it all worthwhile.

However, we are talking about big telescopes, and it is difficult for a beginner to tackle a mirror over 8 inches aperture, let alone 16 inches. It takes time and talent, and the willingness to master a new set of skills. This usually means making a series of mirrors starting at 6 or 8 inches and working up in size to a good 12 inch. Even commercial optical shops sometimes have trouble figuring a big mirror, and they're pros. We recommend that unless you have made several good mirrors around 12 inches you are better off to buy a large mirror, or more precisely, to buy a *good large* mirror.

However, we have included **Appendix B**, on grinding, polishing, and figuring large telescope mirrors because we want you to know what's involved in making a mirror. If you decide to grind a mirror, more power to you. You'll probably get hooked, become a mirror maker, and hardly ever look through an eyepiece again. Don't say we didn't warn you.

4.2.4 Choice 4: What Glass?

The standard substrate for telescope mirrors is Pyrex glass. Pyrex is the brand-name for Corning's low-expansion borosilicate glass. This material has a coefficient of thermal expansion about ⅓ that of ordinary soda-lime window glass. Today others make glass that is equivalent to Pyrex. If an optical shop offers you a mirror made on a low-expansion borosilicate glass that you have never heard of, find out if it is equivalent to Pyrex. There is no reason to think that this material is inferior, and it may save you money.

The low expansion.property is important because a temperature difference of a fraction of a degree across a blank can bend it far more than a ¼-wave. Soda-lime glass's three times greater thermal expansion than Pyrex makes it all but impossible for an optician to figure in disks over 10 inches across. For mirrors over 30 inches across, it is extremely difficult to produce a truly good surface even on Pyrex.

Beyond Pyrex, there are a variety of zero-expansion glasses and glass-like ceramics formulated to have nearly zero thermal expansion. Germany, Japan, and Russia make these materials for export to the United States. These mirror materials are far easier for an optician to figure, and therefore greatly preferred in optical shops. Unfortunately, zero-expansion materials are often considerably more expensive than Pyrex. If you can afford a mirror made of fused quartz, Zerodur, *ULE*, or another zero-expansion material, by all means do so. But be prepared to pay somewhat more than you would for Pyrex.

4.2.5 Choice 5: Cut Sheet or Molded Blanks?

Cutting mirror blanks out of sheet Pyrex is cheaper than molding separate blanks, but molded blanks are better. The difference is the uniformity of the material in the glass disk.

To make a molded blank, red-hot Pyrex is poured into a mold. Because the molten material flows in all directions, the blanks are uniform in their mechanical properties. However, because the mold is not as hot as the molten glass, the glass initially cools with stresses locked in. To remove these stresses the glass is heated until it softens ever so slightly in a process called annealing. Molded blanks can be made in any thickness, but thin blanks are less expensive and lighter than thick blanks.

Sheet Pyrex gets rolled out like cookie dough from a house-sized tank at the glass factory, so it tends to have slightly different mechanical properties in length and width. Annealing does not remove these differences because they are not caused by frozen-in stress.

Opticians who work on sheet Pyrex sometimes complain about it, saying that the difference along the two axes can produce astigmatism that is time consuming (and therefore expensive) for them to remove. We have seen excellent mirrors made with sheet Pyrex, so we know it can be done.

The bottom line is this: if the optical shop does its job well, your mirror will be free of astigmatism. We have seen no evidence that mirrors made from sheet Pyrex are any worse or any better than mirrors made with molded Pyrex. If you think you will feel more comfortable with a molded blank, then select a firm that uses molded blanks for its mirrors.

4.2.6 Choice 6: Fine or Precision Annealed?

If glass cools quickly or unevenly when it is manufactured, stresses become locked in. Over many years, the glass deforms to relieve the stress, distorting any optical surface that has been ground into the surface.

Annealing is a heat treatment applied to telescope mirror blanks to remove internal stress. At elevated temperatures, the glass flows until the stresses are gone. After it has been annealed, the blank is stable and can hold an optical figure for centuries. Most mirror manufacturers fine-anneal their Pyrex glass before grinding. Some manufacturers "precision" anneal their mirror blanks. This means it stays in the annealing oven twice as long as the fine annealed blanks. The extra time gives the glass more time to release stresses. In theory, a precision annealed mirror should hold its figure better and longer.

Frankly, we don't know if precision annealing is any better than fine annealing for ordinary telescope mirrors. We have never seen convincing proof that any mirror has changed its figure in a telescope, even after many years. Other factors being equal, we would prefer a precision annealed mirror over a fine annealed mirror, but then again neither of us would lose sleep over it. The bottom line is that it's probably more important to get a mirror with a good figure and to mount it properly in a good mirror cell than it is to fret about the type of annealing.

4.2.7 Choice 7: How Thick?

Pyrex for large Dobsonian mirrors comes in difference thicknesses. In the world of large mirrors, "thin" means 2 inches thick and "ultra-thin" means 1½ inches thick. There are advantages and disadvantages to both. The 2-inch-thick mirrors are twice as stiff as the 1½-inch-thick mirrors. This is due to a law of physics: stiffness is proportional to the cube of the thickness. We recommend getting the 2-inch mirror. After all, flexure could mean bad images and good image quality is what it's all about. The extra weight in the bottom end will help to balance the scope anyway.

On the other hand, ultra-thin mirrors cool quicker to the temperature of the outside air. This means you get good images earlier than the guy with the 2-inch mirror. Ultra-thin mirrors can perform well, but to do so they must be mounted properly in a flotation cell. The 2-inch mirror is more forgiving because it is inherently twice as stiff.

Why is there such a limited range of mirror thicknesses? The main reason is availability of raw Pyrex glass. Pyrex comes in sheets 1⅛ and 1⅝ inches thick, and some shops can get blanks 2⅛ inches thick. Optical shops cut their mirror blanks from the glass sheet, and the extra ⅛ inch gets ground off. The amateur telescope market is much too small for the big glass factories to tool up to make thicker sheet Pyrex. Anything over 2⅛ inches thick has to be custom molded and cast at great expense.

Forget about full-thickness mirrors with the standard 6-to-1 diameter-to-thickness ratio. A full-thickness 25-inch mirror is over 4 inches thick and weighs 150 pounds. This book describes how to make a *portable* large-aperture telescope.

If you are lucky enough to have a permanent observatory for your big scope, then you may be interested in using a full-thickness primary. For large portable telescopes, full-thickness mirrors are not practical.

4.2.8 Choice 8: Blanchard Ground?

Specify that the back of the blank is to be Blanchard ground. A Blanchard grinder is a large machine for finishing optical surfaces. For ease in mounting, you want a mirror that has been ground flat on the back side. The flat back helps the flotation cell work correctly. A surface that is Blanchard-ground looks smooth and milky, with traces of a circular grinding pattern. If the job has been done right, the entire back surface is flat to better than 0.001 inch. That's what you want.

4.2.9 Choice 9: What Coatings?

All coatings fail. Some fail earlier than others. Humidity and air pollution are the enemies of optical coatings. Where you live, how much money you are willing to spend, and how much down-time you will put up with—these all influence your choice of mirror coatings.

The problem is that much of the United States has air that is harmful to optical coatings. If you live downwind from an urban area, air pollution will claim the coating in a couple of years. If you live along the seacoast, salt and water in the air will ruin it. If you live east of the Mississippi River, the high summertime humidity and acid dew will eat holes in the coatings. Even out west, parts of Arizona suffer from severe sulfur dioxide pollution.

Your basic options in coatings are: 1) standard overcoated aluminum that reflects 88% and lasts about ten years; 2) enhanced aluminum that reflects 96% and lasts five years; and 3) overcoated silver that reflects as much as 98% and is likely to last a year or less.

If you live a place with chemical-laden air, don't even consider enhanced coatings. Stick with standard silicon monoxide overcoated aluminum on the primary and the diagonal. To get a decade of service from a standard overcoated aluminum coating take good care of it and *never* let it get covered with acid dew.

If you live in a place with reasonably clean air, you may feel it is worthwhile to get enhanced aluminum coating on the secondary. You'll need to have the secondary recoated in five years, but you'll gain about 8% more light. The secondary mirror is a snap to ship and is cheap to recoat. It's a trade-off, and it's what we recommend.

4.2.10 Choice 10: Optical Quality

Everybody wants great optics, but nobody's willing to pay. Well, that's what most of the opticians who make telescope mirrors claim, and in our modern discount-oriented society, it's probably true. If you place two mirrors side by side, you can't tell the difference between an optically perfect mirror and a mirror that is terribly

Table 4.3
The Reflectivity of Typical Optical Coatings*

Wavelength (Angstroms)	Overcoated Aluminum	Enhanced Aluminum	All-Dielectric Coating
4000	84	93	39
4200	85	93	57
4400	86	94	80
4600	87	95	92
4800	88	96	96
5000	88	96	99
5200	88	96	99
5400	88	96	99
5600	89	96	98
5800	89	96	97
6000	89	96	97
6200	89	95	96
6400	88	95	84
6600	88	94	70
6800	87	93	50
7000	87	92	18
Reflectivity given as percentage at normal incidence.			

flawed. (Look at what happened with Hubble—not even NASA knew they had a problem until they tried it out on real stars!) For that matter, you cannot easily tell the difference between an "okay" mirror and a really great mirror without careful testing in the telescope on nights when the air is steady.

Over the short term, you could be very happy with a mirror that's no better than "okay" because there are not many nights when it makes a big difference. But over the long term, you'll learn to live for those nights when the air is perfect and a really great mirror gives transcendent images.

So—how can you obtain "a really great mirror?" The best bet is to place an order with a firm that has a good reputation and then hope. Before you order, discuss the impending mirror with the firm. They can tell you the aperture, focal length, thickness, what material it will be made from, and give you an estimated delivery date. When you mail your order, specify exactly what you expect to receive. If you do a thorough job, there should be little room left for misunderstanding—except for the most important thing of all: the mirror's optical quality.

Optical quality is hard to specify and difficult to measure. Long ago the optical industry and the military worked out procurement based on the customer performing specified acceptance tests that the product had to meet. You probably don't have equipment on hand to perform an acceptance test, and even if you did, you and the optical shop have no agreed-upon standard for mirror quality on which to base acceptance. Besides, without a mega-billion dollar procurement budget, to them you're just "a perfectionist nut" about optics.

What you would like to know is how the optical shop measures a mirror to determine when it is finished. Ideally, after the usual indoor tests, the optician would place each mirror in a telescope and test it on the stars. It's the best test because that's what the mirror will spend the rest of its life doing, and it's also how you and others will ultimately judge the mirror. But the simple fact is that optical shops don't have time to test mirrors this way, and steady nights that are needed to test truly fine mirrors are rare. In short, it's just not practical.

The best shop alternative is an interferometric test, because it is objective. The optician sets up the mirror in front of the test unit so that interference fringes cross the face of the mirror. With a video camera, the optician grabs a snapshot of the fringes. A standard software package analyzes the fringe pattern and prints the results. (It's like getting your auto's emission test—neither you nor the technician know how the test will turn out.) The printout shows a million-times exaggerated profile of the mirror, and the peak-to-valley and r.m.s. error are calculated and displayed. If it's better than $\frac{1}{4}$ wave peak-to-valley on the wave front and $\frac{1}{7}$ wave r.m.s. on the wavefront, you've got a gem.

Interferometric testing is a bit problematic for optical shops that make mirrors-for amateurs. Testers made by firms like Zygo are an enormous investment for a small company, and from the optician's point of view, an interferometer is a device that can change a perfectly fine $3,500 mirror into an unsalable piece of junk if the test shows it to be 0.26 waves peak-to-valley. Furthermore, the optician knows that his competitor claims—on rather shaky grounds—to be making "$\frac{1}{10}$-wave" optics. As far as small optical firms are concerned, interferometric testing is a fancy way to go out of business.

Next best is to select a shop that null tests its mirrors. There are lots of different null tests and there are plenty of ways to interpret and misinterpret null tests, so null testing *by itself* doesn't mean a whole lot. But if the firm has a good reputation and you have seen other good mirrors from them, then they're doing a good job and null testing is a good way to do the job.

If you talk to several optical firms, you will hear a lot of conflicting claims. In this respect, buying a mirror is no different than purchasing a car. Listen politely to all of it, and base your decision to some extent on how you think the people you talked to would treat you if, after you got your mirror, you called up with a quality problem. It's important that they be willing to listen to you, too.

What to do? Well, for starters make it clear to the optical shop that the first thing you will do when you receive the mirror is to star test it. Since some mirrors come out better than others, they might just send you the best one in the current batch. This alone might be the easiest way to get the best mirror they know how to make. Someone else will get the less-good mirror you might otherwise have gotten.

Next, if one firm can perform an interference test and another cannot, prefer the firm with the interference test *regardless* of their respective claims of surface accuracy. It's better to have a ¼-wave mirror that really is ¼-wave than to have a supposedly ¹⁄₁₀-wave mirror that's really ½-wave. Interferometers tell the truth.

The ultimate way to assure that you get the optical quality promised is to have an interferogram done by an independent testing company. Have the primary sent directly from the manufacturer to the testing company. If they are quality conscious, they will readily comply. Have the interferogram results sent to you and the manufacturer. If you are not happy with the test results, have them send the mirror back to the optics company for refiguring or a refund. You should pay for the interferogram and all shipping involved.

Finally, recognize that much as you want a truly great mirror, roughly 98% of the time you wouldn't be able to tell it from an "okay" mirror. Interference testing and placing stringent conditions on how good your mirror must be and what tests it must pass inevitably drive up the price. Your best bet may be to specify that the mirror be diffraction limited and stand aside to let an optician you trust do the job right.

4.3 Thinking About Eyepieces

Having chosen the primary mirror, you must next decide on a secondary mirror—and that forces you to consider the type of eyepieces you expect to use. The chain of logic that links the secondary mirror to the eyepieces begins with the size of the fully dilated pupil of a young human: 7 millimeters. To enter your eye, therefore, the beam of light from the eyepiece—called the "exit pupil"—must be the same size or smaller than the pupil of your eye. If the exit pupil is larger than this some light will be lost. As we become older, wiser, and more experienced observers, our pupils of don't dilate as fully and our night vision grows dimmer. (All the more incentive to build your telescope quickly!) By the time we are 50, our pupils may open no more than 5 millimeters. But it turns out that's acceptable. Skilled observers report that they can see the very faintest objects when the exit pupil of the telescope is 4 millimeters. You get a brighter image when the exit pupil is 7 millimeters, but the retina detects fainter objects when the image is bigger. An eyepiece that yields a 4 millimeter exit pupil gives a big, pleasing image—the kind of image that gets those oohs and ahhs—and the exit pupil fits nicely in the eye of a middle-aged observer. So that defines the target to aim for: a 4 millimeter exit pupil. You can easily calculate the exit pupil from any telescope and eyepiece: it equals the focal length of the eyepiece divided by the focal ratio of the telescope.

$$\text{exit pupil} = \frac{\text{focal length of eyepiece}}{\text{f-number of primary}}$$

With an *f*/5 mirror, an eyepiece with a focal length of 20 millimeters produces an exit pupil of 4 millimeters. So, to figure out how big the fully illuminated area must be, we just ask how big is the field of view of a 20 millimeter eyepiece. (Remember: we're designing our telescope around the optics, not the other way around. We first pick the primary, then select the eyepiece, and then select the secondary.) With some eyepieces it's easy to figure: you just measure the diameter of the field stop and that's the diameter of the field of view. With typical Erfle, König, and Wide Angle designs—eyepieces in which the focal plane lies outside the eyepiece—the field stop is about 20 millimeters across.

With fancy modern eyepieces, it's a little trickier because the focal plane is located inside the eyepiece, so we get the answer in a roundabout way by figuring out the magnification of a 20 mm eyepiece, getting the angular field of view, and then converting that to the actual linear diameter. With a 20-inch *f*/5 mirror, the focal length is 100 inches, or 2540 millimeters (1 inch equals 25.4 mm). Next, figure out the magnification:

$$\text{magnification} = \frac{\text{focal length of primary}}{\text{focal length of eyepiece}}.$$

You divide 2540 mm by 20 mm to get a magnification of 127x. Finally, figure out the field of view by dividing the apparent field of view (all manufacturers list the apparent field of view in their literature; it's kind of a boast) by the magnification:

$$\text{field of view} = \frac{\text{apparent field of view}}{\text{magnification}}.$$

Using a 20-millimeter Nagler eyepiece as an example, that's 80° apparent field of view divided by 127, or about 0.63°. Scaling from the ½° diameter size of the Moon, this eyepiece views a field 28 millimeters in diameter. If you are anxious not to lose a single photon, you may decide that you need a 28 mm diameter fully illuminated field of view to make full use of your best wide-angle Naglers. But a field of view that large means a big, heavy and expensive secondary mirror. You don't need a fully illuminated field anywhere near that large for visual astronomy. That's because the illumination does not drop to zero at the edge of the fully illuminated field of view. It drops to 99% and then 98%, and so on, quite gradually. The edge of the 28 millimeter field will be at least 80% illuminated, which is about the smallest light loss the eye can detect in the field of an eyepiece—and remember that the eyepiece we chose for this example has about as wide a field as anyone would ever use. For most observing with a large telescope, a fully illuminated field of view 20 mm across would be more than adequate. On a modest telescope in the 16-inches and under range of aperture, a fully illuminated field of view 12 mm across is entirely adequate. The bottom line, about which you can argue back and forth all day, is that the fully illuminated field of view needed for a big Dobsonian almost always works out to be between 12 millimeters and 20 millimeters across.

4.4 Thinking About the Focuser

Although it is not part of the optical system, the focuser plays a significant role because its height partially determines the spacing between the secondary mirror and the focal plane. The focuser plays a second significant role because its tube blocks stray light that might otherwise reach the focal plane. We suggest that you look for a focuser that stands between 2 and 3 inches high when it is fully racked in, the type often referred to as a "low-profile" focuser.

Any focuser is a compromise between too short and too tall. If you select a focuser that is very short, one of the "ultra-low-profile" units, then you can install a smaller secondary mirror. However, a very short tube may allow light to sneak over the top and bottom edges of the secondary cage and reach the eyepiece. If you select a focuser that is very tall, one of "standard height" units, you will need a slightly larger secondary mirror and the too-long tube can cut off light that should reach the edge of the field of view.

For most observers, a focuser between 2 and 3 inches tall is usually a good compromise. It means buying a slightly larger secondary mirror, but it also means that you will be less bothered by stray light. For additional stray light protection, you can install a baffle at the base of the focuser and another partway up the focuser tube, as described in **Section 12.6.2.**

4.5 Sizing the Secondary Mirror

The secondary mirror must fully illuminate the center of the focal plane, and should also fully illuminate the region around it. Here is the formula for the minor axis of the secondary mirror:

$$a = d + \frac{(D-d)L}{F}$$

where:

a = minor axis of secondary
d = fully illuminated field desired
D = diameter of the primary mirror
L= distance from secondary to focal plane
F = focal length of primary mirror.

You can see why it is important to consider what eyepieces you will use; the field of view impacts quite directly on the size of the secondary mirror. At this early stage of planning, you do not know L precisely, but you can get a good handle on it by calculating the inside diameter of the secondary cage, dividing it by two, and then adding 3 inches for the height of a typical, low-profile focuser.

$$T = D + \frac{F}{100}$$

T is the inside diameter of the secondary cage, D is the diameter of the primary mirror, and F is its focal length. Let's run through the numbers for a typical 20-inch $f/5$ telescope so you can see the entire process.

Step 1: Determine D and F. Since we're talking about a 20-inch $f/5$ mirror for our example, D is 20 inches and F is 100 inches. Be sure to check the actual focal length of your mirror when it arrives—either on the bill of sale or written or scratched on the back of the mirror. Even if the exact focal length of the mirror were 99.1 or 102.3 inches, it would still be called a 20-inch *f*/5 mirror.

Step 2: Determine *L*. This is the distance from the secondary mirror to the focal plane. The inside diameter of the secondary cage is 21 inches (see **Section 6.1.2**), so the distance from the secondary mirror to the inside of the tube is 10.5 inches. Add to that 0.25 inches for the thickness of the focuser board, 2.5 inches for the racked-in height of the focuser you wanted to buy, and 0.25 inches of focus travel out, for a total of 13.5 inches.

Step 3: Determine *d*. As an avid deep-sky observer, you crave a fully illuminated field 20 mm (0.78 inches) across to match the wonderful Nagler your understanding spouse gave you for Christmas. Even though you know that the useful field will be much larger than 20 mm, you can't bear the thought of losing a single photon.

Step 4: Plug in the numbers. Pull out your pocket calculator if your math has gotten rusty. Bear in mind that even if your calculator shows eight decimal places, the numbers you're putting into the equations are good to two decimal places at best.

$$a = 0.78 + \frac{(20 - 0.78)13.5}{100}$$

$$a = 3.37$$

Step 5: Order a secondary mirror. The standard stock sizes for secondary mirrors are: 2.6, 3.1, 3.5, 4.0, 4.5, and 5.0 inches. In this case, where you have calculated that you need a 3.37-inch secondary, the decision could go either to 3.1 inches or 3.5 inches.

How do you decide? Well, in this case you opted for a 20 mm diameter field size, which is actually quite large. You could go down to a 3.1-inch secondary without losing anything but a small amount of light at the edge of your lowest-power eyepieces. This decision makes a lot of sense when you compare the price of a 3.1-inch with price of a 3.5-inch secondary mirror. The difference is enough to buy another eyepiece.

We should make it clear that the decision is not always so easy because the trade-offs are complex. You can make the secondary smaller by specifying a short focuser, but you will also have to pay more attention to baffling stray light with a short focuser. Another thing to consider is that errors on diagonals are usually confined to the edge. If you plan to observe from dark-sky sites, stray light is usually not a major problem.

As you play with the equations, you find that you can squeeze down the size of the secondary by a variety of methods, but the whole point of building a big telescope is having wide, bright views of the universe. In general, we urge you to resist cheaping out with a secondary that is too small. Save the dollars elsewhere and design your telescope with optimum optics.

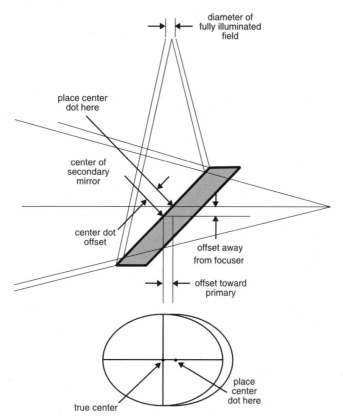

Fig. 4.4 *There are two distinct offsets: one away from the focuser and the other toward the primary mirror. The two offset distances are equal. The combined effect of these two offsets is that you should place the center dot at a distance of $\sqrt{2}$ times the offset from the true cen-*

4.6 To Offset or Not to Offset?

The secondary mirror in a Newtonian telescope should be offset slightly from the optical axis of the primary mirror. However, the offset distance is small and you can ignore it if you don't feel like bothering with it. Your telescope will still work well.

On the other hand, if you like to do things "by the book," you will want to offset the secondary. It's your choice. Mechanically speaking, it is simpler to ignore the offset and center the secondary, but optically speaking, the secondary offset ensures that the fully illuminated field of view is centered on the optical axis.

Consider the limiting case: a secondary mirror with a minor axis just big enough to reflect the cone of light coming from a single star to the eyepiece. If you place the secondary exactly on the optical axis, the part of the secondary that is closer to the eyepiece will intercept the converging cone closer to focus, where the

Table 4.4a
Secondary Mirror Size and Offset

Diameter of 100% illumination: 0.5 inch

Height of focuser: 2 inches

Primary	Minor Axis of Secondary Mirror in Inches				
	f/4.0	*f*/4.5	*f*/5.0	*f*/5.5	*f*/6.0
14.5	2.863	2.608	2.404	2.238	2.099
15.0	2.929	2.667	2.457	2.286	2.143
16.0	3.060	2.784	2.563	2.383	2.232
17.5	3.256	2.960	2.722	2.528	2.366
18.0	3.322	3.018	2.775	2.576	2.410
20.0	3.583	3.252	2.986	2.769	2.588
22.0	3.845	3.485	3.197	2.962	2.766
24.0	4.106	3.718	3.408	3.154	2.943
25.0	4.236	3.835	3.513	3.251	3.032
30.0	4.888	4.417	4.040	3.732	3.475
32.0	5.149	4.650	4.250	3.924	3.652
36.0	5.670	5.115	4.671	4.308	4.006
40.0	6.190	5.580	5.092	4.692	4.359

Primary	Secondary Mirror Offset in Inches				
	f/4.0	*f*/4.5	*f*/5.0	*f*/5.5	*f*/6.0
14.5	0.173	0.140	0.116	0.098	0.084
15.0	0.177	0.143	0.119	0.100	0.086
16.0	0.185	0.150	0.124	0.105	0.090
17.5	0.198	0.160	0.132	0.112	0.096
18.0	0.202	0.163	0.135	0.114	0.098
20.0	0.218	0.176	0.146	0.123	0.105
22.0	0.235	0.189	0.156	0.132	0.113
24.0	0.251	0.202	0.167	0.140	0.120
25.0	0.259	0.209	0.172	0.145	0.124
30.0	0.300	0.241	0.199	0.167	0.142
32.0	0.317	0.254	0.209	0.176	0.150
36.0	0.349	0.280	0.230	0.193	0.165
40.0	0.382	0.306	0.251	0.211	0.179

Table 4.4b
Secondary Mirror Size and Offset

Diameter of 100% illumination: 0.5 inch

Height of focuser: 3 inches

Primary	Minor Axis of Secondary Mirror in Inches				
	f/4.0	*f*/4.5	*f*/5.0	*f*/5.5	*f*/6.0
14.5	3.104	2.823	2.598	2.413	2.260
15.0	3.170	2.882	2.651	2.462	2.304
16.0	3.302	2.999	2.757	2.559	2.394
17.5	3.499	3.175	2.916	2.704	2.528
18.0	3.565	3.234	2.969	2.753	· 2.572
20.0	3.827	3.468	3.181	2.946	2.751
22.0	4.089	3.702	3.393	3.140	2.929
24.0	4.351	3.936	3.604	3.332	3.106
25.0	4.481	4.052	3.710	3.429	3.195
30.0	5.134	4.635	4.237	3.910	3.638
32.0	5.395	4.868	4.447	4.103	3.816
36.0	5.916	5.334	4.868	4.487	4.170
40.0	6.437	5.800	5.289	4.872	4.524

Primary	Secondary Mirror Offset in Inches				
	f/4.0	*f*/4.5	*f*/5.0	*f*/5.5	*f*/6.0
14.5	0.187	0.151	0.125	0.106	0.091
15.0	0.192	0.155	0.128	0.108	0.093
16.0	0.200	0.161	0.134	0.113	0.097
17.5	0.212	0.171	0.142	0.119	0.102
18.0	0.217	0.175	0.144	0.122	0.104
20.0	0.233	0.188	0.155	0.131	0.112
22.0	0.250	0.201	0.166	0.139	0.119
24.0	0.266	0.214	0.176	0.148	0.127
25.0	0.274	0.221	0.182	0.153	0.130
30.0	0.316	0.253	0.208	0.175	0.149
32.0	0.332	0.266	0.219	0.184	0.157
36.0	0.365	0.292	0.240	0.201	0.171
40.0	0.397	0.318	0.261	0.219	0.186

Table 4.4c
Secondary Mirror Size and Offset

Diameter of 100% illumination: 0.5 inch

Height of focuser: 4 inch

Primary	Minor-Axis of Secondary Mirror in Inches				
	f/4.0	*f*/4.5	*f*/5.0	*f*/5.5	*f*/6.0
14.5	3.346	3.037	2.791	2.589	3.346
15.0	3.412	3.097	2.844	2.638	3.412
16.0	3.544	3.215	2.951	2.735	3.544
17.5	3.742	3.391	3.111	2.881	3.742
18.0	3.808	3.450	3.164	2.930	3.808
20.0	4.071	3.685	3.376	3.124	4.071
22.0	4.333	3.919	3.588	3.317	4.333
24.0	4.595	4.153	3.800	3.510	4.595
25.0	4.726	4.270	3.905	3.607	4.726
30.0	5.380	4.854	4.433	4.089	5.380
32.0	5.641	5.087	4.644	4.282	5.641
36.0	6.163	5.553	5.066	4.667	6.163
40.0	6.684	6.019	5.487	5.051	6.684

Primary	Secondary Mirror Offset in Inches				
	f/4.0	*f*/4.5	*f*/5.0	*f*/5.5	*f*/6.0
14.5	0.202	0.163	0.135	0.114	0.097
15.0	0.206	0.166	0.137	0.116	0.099
16.0	0.215	0.173	0.143	0.120	0.103
17.5	0.227	0.183	0.151	0.127	0.109
18.0	0.231	0.186	0.154	0.129	0.111
20.0	0.248	0.200	0.165	0.138	0.118
22.0	0.265	0.213	0.175	0.147	0.126
24.0	0.281	0.226	0.186	0.156	0.133
25.0	0.289	0.232	0.191	0.161	0.137
30.0	0.331	0.265	0.218	0.183	0.156
32.0	0.347	0.278	0.229	0.192	0.163
36.0	0.380	0.304	0.250	0.209	0.178
40.0	0.413	0.330	0.271	0.227	0.193

Table 4.4d
Secondary Mirror Size and Offset

Diameter of 100% illumination: 0.78 inch

Height of focuser: 2 inches

Primary	Minor Axis of Secondary Mirror in Inches				
	f/4.0	*f*/4.5	*f*/5.0	*f*/5.5	*f*/6.0
14.5	3.096	2.846	2.646	2.483	2.347
15.0	3.162	2.905	2.700	2.532	2.392
16.0	3.294	3.023	2.806	2.629	2.481
17.5	3.491	3.199	2.966	2.774	2.615
18.0	3.557	3.258	3.019	2.823	2.660
20.0	3.819	3.492	3.231	3.017	2.838
22.0	4.081	3.726	3.442	3.210	3.016
24.0	4.343	3.960	3.653	3.403	3.194
25.0	4.474	4.077	3.759	3.499	3.283
30.0	5.126	4.660	4.286	3.981	3.726
32.0	5.387	4.893	4.497	4.173	3.904
36.0	5.909	5.359	4.918	4.558	4.258
40.0	6.430	5.824	5.339	4.943	4.612

Primary	Secondary Mirror Offset in Inches				
	f/4.0	*f*/4.5	*f*/5.0	*f*/5.5	*f*/6.0
14.5	0.183	0.150	0.125	0.107	0.093
15.0	0.187	0.153	0.128	0.109	0.094
16.0	0.196	0.160	0.133	0.114	0.098
17.5	0.208	0.170	0.142	0.120	0.104
18.0	0.213	0.173	0.144	0.123	0.106
20.0	0.229	0.186	0.155	0.132	0.114
22.0	0.246	0.200	0.166	0.141	0.121
24.0	0.263	0.213	0.177	0.150	0.129
25.0	0.271	0.219	0.182	0.154	0.133
30.0	0.312	0.252	0.209	0.176	0.151
32.0	0.329	0.265	0.219	0.185	0.159
36.0	0.361	0.291	0.241	0.203	0.174
40.0	0.394	0.317	0.262	0.220	0.188

Table 4.4e
Secondary Mirror Size and Offset

Diameter of 100% illumination: 0.78 inch

Height of focuser: 3 inches

	Minor Axis of Secondary Mirror in Inches				
Primary	*f*/4.0	*f*/4.5	*f*/5.0	*f*/5.5	*f*/6.0
14.5	3.332	3.056	2.836	2.655	2.504
15.0	3.399	3.116	2.889	2.704	2.550
16.0	3.531	3.234	2.996	2.802	2.640
17.5	3.730	3.411	3.157	2.948	2.774
18.0	3.796	3.470	3.210	2.997	2.819
20.0	4.059	3.706	3.423	3.191	2.998
22.0	4.322	3.940	3.635	3.385	3.177
24.0	4.585	4.175	3.847	3.579	3.355
25.0	4.716	4.292	3.953	3.675	3.444
30.0	5.370	4.876	4.481	4.158	3.889
32.0	5.631	5.110	4.692	4.351	4.066
36.0	6.153	5.576	5.114	4.736	4.421
40.0	6.675	6.042	5.535	5.121	4.776

	Secondary Mirror Offset in Inches				
Primary	*f*/4.0	*f*/4.5	*f*/5.0	*f*/5.5	*f*/6.0
14.5	0.197	0.161	0.134	0.114	0.099
15.0	0.201	0.164	0.137	0.117	0.101
16.0	0.210	0.171	0.143	0.121	0.105
17.5	0.223	0.181	0.151	0.128	0.110
18.0	0.227	0.184	0.154	0.130	0.112
20.0	0.244	0.198	0.164	0.139	0.120
22.0	0.261	0.211	0.175	0.148	0.128
24.0	0.277	0.224	0.186	0.157	0.135
25.0	0.286	0.231	0.191	0.162	0.139
30.0	0.327	0.264	0.218	0.184	0.158
32.0	0.343	0.277	0.229	0.193	0.165
36.0	0.376	0.303	0.250	0.211	0.180
40.0	0.409	0.329	0.271	0.228	0.195

Table 4.4f
Secondary Mirror Size and Offset

Diameter of 100% illumination: 0.78 inch

Height of focuser: 4 inches

Primary	Minor Axis of Secondary Mirror in Inches				
	f/4.0	*f*/4.5	*f*/5.0	*f*/5.5	*f*/6.0
14.5	3.569	3.267	3.025	2.827	2.662
15.0	3.636	3.326	3.079	2.876	2.708
16.0	3.769	3.446	3.187	2.975	2.798
17.5	3.969	3.624	3.348	3.122	2.934
18.0	4.035	3.683	3.401	3.171	2.979
20.0	4.300	3.919	3.615	3.366	3.158
22.0	4.563	4.155	3.828	3.561	3.338
24.0	4.827	4.390	4.040	3.755	3.516
25.0	4.958	4.507	4.147	3.852	3.606
30.0	5.613	5.093	4.676	4.335	4.051
32.0	5.875	5.326	4.887	4.528	4.229
36.0	6.398	5.793	5.310	4.914	4.584
40.0	6.920	6.260	5.732	5.299	4.939

Primary	Secondary Mirror Offset in Inches				
	f/4.0	*f*/4.5	*f*/5.0	*f*/5.5	*f*/6.0
14.5	0.211	0.172	0.143	0.122	0.105
15.0	0.215	0.175	0.146	0.124	0.107
16.0	0.224	0.182	0.152	0.129	0.111
17.5	0.237	0.192	0.160	0.136	0.117
18.0	0.241	0.196	0.163	0.138	0.119
20.0	0.258	0.209	0.174	0.147	0.126
22.0	0.275	0.223	0.185	0.156	0.134
24.0	0.292	0.236	0.195	0.165	0.142
25.0	0.300	0.243	0.201	0.170	0.146
30.0	0.342	0.276	0.228	0.192	0.164
32.0	0.358	0.289	0.238	0.201	0.172
36.0	0.391	0.315	0.260	0.219	0.187
40.0	0.424	0.341	0.281	0.236	0.202

cone is smaller. The outermost ray of light will strike the mirror somewhat inside its rim. The part of the secondary that is farther from the eyepiece intercepts the cone of light farther from focus, where the converging light cone is larger—in fact, enough larger that light from the outer edge of the mirror misses the edge of the secondary entirely. In this case, failing to offset the secondary causes some light to miss the reflective surface of the diagonal.

To reflect the cone from the entire mirror surface to the eyepiece, you can simply "slip" the mirror in its own plane away from the eyepiece and downward to the primary mirror by a small amount. The far side of the secondary now catches all the light from the outer edge of the light cone, and there is no "wasted" part of the secondary on the side closer to the eyepiece.

Now consider a telescope with a generously sized secondary mirror, big enough to fully illuminate a 20 mm (0.78-inch) diameter circle. If you mount the secondary mirror in the exact center of the tube, what happens? On the optical axis and for some distance around it, the focal plane is fully illuminated.

However, because the secondary mirror is centered, the rays of light on the eyepiece side of the secondary strike the mirror when they are closer to the optical axis than do the rays on the opposite side of the secndary mirror. As a result, when the secondary mirror is centered, the center of the fully illuminated region at focus is offset from the optical axis.

For a typical 20-inch *f*/5 optical system with a 20mm fully illuminated field of view, the offset is about ⅙ of an inch—which seems like a pretty small distance, so why bother? The answer is that unless you offset the secondary mirror, the center 20mm fully illuminated field of view will be 4 millimeters off the optical axis. By offsetting the secondary mirror, you are centering the fully illuminated field of view on the optical axis.

4.7 Testing Telescope Optics

Once you have installed a mirror in the telescope, you can do the poor man's optical test, the "star test." It is also the best test. Star testing gives you a wealth of qualitative information about the surface of a mirror, which is exactly what you want. You can determine if the figure is smooth, whether astigmatism is present, and whether it is under- or over-corrected. These qualities matter when you observe, not some quantitative fraction-of-a-wave rating.

To perform a reliable star test, you need a few nights of good seeing. The mirror must be given time to cool so it is the same temperature as the air. The secondary mirror must be of good quality and mounted properly in its cell. The telescope must be properly collimated. The mirror must be resting comfortably on the sling support and against the flotation pads of its cell.

Fortunately, the star test itself will tell you if your telescope does not meet these conditions. If the seeing is poor, you see it in the star image. If the mirror is warm, or stressed, or hanging wrong in the sling, you see the effects. In fact, anything and everything that degrades optical performance shows up in the star test. The

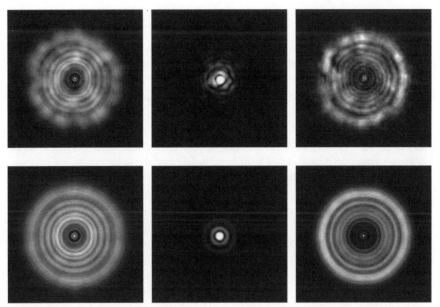

Fig. 4.5 *This large mirror has an error mild enough to make it difficult to see most nights. Experienced star testers might peer through the turbulence on the top row long enough to give them an idea of the error, but most people will have to wait for the tranquil conditions of the bottom row to allow them to see that this is a slightly overcorrected mirror. Computer simulation by Harold R. Suiter.*

key is knowing how to interpret what you see.

Luckily, a very thorough and comprehensive book, *Star Testing Astronomical Telescopes*, describes the procedure in detail and suggests how to interpret what you see. The author, Harold R. Suiter, filled the book with hundreds of realistic computer-generated star images showing just about everything that can affect optical performance. The book is exceptionally valuable because you learn to recognize many problems—poor seeing, sling stress, tube currents, and poor collimation—that you can easily correct. The star test also allows you to evaluate the optical quality of your mirror.

Imagine that you are out with your telescope. The night is dark and crisp and the stars are twinkling. Great night—but *not* for star testing. Big reflectors are extremely sensitive to seeing conditions. Twinkling is a sign of turbulent air and poor seeing. With a good 12mm eyepiece, examine a moderately bright star (Polaris is good because it moves very little). In focus you see a hopeless mess. Rack outside of focus and you'll see an out-of-focus doughnut crossed by a rapid flow of what?—air cells! You're looking directly at silhouettes of turbulent parcels of air a few miles above your telescope. The air parcels act like lenses to break up and distort your star image. If the seeing is poor, postpone testing for a better night. Instead, put in a low-power eyepiece and spend the evening searching out faint nebulae.

Another night, a cool night following a hot day. The stars are rock steady. You have just set up your telescope but the star images are squirming fuzz balls. What gives?—the seeing ought to be great. Rack the image out of focus and you see the out-of-focus star filled with moving blobs. You're now looking through tube currents, warm parcels of air rising from your still-warm mirror. Relax—the mirror will cool and you may yet enjoy a fine night of observing.

On a night when the seeing is good and the telescope is in equilibrium with the air, put in an 8mm eyepiece and point your scope at a moderately bright star. Slowly rack the eyepiece in and out so you see the star image expanded to a doughnut. Examine it on both sides of focus. You will certainly see some stuff flowing, and some slowly moving blobs. If you put your hand in front of the tube, you'll see it in silhouette with plumes of warm air rising from it. Hold a popsicle in front and you'll see descending plumes of cold air.

Mentally subtract the air currents. Concentrate on the what the starlight does as it goes through focus. The doughnut contracts to a tiny point and expands again. The doughnut has a dark center, the silhouette of the secondary mirror, and rings like a tree. The further you go from focus, the more rings you see. The rings are not terrifically apparent, but they are definite. On the opposite side of focus, the doughnut patterns look identical. At focus the image almost seems to snap together as a tiny disk and faint diffraction rings that form and break apart as little waves of air turbulence pass. This is what you want—a night of nearly great seeing and a flawless mirror.

Be sure to look at the in-focus images as well. Consider how well the mirror has performed since you got it. A bad mirror never gets better, but a good mirror can be hounded by optical difficulties that have nothing to do with its shape. If you've witnessed instants of breathtaking performance between hours of so-so imaging of the planets, you have the seeing problems that plague big mirrors. But your mirror is probably good. If a planet seems to have a preferred "corner" of the eyepiece, perhaps you are partially counterbalancing a misalignment aberration with an off-axis eyepiece aberration. You'll have to realign before you can judge the mirror.

If you feel that there is a problem with the telescope, but are not sure what, think hard about what you see. People tend to blame the first thing they think of and the first thing they think of is the main mirror. If the images appears washed out by light, maybe some baffling improvements would help. It it unlikely that anything as large in scale as the mirror's figure is contributing to a stray light problem, although polish quality could be a factor.

If you are diagnosing some complicated optical problem that appears different depending on where you have the image centered, think about your eyepieces. Are you still using those Modified Achromatics you inherited from your 6-inch *f*/8? An *f*/4 telescope demands premium-quality oculars. The light cone is too steep to be handled by less complicated eyepieces. A Huygens or Ramsden eyepiece will adequately serve in a *f*/15 system. Kellner or Modified Achromatics will work in instruments down to about *f*/6. For light cones as fast as *f*/4, only recent, advanced designs will work and you're compromising even then. Replace your eyepieces from the bottom up, high-powers and Barlows first. For now, borrow a quality eyepiece.

Practice star testing. When you attend a star party look through as many telescopes as you can. You'll feel sneaky pretending to "ooh" and "ahh" at the Orion Nebula when, in fact, you are actually star testing. In one telescope, the central spot is off center and you see coma when you focus on the star. "This guy needs to collimate!" you think, wondering why he seems oblivious to such an easily corrected problem.

Through another telescope, the images look mushy at focus. The inside edge of the doughnut is bright on one side of focus and the outside is bright on the other. Racking out further, it is hard to miss the uneven appearance of the doughnut. "Spherical aberration," you think, "and zones . . . Big contrast killers. You can bet this guy doesn't put much stock in planetary observing." As you step away, the whine of motors tells you he's already off to another object.

Later you see a telescope that shows a thin bright ring on the doughnut outside focus and a faint spray of light inside focus, a classic case of turned-down edge. The image looks good at focus. You ask politely if he made the mirror himself, knowing he will proudly answer, "Yes." You appreciate the work he has put into his big mirror. You can't feel superior—he's made a good mirror with a minor problem. Would that you could do so well yourself.

When you start to star test, it's important to remember that the test is very sensitive. With it you can detect minor problems that would hardly ever bother a serious observer. But you will also see major problems—and learn that many people seem to look right past these problems as if they didn't exist. Well, if those people are happy looking through telescopes like that, let them be happy. For yourself, though, you want a great mirror housed in an easy-to-use telescope. You are willing to maintain your telescope in tip-top condition. And the star test is just one more tool to insure that you will see all the universe has to offer.

4.8 What To Do If You Don't Like a Mirror

First of all, don't freak out. The stars will be there tomorrow and forever. You have a lifetime to observe so don't get upset over a "bad" optic. Unless this is your first attempt at making a telescope, you already know that a telescope is never finished. If the mirror needs to be refigured, so be it.

But—before you make that bad-news phone call to the mirror company, be sure you have done your homework. The most frequent complaints about big mirrors are, "I get fuzzball images," and, "The stars are not pinpoints." Those aren't mirror complaints, they're seeing complaints. Don't look one time and complain; test the mirror properly. Collimate the telescope. Make sure the mirror is properly seated on the pads, hanging in the sling, and that it has reached thermal equilibrium. Test when the seeing is good. Ask some knowledgeable friends for their assessment. Being less involved, they can provide a more objective verdict.

If the problem is the mirror and it's really obvious, you're in luck. Chances are the mirror makers will agree with you and fix it. If the problem is hard to see and you have trouble convincing yourself, you'll have trouble convincing the people at the optical shop, too. Make careful notes, and document your test procedures.

When you are certain there is a problem, keep a level head. You'll get faster results and a better job if you work *with* the guys at the optical shop and not against them. Belligerent customers never get as good a product or service as friendly customers. On the other hand, be firm. The mirror has a problem and it won't go away by itself.

The people at the shop will almost certainly ask how you tested the mirror, how the mirror cell is constructed, and whether the mirror is collimated. You may feel that they are questioning your integrity, but they are not. Put yourself in their position: it will cost you a lot of money to ship the mirror back and take them a lot of time to retest the mirror, so they need to be convinced there really is a problem.

This is only fair: they have probably heard lots and lots of unjustified complaints and they want to make sure that you know what you are talking about. Explain what you have done and what you have observed. If they have suggestions for you to try, agree to carry them out. A little cooperation and respect will go a long way.

Most commercial optical companies want you to be satisfied, especially those that make big mirrors. They know their livelihood depends almost entirely on their reputation. Big-aperture scopes get a lot of attention and an unhappy customer is their worst nightmare. Once they are convinced that you have done your testing properly, most shops will gladly look at your mirror again and, if necessary, correct the problem. However, you should pay the shipping charges.

Opticians are only human. Parabolizing a big mirror is an art. It depends on the skill and attitude of the optician. Everybody has a bad day now and then. Maybe they could have done a better job, but there's no need to ram it down their throats. If you act patient and understanding, you have a right to expect prompt attention, a better surface on the mirror, and a new coating.

We have seen many commercial telescope mirrors, and many of them delivered good or very good images. Most of those that were bad were made years and years ago, before opticians had much experience with making big, thin mirrors. Today, if you order a large mirror from a reputable company you can be reasonably sure that you will receive a fine optic that will give you years and years of great observing.

Chapter 5
The Primary Mirror Cell

You paid good money to get a mirror of the highest possible quality, and you expect great performance. But optical quality means nothing if you don't properly support a thin mirror. With a well made mirror cell, a thin mirror is capable of delivering the same optical performance as a far more expensive full-thickness mirror. It's up to you to do the job right by mounting your mirror in a cell that provides the necessary support. This chapter tells you how.

Without support, all mirrors bend under their own weight. The amount of flexure is small, often just a few wavelengths of light, but a telescope mirror must hold its shape to within ¼-wavelength of light—or about five millionths of an inch. The classic solution is to make the mirror so thick that it bends less than that tiny amount—hence the traditional recommendation that you get a "full-thickness" mirror with a 6-to-1 ratio of diameter to thickness. The idea behind the 6-to-1 ratio is to make the mirror so thick that it won't sag under its own weight. This works fairly well for mirrors up to 12 inches or so, mirrors for telescopes that were considered large a few decades ago. Larger mirrors cannot be made thick enough to perform well without some kind of support. Big mirrors, both thin and thick, are always flexible; the only question is *how* flexible.

Dobsonian builders prefer thin mirrors. They weigh less and cool quicker, so the telescope is more portable and it performs well soon after dark. When you decide to buy a thin mirror, however, you are committed to constructing a mirror cell to provide proper support. Two decades of Dobsonians have proven that properly supported thin mirrors match or exceed the performance of full-thickness mirrors in big, portable telescopes.

Opticians favor full-thickness mirrors. This is entirely understandable because thick mirrors are easier to work on. They bend less under the pressure of polishing tools and on the test stand they flex less under their own weight, so it's easier for the optician to figure and test a thick mirror. A bonus for the optician is that full-thickness mirrors fetch a higher price.

As a telescope builder, you may be reluctant to part with several thousand dollars more for a mirror that seems to take all night to cool. If you had the extra dollars to spend, you would rather order an even bigger thin mirror!

So the logic is inexorable. Telescope makers know that opticians can make

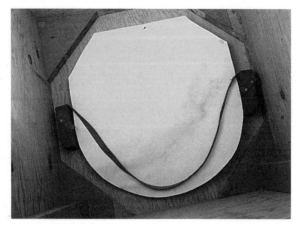

Fig 5.1 *Classical Dobsonians had an elegant sling-and-flotation system behind their port-hole-glass mirrors. The view shows the cell for a 24-inch mirror with the sling hanging loose ready to receive the mirror. The back of the mirror rests on an 18-point flotation mechanism hidden behind the paper.*

big, thin mirrors that perform extremely well, and we will show you that first-rate mirror cells are not hard to build. The question is not "Should I get a thin mirror?" but "How can I build a mirror cell that will help my mirror deliver all the optical performance it can?" That's what this chapter is all about.

5.1 The Flotation System and the Sling

Everybody loves water beds. When you lie on a water bed, you're floating. Every part of your body is supported uniformly from below. Telescope makers support mirrors the same way water beds support the human body, with a mechanical *flotation system*. The flotation system consists of levers and triangles that distribute an equal fraction of the mirror's weight to each of a large number of pads. The levers and triangles ultimately rest on three large bolts attached to the tailgate of the telescope. Even though the flotation system is made of hard steel or aluminum parts, it supports every part of the mirror equally, just as the water bed supports you.

The flotation system is only half of the mirror cell. The other half is the sling. When the telescope is pointing straight up, the mirror rests on its back on the flotation system. When the telescope points at the horizon, however, the mirror is standing on edge and the sling supports it around the lower edge. The sling acts like a hammock, and what is more comfortable than a hammock on a hot summer day?

The sling is a flexible band of metal or fabric around the bottom rim of the mirror. The sling presses up on each section of the mirror's lower rim exactly as much as the mirror's weight presses down on the sling. The mirror cell thus provides two kinds of support: flotation supporting the mirror from the back, and a

Fig. 5.2 *Folding back the sheet of paper reveals the bars and triangles of a simple yet effective 18-point flotation support made of plywood. The large triangles are stapled to the triangular sheet of cardboard behind them. In this design, however, there is no way for air to circulate around the mirror.*

sling supporting the mirror around the rim.

As the telescope is pointed higher or lower in the sky, the proportion of the mirror's weight carried by the sling and flotation system changes. When the telescope is pointed at the zenith, the mirror's entire weight rests on the flotation system. Point the telescope at the horizon and the entire weight of the mirror hangs in the sling. Between the zenith and horizon, the flotation system and sling share the job of supporting the mirror.

5.1.1 Flotation in Theory

Small, thick mirrors such as the classic 1-inch-thick 6-inch *f*/8 mirror flex so little under their own weight that most amateur telescope makers mount them in a cell with three support pads. Three is a "magic number" because three points define a plane, and thus the orientation of the mirror. To adjust the tilt and tip of the mirror in the telescope, you adjust the three support points behind the mirror.

Recall from **Chapter 3** that bending in a simple beam of constant thickness depends on the cube of the length of the span, that is:

$$\text{Telescope mirror deformation is proportional to } \frac{\text{Force} \times \text{Length}^3}{\text{Thickness}^4}.$$

Suppose that instead of a 1-inch-thick mirror of 6 inches aperture, you place a 1-inch-thick mirror of 12 inches aperture on three points. What happens? The length of the span doubles and the force on the span (that is, the weight of the mirror) quadruples, so the 12-inch mirror shows *32 times as much flexure* as the 6-inch did. It's quite likely that so flexible a mirror will deform more than a ¼-wave, producing images that are less than fully pleasing.

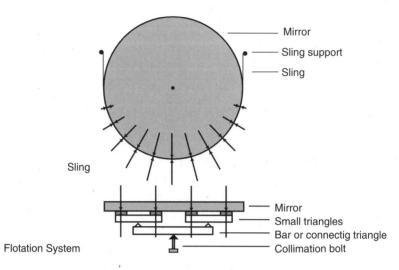

Fig. 5.3 *The sling and flotation cell support the mirror around the rim and from behind by pressing against the mirror with just the right amount of force. These simple devices make large, thin mirrors and giant telescopes a viable option for amateur astronomers.*

The aim of the flotation system is to divide the mirror into many small spans each supported equally. Instead of a single span over a wide gap, you can think of the mirror as a set of small spans over short gaps. Because the forces floating each section of the mirror are equal, the small spans do not bend their neighbors. If the 12-inch thin mirror in the example rests on 9 points, then the average span between supports is about 4 inches. Because it is supported over lots of short spans instead of one long span, a 12-inch thin mirror performs quite well on a flotation system.

The mirror of the 10-meter Keck telescope is very thin and also segmented, so it is supported on a complex set of levers and triangles. The support assemblies are called "wiffletrees." If you look up "wiffletree" in the dictionary, you'll find it's the swing bar to which the traces of a harness are attached on a horse-drawn wagon. The wiffletree equalizes the pulling forces from both horses in a team.

At this juncture, it is tempting to ask why not simply rest the mirror on a solid steel plate. Wouldn't this provide plenty of support? Sadly, the answer is no. If you place a rigid object on another rigid object, they contact each other at exactly three points. Only when the surfaces deform can the two objects contact at more points. You can support an uncooked pizza on a flat steel plate because the pizza sags until it rests on the steel, but that's no way to support a telescope mirror.

With the back side of the mirror supported, let's consider the support around the edge of the mirror. You can't just grab a thin mirror and bolt it to the mirror cell. Neither can you let it rest on retaining clips at the edge; the optic will bend from strong local stress over the clips. What's needed is a uniform support.

To support the large, thin mirrors in his telescopes, John Dobson adopted the mirror sling. There is nothing fancy or high-tech about the sling. Testing mirrors in a sling is a time-honored practice in optical shops. With alt-azimuth telescopes like the Dobsonian, the same end of the mirror always points down so the sling gives uniform and excellent support.

Sling supports do not work in equatorially mounted telescopes because when the telescope is pointed at different parts of the sky, gravity pulls on the mirror in different directions. What happens when you turn a hammock upside-down? Oops—you don't want that happening to your mirror! But the tube of a Dobsonian never turns over. If it were not for John Dobson's putting slings into his mirror cells in the 1970s, you probably wouldn't be building your dream telescope today.

It is important that you be aware that the flotation-plus-sling cell does not provide perfect support. The front surface of the glass disk is concave, so when the mirror hangs heavily in the sling, as it does when the elevation angle is less than 15 degrees, the slightly hollowed-out front surface of the mirror slumps forward and assumes a "potato chip" shape. This distorts the images of stars low in the sky, making them look slightly astigmatic. This is not a major problem, but we don't want you to be surprised when it happens.

One way to reduce potato-chipping is to use a thicker mirror, but that reintroduces problems of weight and long cool-down times that you also want to avoid. Another remedy is to wait until the objects you want to see rise more than 15 degrees above the horizon. The sky is murky and turbulent near the horizon, and often light polluted as well. With a large-aperture telescope, you won't be lacking for deep-sky objects to observe at higher elevations.

5.1.2 Flotation in Practice

In this section, we'll sketch how theory becomes practice in the modern-style Dobsonian. Remember that the mirror cell comprises three principal structures: the flotation system, the sling, and the tailgate. The flotation system supports the mirror from behind. The sling supports the mirror from the bottom. The tailgate connects the sling and the flotation system to the mirror box. In addition, the tailgate holds the side pins and mirror clips which are safety devices that prevent the mirror from bouncing out of the cell when you're transporting the telescope. And last but not least, the tailgate holds a muffin fan to help cool the mirror to ambient temperature.

The flotation system supports the mirror on a hierarchy of *bars, triangles,* and *pads*. The pads, or points, directly support the mirror. A balance bar—such as a teeter-totter or a seesaw—with arms of equal length pushes back with an equal force on each end when loaded; similarly, a balanced triangular plate supported at its center pushes back with an equal force at each corner when it is loaded.

The simplest cell has a pad atop each of three collimation bolts. Because they are equally spaced, each pad carries the same fraction of the mirror's weight. For small mirrors, a *three-point support* is very common. Adjusting the bolts allows you to collimate the mirror.

Table 5.1
Recommended Mirror Cell Flotation Points

Aperture (inches)	Thickness		
	1.5 inches	2 inches	"Full"
12.5	9 or 18	9	9
14.5 and 15	18	9	9
16	18	9 or 18	9
17.5 and 18	18	9 or 18	9
20	27	18	18
25	n/r	18 or 27	18
30	n/r	27	27
32	n/r	27	27
36	n/r	27	27
40	n/r	27	27

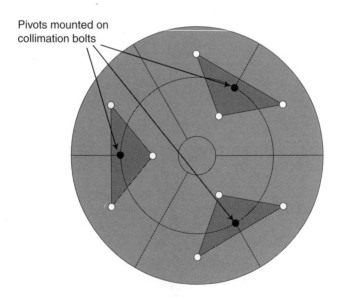

Pivots mounted on collimation bolts

Fig. 5.4 In a nine-point flotation cell, the mirror is divided into nine segments each having the same weight. Each segment is supported by a contact point at the apex of one of three triangles, arranged so that the support at the contact points is all the same. The black circles denote the collimation bolts that the triangles pivot on.

Pivots mounted on
collimation bolts

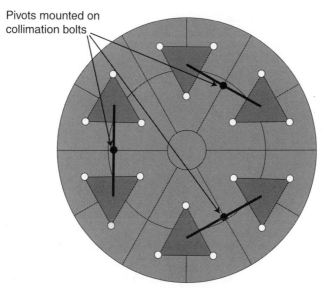

Fig. 5.5 Dobsonian telescopes in the mid-size range usually have an 18-point flotation cell. In this design, each collimation bolt supports a bar that carries two triangles. The lengths of the bars and sizes of the triangles are designed so that each tip of each triangle carries the same share of the mirror's weight. The black circles denote each collimation bolt that the pivot bar rests on.

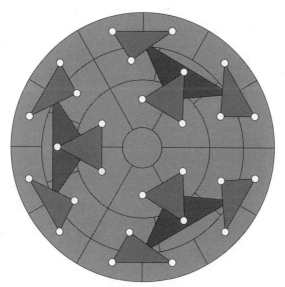

Fig 5.6 This 27-point flotation cell is entirely adequate to support the largest amateur telescope mirror. With nine small triangles balanced on the points of three large triangles, this design is both practical to construct and robust enough to use and maintain. The three collimation bolts that support each of the three big connecting triangles are hidden under the apices of the three inner triangles.

If you place a triangle atop each of the three collimation bolts in a mirror cell, and place a pad at the corner of each triangle, there are nine points available to support a mirror, and the assembly is called a *nine-point flotation system*. The trick is to choose positions of the collimation bolts and dimensions for the triangles so that each pad supports an equal share of the mirror's weight.

If you place a bar atop each collimation bolt, you have a six-point cell. The ends of the bars spread the load more evenly.

Pair up two triangles with one on either end of a short length of stainless steel bar. Let the two triangles balance on the ends of the bar and let the bar now balance on the collimation bolt. Three of these pairs, one pair for each of three collimation bolts and *presto*, you've got an *18-point flotation system*.

Instead of three balance bars, imagine using three large triangles each balanced on the three collimation bolts. Balance a smaller triangle on each corner of the connecting triangles, put a pad on each corner of the small triangles, and you've created a *27-point flotation system*—three collimation bolts times three connecting triangles times three small triangles equals 27 points.

Can you add more balance bars or triangles? Of course. You can put a tiny bar in place of each pad in a 18-point system for 36 points, or add a tiny triangle where each of the 27 points was to make an 81-point flotation system, but at some point it becomes overkill. Friction, the weight of the triangles, and mechanical instability keep the growing proliferation of bars and triangles from supporting the mirror any better than a smaller number of points.

The *sling* is unbelievably simple. It's nothing more than a length of tough nylon webbing—the stuff of automobile seat belts—running under the lower half of the mirror. Supporting the sling are two bolts with slots cut down their lengths so that you can adjust the tension of the sling.

Both the flotation system and the sling are attached to the mirror box of the telescope through a strong ladder-like structure called the *tailgate*. The Kriege tailgate looks very different from that which Dobson used in his early telescopes, but the principles are identical. The tailgate in the modern-style Dobsonian is an open structure welded from square steel tubing, bolted into the mirror box. Tailgates are covered in detail in **Section 5.3**.

The mirror in a modern-type Dobsonian is held in place by gravity. When you transport the telescope, therefore, the mirror bounces around. To keep it safe, three *side pins* attached to the tailgate prevent the primary from moving sideways or bouncing out of the cell. The side pins are plastic or wood cylinders spaced equally around the mirror. They do not touch it: there is a ⅛ to ¼-inch clearance between the side of the mirror and the pin. The side pins never touch the mirror when you are observing.

Mounted at the top of each side pin assembly is a *mirror clip*. The clips are another safety device to prevent the primary from tumbling out of the mirror box in the event of a catastrophe like the whole scope tipping over in a high wind. Lastly, a 12-volt *muffin fan* is mounted on the tailgate frame to provide air circulation to cool the mirror on nights when the air is still.

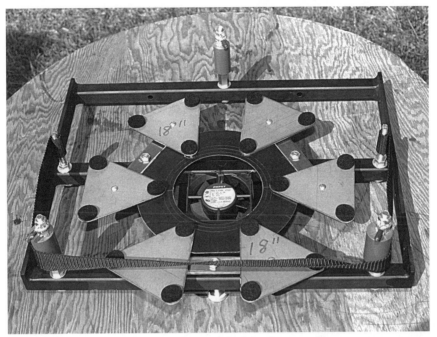

Fig. 5.7 *This 18-point mirror cell provides uniform support from behind and uniform support around the rim. Note the inner ring of flexible Kydex plastic that keeps the six triangles in the proper orientations. The sling is placed inside the two lower side pins during use.*

5.2 Building the Flotation System

Designing a flotation system requires three key inputs: the diameter of the mirror, the thickness of the mirror, and the number of pads to be provided. If your mirror is a standard size, you can use the numbers in the **Table 5.4**, cut and finish the mechanical parts, and assemble it.

If your mirror is an odd size, you can simply scale down the dimensions of the next larger size to your mirror's diameter. If you have a 22⅝-inch mirror that is 1⅞ inches thick, use the cell design recommended for a 25-inch mirror that is 2 inches thick. Multiply each of the recommended dimensions by 22.625/25, or 0.905, then proceed as you would making one of the standard cell sizes.

We have already covered selecting and ordering a mirror, but the question remains: How many points? For large-aperture thin mirrors, a handy rule of thumb is that the number of points should be 1.5 times the width-to-thickness ratio. For example, an 18-inch primary that is two inches thick has a nine to one width-to-thickness ratio. Multiply nine times 1.5 to get 13.5 points. Since there is no way to make a flotation system with 13.5 points, go with an 18-point design. With a full-thickness 18-inch mirror, you could use a 9-point cell.

Table 5.2
Flotation System Hierarchy

Points	Bars and Triangles	Pads
3	none	1 each collimation bolt.
9	3 triangles.	1 each corner.
18	3 bars, 6 triangles.	1 each triangle corner.
27	3 connecting triangles, nine contact triangles.	1 each contact triangle corner.
36	3 bars, 6 triangles and 18 small bars.	1 each end of each small bar.

5.2.1 Flotation System Design

The cell always builds on the three collimation bolts. **Table 5.2** shows how cells are constructed. Unless you've got a bad case of "point fever," do not exceed 27 points. Building a cell with more than 27 points greatly complicates the construction of an amateur telescope and probably adds nothing to its performance. Potato chipping, cool-down time, and a host of other practical issues are more important than some theoretical advantage you might obtain from putting 36, 54, or 81 pads under your mirror. Besides, do you really want to spend your life in your shop cutting out little triangles?

With a plan of bars and triangles in hand, put some actual dimensions to it. Although designs for flotation systems have been available for decades, a sharp fellow named Dave Chandler came up with a program to compute mirror cell dimensions. Dave's program divides the mirror into segments of equal weight, then places the support pads at the center of each segment. For a 9-point flotation system, he divides the mirror into an inner and outer zone with six segments in the outer zone and three in the inner zone. The outer zone has exactly twice the mass of the inner zone. Each of the three triangles supports two outer segments and one inner segment.

For more points, he builds even more complex divisions into zones. For the 18-point cell, the mirror is divided into and inner and outer zones, with the outer zone having 12 segments and the inner zone having six.

For a 27-point cell, Dave divides the mirror into inner, middle, and outer zones. The inner zone is divided into six, the middle zone into nine, and the outer zone into twelve segments. Each of three inner triangles supports two inner zone segments and one middle zone segment. Six outer triangles each support one middle zone segment and two outer zone segments.

On the basis of your mirror's size and thickness, decide how many points the cell needs. For big Dobsonians, the choice is almost always between 18 and 27 points. Look up or scale the dimensions from the appropriate table. And that's all there is to designing the flotation system for a large, thin mirror.

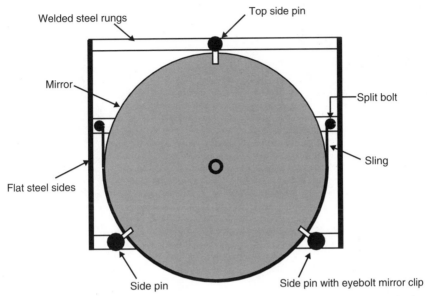

Welded steel rungs

Top side pin

Mirror

Split bolt

Sling

Flat steel sides

Side pin

Side pin with eyebolt mirror clip

Fig. 5.8 The mirror sling is simple and elegant: two split bolts support the length of nylon webbing in which the mirror rests. Side pins do not touch the mirror, but stand ready to prevent the mirror from bouncing sideways more than ¼-inch during transport.

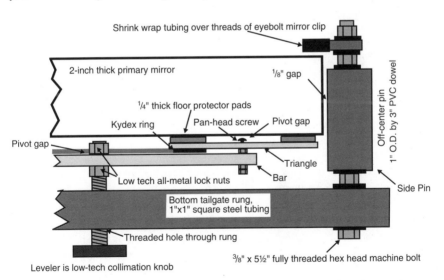

Shrink wrap tubing over threads of eyebolt mirror clip

2-inch thick primary mirror

⅛" gap

Off-center pin
1" O.D. by 3" PVC dowel

¼" thick floor protector pads

Kydex ring

Pan-head screw

Pivot gap

Pivot gap

Triangle

Low tech all-metal lock nuts

Bar

Side Pin

Bottom tailgate rung,
1"x1" square steel tubing

Threaded hole through rung

³/₈" x 5½" fully threaded hex head machine bolt

Leveler is low-tech collimation knob

Fig. 5.9 This cross section shows how a low-tech flotation cell is made from readily available components. The hardest parts to make are the bars and triangles, which should be cut from stainless steel if at all possible. Components such as the pads, bolts and machine screws are standard items in every hardware store.

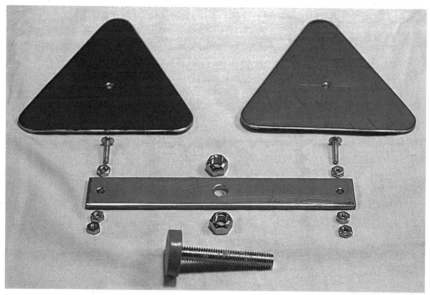

Fig. 5.10 *Here are the components for one "branch" of an 18-point flotation cell. The large knob is an off-the-shelf leveling screw, and the small bolts and nuts are standard hardware-store fare. The bar and triangles are cut from stainless steel or aluminum stock.*

5.2.1.1 Material for Bars and Triangles

Stainless steel is the best material for the bars and triangles. It won't rust and it doesn't need painting. If stainless steel is not available, use mild steel and be sure to paint it with a hard, rust-resistant finish.

Many telescope makers have difficulty cutting metals. If cutting stainless steel is a problem for you, ask the folks at a machine shop or welding shop to cut the parts for you. Make a paper template of a triangle and a bar. For a typical large-aperture scope, ask them to cut the triangles from ⅛-inch stainless steel and the bars from ¼-inch stainless. A few extra dollars for the person running the metal cutter usually increases the quality of the work. If for some reason you must settle for mild steel rather than stainless, remember that you will have to paint it.

If you cannot work with steel, make the parts from aluminum. Aluminum is much softer than stainless steel. If you have nothing but wood-working tools, you can still cut aluminum by putting a metal-cutting blade in a hand-held saber saw. If you can get access to one, a band saw is faster and better. If you make the parts from aluminum, use double the thickness recommended for stainless steel.

If you are a confirmed woodworker, you *can* make the parts from plywood—but select a high quality material like Baltic Birch and make sure the plywood is thick enough to support the weight of a mirror. Wood is user-friendly and most telescope makers are comfortable with it, but for the mirror cell, push yourself a bit. For this purpose, metal is better.

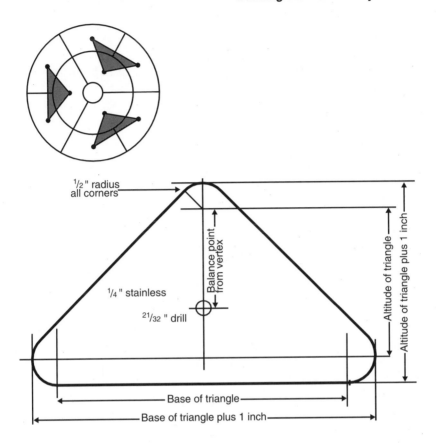

Table 5.3
9-Point Flotation System Dimensions*

Size of mirror	12.5	14.5	15	16	17.5	18
Radius of outer support points	4.85	5.63	5.83	6.22	6.81	7.01
Radius of inner support points	2.02	2.35	2.44	2.61	2.86	2.94
Base of triangle (isosceles)	4.85	5.63	5.83	6.22	6.81	7.01
Altitude of triangle	2.18	2.52	2.61	2.80	3.06	3.15
Balance point from vertex	1.45	1.68	1.74	1.86	2.04	2.10
Radius to balance point	3.47	4.03	4.18	4.47	4.90	5.04
Thickness of steel triangles	⅛	⅛	⅛	⅛	⅛	⅛
All dimensions in inches.						
*Support mirrors 10 inches and under on 3 points.						

Table 5.4
Parts List for 9- and 18-Point Mirror Cell and Tailgate
9- Point: mirrors 12.5 to 18" aperture and 18-Point: 18 to 25" aperture*

9-Point Number	18-Point Number	Description
2	2	¼-inch x 1¼-inch steel flat stock for tailgate side rails
3	3	1-inch x 1-inch heavy wall square steel tubing for tailgate rungs
3	6	⅛-inch thick steel triangles (or substitute ¼-inch thick aluminum)
none	3	¼-inch thick steel bars (½-inch thick aluminum)
9	18	Floor protector pads for mirror cell points, ~¼-inch thick and relatively incompressible
none	6	8-32 x ¾-inch pan-head screws that hold triangles to bars
none	18	8-32 nuts for pan-head screws
9	9	⅜-16 all-metal lock nuts, two per triangle on 9-point cell, or two per bar on 18-point cell, plus one above each mirror clip
3	3	⅜-16 x 2½-inch leveling jacks or collimation bolts; or weld a flat washer to a hex bolt
2	2	⅜-16 x 4-inch hex-head bolts, fully threaded and split to hold sling
3	3	⅜-16 x 5½-inch hex-head bolts, must be fully threaded for side pins
3	3	Eyebolts to serve as mirror clips; eye must be ⅜-inch inside. Obtain from turnbuckles
9	9	⅜-inch lock washers to put above side pins and above and below mirror clips
8	8	⅜-16 nuts, one to retain each split bolt, one to retain each side pin bolt, and one at top of each side pin
8	8	⁵⁄₁₆-inch washers (⅜-inch hole size); one for each split bolt, the rest for the side pins
4	4	1-inch #10 sheet metal screws to mount fan on fan board
4	4	8-32 x ¾-inch round-head screws to mount fan board on tailgate rungs
3	3	Short lengths of shrink wrap for mirror clips
1	1	1½-inch nylon strapping, or seat belt from an old car
3	3	1-inch O.D. x 3-inch long PVC dowels for side pins; wood pins okay instead of PVC
1	1	Double-sided tape
1	1	¼-inch thick plywood or plastic fan mounting board
1	1	12-volt muffin (cooling) fan

* If you use different thickness steel or opt to go with aluminum for your triangles and bars, then use longer fasteners, bolts, and off-center pins.

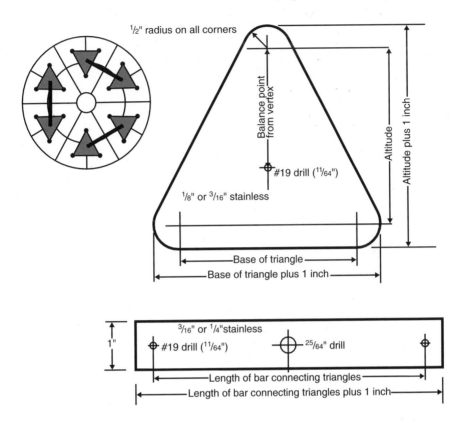

½" radius on all corners

Balance point from vertex

#19 drill (¹¹/₆₄")

⅛" or ³/₁₆" stainless

Altitude

Altitude plus 1 inch

Base of triangle

Base of triangle plus 1 inch

³/₁₆" or ¼"stainless

1"

#19 drill (¹¹/₆₄")

²⁵/₆₄" drill

Length of bar connecting triangles

Length of bar connecting triangles plus 1 inch

Table 5.5
18-Point Flotation System Dimensions

Size of mirror	16	17.5	18	20	22	24	25	30
Radius of outer support points	6.44	7.05	7.26	8.07	8.89	9.71	10.12	12.08
Radius of inner support points	3.01	3.30	3.40	3.79	4.18	4.58	4.78	5.64
Base of triangle (isosceles)	3.34	3.65	3.76	4.18	4.60	5.03	5.24	6.25
Altitude of triangle	3.21	3.51	3.61	4.01	4.41	4.81	5.00	6.02
Balance point from vertex	2.14	2.34	2.41	2.67	2.94	3.20	3.33	4.02
Length of bar connecting triangles	5.15	5.64	5.81	6.46	7.12	7.78	8.11	9.66
Radius to center point of bar	4.46	4.89	5.03	5.60	6.16	6.74	7.02	8.36
Thickness of steel triangles	⅛	⅛	⅛	⅛	⅛	⅛	⅛	³/₁₆
Thickness of steel bars*	³/₁₆	³/₁₆	³/₁₆	³/₁₆	³/₁₆	³/₁₆	³/₁₆	¼
All dimensions in inches.								
* Bars are 1 inch wide, except for 30-inch mirrors, which are 1.25 inches wide.								

Table 5.6
Parts List for 27-point Mirror Cell and Tailgate
For mirrors 30 to 40 inches aperture*

Number	Description
2	¼-inch x 2-inch steel flat stock for tailgate side rails
3	1½-inch x 1½-inch steel tubing for tailgate rungs (heavy wall preferably)
6	3/16-inch thick steel scalene outer triangles (or substitute ⅜-inch thick aluminum)
3	3/16-inch thick steel isosceles inner triangles (or substitute ⅜-inch thick aluminum)
3	⅜-inch thick steel connecting triangles (or substitute ¾-inch thick aluminum)
27	Floor protector pads for mirror cell points, ¼-inch thick and relatively incompressible
9	Pan-head screws 8-32 x 1-inch long to hold triangles to connecting triangle
27	8-32 nuts for pan-head screws
6	½-20 all-metal lock nuts, one on either side of connecting triangle
3	½-20 x 4-inch custom-made collimation bolts. Weld large flat washer to ½-20 by 4-inch fully threaded bolt. Place rubber floor protector over washer for an attractive knob
2	½-20 x 5-inch hex-head bolts, must be fully threaded and split to hold sling
3	½-20 x 6½-inch hex-head bolts, fully threaded for side pins (or substitute threaded rod)
3	Eyebolts for mirror clips; eye must be ½-inch diameter (obtain from turnbuckles)
9	½-inch lock washers to put above side pins and above and below mirror clips
8	½-inch x 20 nuts
8	7/16-inch washers (½-inch actual hole size) one for each split bolt, the rest for side pins
8	#10 x 1-inch sheet metal screws to hold fans to fan boards 8, 8-32 x ¾-inch round-head screws to mount fan board on tailgate rungs
3	Short lengths of shrink wrap for mirror clips
1	2-inch nylon strapping, or a seat belt from an old car
3	1¼-inch O.D. x 3-inch long PVC dowels for side pins; wood pins okay instead of PVC
1	Double-sided tape
1	¼-inch thick plywood or plastic fan mounting board
2	12-volt muffin fans. Why two fans? Lots of glass and thermal inertia.

* If you use different thickness steel or opt for aluminum in your triangles and bars, then use longer fasteners, bolts, and off-center pins.

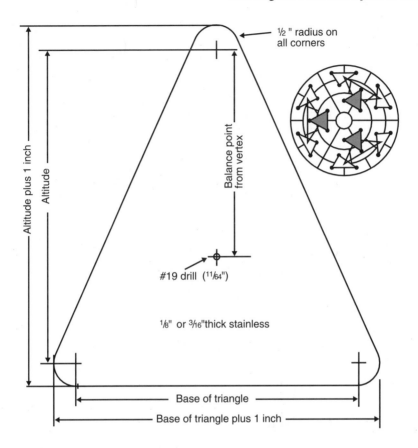

Altitude plus 1 inch

Altitude

Balance point from vertex

½ " radius on all corners

#19 drill (¹¹⁄₆₄")

⅛" or ³⁄₁₆"thick stainless

Base of triangle

Base of triangle plus 1 inch

Table 5.7
27-Point Inner Triangle Dimensions

Size of mirror	24	25	30	32	36	40
Inner triangle is isosceles pointing outward						
Base	3.75	3.91	4.84	5.20	5.92	6.67
Altitude	4.23	4.40	5.34	5.70	6.44	7.18
Balance point from vertex	2.82	2.94	3.56	3.80	4.29	4.79
Thickness	⅛	⅛	⅛	⅛	³⁄₁₆	³⁄₁₆
All dimension in inches.						

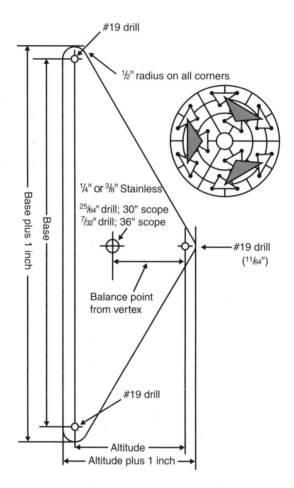

#19 drill

½" radius on all corners

¼" or ⅜" Stainless

²⁵⁄₆₄" drill; 30" scope
⁷⁄₃₂" drill; 36" scope

#19 drill
(¹¹⁄₆₄")

Balance point
from vertex

#19 drill

Altitude

Altitude plus 1 inch

Base plus 1 inch

Base

Table 5.8
27-Point Connecting Triangle Dimensions

Size of mirror	24	25	30	32	36	40
Connecting triangle is isosceles pointing inward						
Base	9.98	10.40	12.61	13.48	15.22	16.98
Altitude	3.12	3.25	3.84	4.09	4.57	5.04
Balance point from vertex	2.08	2.16	2.56	2.72	3.04	3.36
Radius to balance point	6.74	7.02	8.54	9.13	10.32	11.53
Thickness	¼	¼	¼	¼	⅜	⅜
All dimensions in inches.						

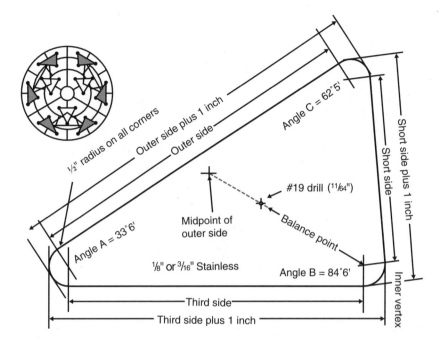

Table 5.9
27-Point Outer Triangle Dimensions

Size of mirror	24	25	30	32	36	40
Outer triangle is scalene						
Outer side	5.45	5.67	6.85	7.32	8.25	9.19
Short side	3.15	3.27	3.83	4.07	4.53	4.97
Third side	4.90	5.11	6.11	6.52	7.32	8.13
Balance point from vertex*	2.06	2.14	2.52	2.68	2.99	3.29
Thickness	⅛	⅛	⅛	⅛	³⁄₁₆	³⁄₁₆
All dimensions in inches.						
* Inner vertex to midpoint of outer side.						

Table 5.10
27-Point Cell Radius of Support Points

Size of mirror	24	25	30	32	36	40
Radius of inner support points	3.75	3.91	4.84	5.20	5.92	6.67
Radius of middle support points	7.47	7.79	9.53	10.21	11.57	12.95
Radius of outer support points	10.52	10.96	13.24	14.14	15.94	17.75

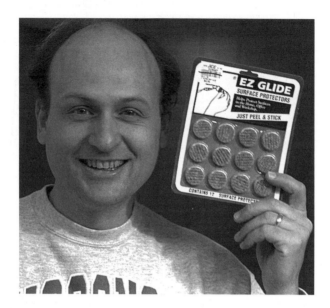

Fig. 5.11 Floor protector pads are the quintessential low-tech solution to the telescope maker's quest for a mirror support "point" that cannot possibly damage the back of your big mirror but is resilient enough to support its weight for many years.

5.2.1.2 Flotation System Pads

Floor protector pads make wonderful mirror cell points. Floor protector pads are used on chair and table legs; they are soft, self-adhesive, bump-absorbing, and squish-resistant even under huge, heavy mirrors. They come in a variety of thicknesses and diameters. They're cheap and you can't beat the ease of installation: just peel off the paper backing and stick them on the corners of the flotation triangles. You can get them in hardware stores. We recommend 1-inch diameter by ¼-inch thick extra-dense pads. Just knowing that your big mirror is sitting on them will make the bumps in the road seem softer. Important: be sure the back of your mirror does not contact any metal fasteners above the bars and triangles. If it does, or if it looks too close for comfort, then install two pads at each support point, one on top the other.

5.2.1.3 Build the Cell Subassemblies

Before you cut any materials, make a full-size cardboard mockup. The type of cardboard doesn't matter so long as it is stiff enough to support its own weight. Cut each of the bars and triangles and with thumbtacks pin them together into a working model of the cell. Place the cell on a cardboard disk the same size as the mirror. Now, stand back and review your mockup. Does each pad really support an equal area? If it looks wrong, recheck the dimensions. This simple reality check can save you a lot of grief down the road.

Next, cut out all the parts. Note that the tables call for triangles ½ inch larger

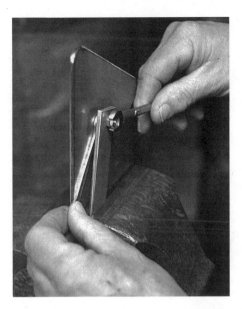

Fig. 5.12 *It may take a few adjustments to set the triangles loose enough to rock into equilibrium but tight enough to remain in place, but once they are made, you'll probably never have to adjust them again.*

on each side than the computations call for because the corners are rounded with a ½-inch radius. The extra width allows you to put a 1-inch diameter floor-protector pad in each corner. If rounded corners are too much to ask, don't fret. Even if it doesn't look as pretty in the daytime, a system with sharp corners will function just as well. At night, who cares?

Although you don't want to put the flotation system together until you have completed the tailgate, it's a big help to assemble the components of the cell so there are no surprises later. The idea is that each small branch must be free to pivot on the next larger branch all the way down to the collimation bolts.

As an example, let's see how to assemble an 18-point flotation system for a 20-inch mirror. The 18-point cell consists of six triangles that ride on three bars, in pairs. When you are done, you'll have three wiffletrees each consisting of a bar connecting two triangles. Fancier cells have correspondingly more complex trees. After drilling all of the holes shown in **Table 5.5**, connect each of the six triangles to the end of a bar with a #6-32 pan-head screw. Pan-head screws have a low profile that doesn't stick above the felt pads, so there is no danger they will contact the glass. You want the mirror to sit *on the pads*, not on the metal screw heads.

Use just one nut above the bar. Use double nuts or all-metal locknuts under the bar so the screw cannot work loose. Attach the screw tightly to the bar, not the triangle. Leave a small gap under the head of the screw so the triangle can pivot slightly on the nut beneath it (see **Figure 5.9**). The triangle has a slightly larger hole than the pan-head screw. If the diameters and spacings are right, each of the triangles should

Fig. 5.13 *A simple ring made from Kydex plastic keeps the complex arrangement of flota-tion-cell triangles and pads properly positioned. To attach the Kydex ring, use double-sided foam tape or small machine screws.*

pivot freely through a small angle. When you press the assembled "tree" against a flat surface, the top triangles should pivot into contact with the surface.

Moving down to the trunk of the tree, each pair of triangles on their bars is connected to the collimation bolts. Each collimation bolt passes through a hole in the center of the bar. Under each of the three balance bars is a ⅜-16 all-metal lock nut. Same as the triangle, the hole in the center of the bar should be about ¹⁄₃₂ inch larger, or about one standard drill size larger than the collimation bolt diameter. The bars of the cell pivot on the beveled upper surface of the lock nuts. To prevent the bars from lifting off the collimating bolts during transport, install an all-metal lock nut above the bars. Do not tighten it. Leave a small gap under it so the bar can pivot lightly (see **Figure 5.9**). Be certain that the bottom of the primary mirror will not contact the tops of these three retaining nuts. You may need to reduce the thickness of the nuts by cutting them in half.

Later on, when the time comes to complete the flotation system, you can mount the partial assemblies on the collimation bolts. To keep the triangles in alignment, make a ring from Kydex or other flexible sheet plastic like Mylar. Cut out a 2-inch-wide ring to cover the apices ("apices" is the plural of "apex," the point on a triangle.) For a 20-inch telescope, the outside diameter of this ring should be about 14 inches. Attach the ring to the triangles with double-sided tape. Kydex is rigid enough to keep the triangles in position but flexible enough to allow them to pivot freely.

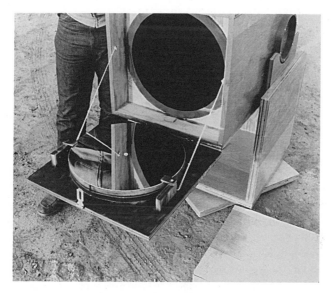

Fig. 5.14 The tailgate in a classic Dobsonian resembled its namesake on a pickup truck or station wagon. The mirror was removed for safe transport. Photo by Bob Kestner.

5.3 The Tailgate

On John Dobson's original versions, the flotation system and sling were mounted on a fold-down board called the tailgate (see **Figure 5.14**). It worked just like the tailgate on a station wagon or pickup truck. This arrangement made it easy to reach the primary mirror for cleaning, and on Dobson's bigger telescopes, made it easy to remove the mirror for travel. Articles published in *Sky and Telescope* and *Telescope Making* magazine brought news of Dobson's tailgate innovation to the telescope building community.

The tailgate served another purpose as well. All the early Dobsonian builders ground and figured their own mirrors on surplus porthole glass, and they tested their mirrors on the stars. The tailgate made it easy to polish on a mirror for a while and then pop the unaluminized glass into the telescope for a quick check of image quality. When an optic tested out to give excellent images it was aluminized.

As word about Dobsonians spread and bigger and bigger mirrors came onto the market, telescope makers continued to build mirror boxes with tailgates. Because the traditional cardboard tube and plywood box design were transported horizontally, you had to remove the primary mirror so it wouldn't come rolling down the tube and join you in the front seat. However, a lot of people got the willies hefting their big mirrors in and out of the tube every time they observed.

In the truss-tube Dobsonian, the tailgate is no longer hinged. Instead, it is a ladder-like frame made of tubular steel, so the mirror box is shallow and open. Rapid thermal equilibrium results because air circulates freely around the mirror. The mirror box is thinner, shorter, and lighter because the steel tailgate strengthens the mirror box. Finally, there is no need *ever* to remove the mirror; the mirror stays safely on its back in the mirror box. In the modern Dobsonian, the tailgate is a win-win situation.

5.3.1 Evolution of the Tailgate

Because the Sidewalk Astronomers were an observing group, there were always plenty of spare hands to set up, move, and knock down big telescopes. The next generation of builders included a lot of lone wolf observers who had to invent ways to transport their telescopes without a lot of help. Swedish telescope builder Ivar Hamberg introduced the open-tube truss concept to large scopes with a two-page article in *Telescope Making*. The truss tube allows the user to break the tube down into three manageable units: the secondary cage, a bundle of truss poles, and the mirror box with the primary mirror in it. Hamberg's innovation meant that the mirror box could be carried around and stored with the mirror on its back, safely, with no danger of breakage.

Next, a clever telescope builder in Columbus, Ohio named Ron Ravneberg realized that the rocker, the ground board, and the mirror box with the primary still in it could be rolled by attaching wheelbarrow handles and wheels to the rocker. Taken together, the truss tube and wheelbarrow handle innovations made one-man observing with really big scopes a reality. Within a few years, most of the Dobsonians between 17½ and 32 inches aperture were constructed this way.

Shortly after these two breakthroughs, in 1988, coauthor David Kriege entered the scene. Kriege wanted to build a 20-inch scope but realized that there would be problems with getting a 2-inch thick mirror to cool at the bottom of a closed tube. In his younger days, his observing had suffered from tube currents in a solid-tube 10-inch *f*/8 Newtonian.

Tube currents are the bane of sharp telescopic images. The bottom end of a closed reflecting telescope is a great "doghouse" for a thick chunk of glass. A doghouse has one opening so that air does not circulate; the dog sleeps comfortably in his doghouse because it holds the heat. On those crisp, clear nights when the air temperature falls all night, the telescope that has only one opening takes hours to cool.

Over the years, many a fine mirror has been unjustly accused of having a poor figure because the observer installed it in the bottom of a closed tube. After the excitement of "first light" wears off, the observer sees fuzzball stars and boiling planets. The morning sun will have risen before the mirror settles down, leaving the true imaging potential of the mirror forever untried. There is an awful lot of thermal inertia in a 25- to 150-pound slab of glass. That's why modern professional observatories are air conditioned during the day: the mirror is cool and ready to go when night arrives. Since you can't air-condition a portable telescope, you need good air flow around the mirror so that it comes to thermal equilibrium rapidly, and provides good images.

5.3.2 The Open-Frame Tailgate

The open-frame tailgate supports the primary mirror on a simple tubular steel framework. The back of the primary mirror is completely exposed; air circulates around the mirror so that the glass reaches ambient air temperature quickly. Couple the open-frame tailgate and a shallow mirror box with an open truss tube, and tube currents become a thing of the past—because the tube no longer exists.

If the air on a given night is very still, you can turn on a little 12-volt fan to move air over the mirror, and it will soon reach equilibrium. If you've got 120-volt power at your dark site, turn on an ordinary window fan and blow as much air through the tailgate as you can. You'll be observing in no time.

Two effects rob a mirror's resolution when it is warmer than the surrounding air. Most obvious (and the one amateur astronomers are familiar with) is that because the outside of the mirror cools before the inside, the mirror is distorted and does not show a good figure until it reaches the same temperature as the surrounding air. The second is less familiar, but it is just as important: a warm mirror heats the air around it. The warm air rises and induces tube currents. If you have money to burn and can afford to purchase a primary made from zero-expansion Zerodur, you can eliminate the first problem. But a warm primary mirror, even one made from Zerodur, will not perform well until it stops heating the air around it.

Letting the mirror come to equilibrium with the air is vitally important, and there's no avoiding it. When your telescope starts out warmer or cooler than the air, you're going to have to wait. This is seldom a problem: you've got to give your eyes time to get dark-adapted anyway.

If you're not already convinced to make your telescope with an open-frame tailgate, here is some food for thought:

1. You get rapid thermal equilibration of the primary mirror and . . .

2. You maintain equilibrium as the air cools during the night. The primary "catches up" faster.

3. The welded tailgate frame is an extremely strong and rigid support for the mirror cell, plus . . .

4. The steel framework reinforces the mirror box and prevents it from getting out of square.

5. The inherent strength of the steel frame lets you eliminate the lower corner of the mirror box for an even lower rocker box profile.

6. You get improved access to the back of the mirror cell and primary. You can adjust the sling quickly and easily from *behind* the telescope.

7. Big plus: there's no danger of dropping a wrench on the mirror!

8. The primary mirror can be left in the mirror box permanently. Reduced chance of breakage or hernia.

9. The frame provides a convenient place to mount accessories like electrical switches, fans and batteries.

10. A low-profile tailgate and mirror cell keeps the primary as close to the bottom of the mirror box as possible. This yields a more compact mirror box and rocker.

11. The three-rung design is simple to construct. Collimation bolts, lateral supports, mirror clips, fan and sling supports are all retained on the rungs.

12. The primary mirror can be washed in the mirror box. That's right, *in the mirror box*. The water runs off the mirror and out the bottom.

13. The back end of your telescope looks impressive. You can show off a fine-looking 18-point mirror cell while other telescope makers salivate.

Well, that's a baker's dozen of good reasons to go for an open-frame tailgate. If you are not yet convinced to go open-frame, build a telescope with a closed bottom, and an always-warm primary and never-ending tube currents. You'll get no sympathy from us.

The tailgate consists of three rungs between two side rails. The rungs are designated the top, middle, and bottom. The rungs are made from 1" square steel tube (1½" square on apertures over 25 inches). The frame holds the mirror box square. The tailgate is bolted to the mirror box by holes drilled in the side rails and the top rung. Because the mirror box is a large, open structure that cannot be reinforced by anything that might block light, the tailgate is vital to the function of the mirror box.

- The *top rung* of the tailgate carries the top side pin with its safety clip. The side pin prevents the mirror from shifting during transport, and the safety clip prevents the mirror from falling out of the cell if you accidentally tip the mirror box too far forward. The top rung thus has one hole for the side pin plus several more for bolting the rung solidly to the mirror box.

- The *middle rung* carries two of the collimation bolts and the two split bolts that support the mirror sling. This means that you must drill four holes in the middle rung: two for collimation bolts and two for the sling split bolts.

- The *bottom rung* holds the bottom collimation bolt, and also carries the two bottom side pins, each with a safety clip. The side pins are located on the low edge of the mirror, 120 degrees apart. These pins, along with the side pin in the top rung, hold the mirror captive during transport. The bottom rung thus has three holes: one for the third collimation bolt and two for the lower side pins.

This arrangement of the three-rung tailgate does an excellent job of supporting and protecting the mirror. You can lay out the tailgate with two collimation bolts on the bottom rung, but please don't. For this arrangement you will need a fourth rung to support the two lower side pins. The three-rung design is much cleaner and more efficient. And it's nice having to squat down to reach only one of the collimation knobs instead of two.

The three-rung cell allows you to round off the lower corner of the mirror box. The rocker sides can thus be shorter, stiffer, and more compact. This may seem trivial now, but when you start building your large-aperture dream scope you'll understand. Everything looks small and manageable on paper, but when it comes time to squeeze the finished product into the back of your Honda, you'll be glad you took off the lower mirror box corner. Remember, the goal is to put your big mirror into a small package.

The three rungs of the tailgate are connected to two strips of steel called *side rails*. The side rails attach the tailgate and mirror cell assembly to the mirror box. The side rails are ¼-inch deeper than the rungs on 25-inch and under telescopes, and on 30-inchers and above, they are ½ inch deeper than the rungs. Deeper side rails provide more metal surface above the rung for the weld and also allow you to drill the tailgate mounting holes a little farther from the end grain of the plywood at the bottom of the mirror box.

The side rails transfer the weight of the heavy primary mirror to the sides of the mirror box that have the altitude bearings. Thus the heaviest component in the telescope is supported close to the side bearings. The shorter the distance this force must travel, the stronger and more rigid will be the assembly.

The 12-volt muffin fan that cools the primary mirror should be mounted between the middle rung and the bottom rung. Install it flush with the bottom of the tailgate, on a piece of ¼-inch plywood, positioned so that the air flow is directed at the back of the mirror. The fan mounting board also serves to mount any electrical switches. You can stash batteries in the corners of the mirror box or between the rungs of the tailgate.

5.3.3 Tailgate Layouts

If you are building a telescope with a standard aperture, you can copy dimensions for the tailgate directly from **Table 5.11**. If you are building a mirror cell for an aperture other than those described here, you'll need to do a little arithmetic to work out the tailgate dimensions.

Start with the dimensions for the flotation system. The crucial dimension is the distance that the collimation bolts lie from the mirror center. To lay out the three rungs in the tailgate, start at the lower collimation bolt in the bottom rung. In a typical 20-inch $f/5$ telescope, this point is exactly 5.60 inches from the center of the mirror. This number comes from the "radius to center point of bar" entry in **Table 5.5**.

You know that an imaginary triangle from the center of the mirror to either

Table 5.11
Tailgate Dimensions (see Figure 5.15)

Aperture of mirror	15	18	20	25	30	36
Mirror box, inside dimensions (A)	16½	20¼	22¼	27½	32¾	39
Rung material	1 x 1	1 x 1	1 x 1	1 x 1	1 x 1 heavy wall	1½ x 1½ heavy wall
Rung length (B)	16	19¾	21¾	27	32¼	38½
Side rail material	¼ x 1¼	¼ x 1¼	¼ x 1¼	¼ x 1¼	¼ x 1½	¼ x 2
Side rail length (C)	13.13	15.63	17.23	21.25	25.41	30.57
Top of rail to center of middle rung (D)	6.06	7.63	8.33	10.25	12.10	14.34
Distance from side rail to collimation bolt, middle rung (E)	4.20	5.55	6.03	7.43	8.73	10.31
Distance from side rail to split bolts, middle rung (F)	0.38	0.50	0.63	0.75	0.75	1.0
Distance from side rail to side pins, bottom rung (G)	1.13	1.63	1.63	2.25	2.75	3.25
Collimation bolt size*	⅜ x 2¾	⅜ x 2¾	⅜ x 2¾	⅜ x 2¾	⅜ x 2¾	½ x 3½
Dimensions of side pin on top rung	0.5 x 3	1 x 3	1 x 3	1 x 3	1 x 3	1¼ x 3
Dimensions of side pins on bottom rung	0.75 x 3	1 x 3	1 x 3	1 x 3	1¼ x 3	1½ x 3
Side pin bolt size	¼ x 6	⅜ x 6	⅜ x 6	⅜ x 6	⅜ x 6	½ x 7
Split bolt size	⅜ x 4	⅜ x 4	⅜ x 4	⅜ x 4	⅜ x 4	½ x 4½
Size of mounting hole in side rails, (number of holes)	5/16, (2)	5/16, (4)	5/16, (4)	5/16, (4)	5/16, (4)	⅜, (6)
Size of mounting hole in top rung, (number of holes)	5/16, (2)	5/16, (2)	5/16, (2)	5/16, (2)	5/16, (4)	⅜, (4)
All dimensions in inches.						
* tap 5/16 hole for ⅜-inch bolt; tap 27/64 hole for ½-inch bolt.						

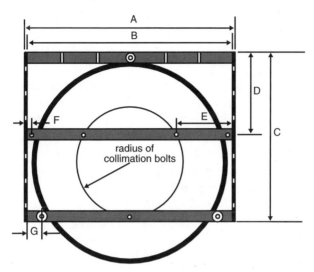

Fig. 5.15 *Two side rails and three rungs comprise the open frame "tailgate" of a modern Dobsonian. When you construct the tailgate, remember that all of the holes in the rungs and rails should be drilled and tapped before you make any welds. For dimensions, refer to **Table 5.11**.*

one of the collimation bolts in the middle rung to the exact center of the middle rung forms a 30°– 60°–90° triangle. (This is because the angle formed by an imaginary line from the collimation bolt in the middle rung to the center of the mirror and a line running the length of the middle rung is 30°.) So now you can determine the distance from the center of the mirror to the center of the middle rung by multiplying 5.60 inches times the sine of 30°. In a right triangle, the length of the side opposite (in your telescope, the distance between the center of the mirror and the middle rung) corresponds to the length of the hypotenuse, which is 5.60 inches times the sine of the angle. Now for the good part: the sine of 30° is 0.5000. Isn't that handy!

So you only have to divide the hypotenuse by two to get the length of the opposite side. For any aperture mirror, you simply multiply the "radius to center point of bar" from **Table 5.5** times 0.5000 and you've got the distance from the center of the middle rung to the center of the mirror. In a typical 20-inch telescope, this is 2.80 inches. If you're making a 9-point or 27-point cell, refer to **Table 5.3** or **5.10** respectively.

Where do the rungs go? The mirror box for a typical 20-inch telescope measures 22.25 inches along the inside edges. Assuming that the center of the mirror box and the center of the mirror are the same, and recalling that the center of the bottom rung is 5.60 inches below the center of the mirror and the center of the middle rung is 2.80 inches above the center of the mirror, you can easily get the spacing of all three rungs.

- The *top rung* is at the top of the mirror box. If you use 1-inch square steel tube for the rungs, then the center of the top rung is ½ an inch from the top inside of the mirror box. The upper edge of the top rung forms the top of the tailgate.

- The *middle rung* is 2.80 inches above the center of the box, so its center lies 11.125 minus 2.80 equals 8.325 inches from the top of the mirror box. The upper edge of the middle rung lies 7.82 inches from the top of the tailgate.

- The *bottom rung* center lies 11.125 plus 5.60 equalling 16.725 inches from the top of the mirror box, and its upper edge lies 16.22 inches from the top of the tailgate.

Now you have the placements for all three rungs, but you still need to figure out where the collimation bolts in the middle rung will be. You've calculated the distance from the center of the mirror to the middle rung which is 2.80 inches and you know that the "radius to center point of bar" is 5.60 inches because all three collimation bolts lie at the end of this radius. Let's use the old Greek discovery: the Pythagorean Theorem. It says for a right triangle, the sum of the squares of the sides is equal to the square of the hypotenuse. This means that the bolt-hole distance squared plus 2.80 squared equals 5.60 squared.

Doing the math: $5.60^2 - 2.80^2 = 31.36 - 7.84 = 23.52$. Haul out your pocket calculator to find the square root of 23.52 and you get 4.85 inches, so the upper collimation bolt holes should each be 4.85 inches from the middle of the middle rung, or 9.70 inches apart.

It is a good idea to check the dimensions by making a full-size drawing for the tailgate and mirror cell. After you have figured out all the correct distances, do not forget to have each of the rungs cut ½₂ inch too short. This little fudge factor guarantees that when the time comes to place the tailgate frame inside the mirror box, it will fit. If the tailgate is ½₂ inch too small, you'll have no trouble shimming it up to fit, but if it's ½₂ inch too large, you won't be able to fit it in. It is better to err on the small side.

5.3.4 Cutting Tailgate Materials

The tailgate of a typical 20-inch telescope requires about 66 inches of 1- by 1-inch square steel tubing for the rungs and about 36 inches of ¼ by 1¼-inch flat steel for the side rails. The cost for this raw stock will run around $20.

Be smart. When you go to the metal shop, have the man cut all the rungs and side rails to exact length when you buy them. The power saws at the shop make accurate and square cuts. Even if it costs few extra dollars to get the steel cut to your specification, the money will have been well spent; there's no way you could do as well by hand with a hack saw.

If you plan to build a 25- or 30-inch telescope, ask for heavy-wall steel tubing for the rungs. If you have a full-thickness 22-, 24-, or 25-inch mirror, or a standard two-inch-thick mirror larger than 30 inches aperture, jump to 1½- by 1½-inch square tubing. This relatively small increase in cross section size yields rungs that are more than three times stiffer.

Even with larger cross sections, the laws of physics work against cells for large mirrors. Longer spans and greater loads make bigger telescopes more flexible. If you were to scale a 20-inch telescope to a 40-inch by doubling the dimensions of every component, the rungs behind the 40-incher would flex *eight times as much*. For big telescopes, go with larger diameter rung cross sections.

If you can't sleep at night worrying about the flexure in the rungs behind your 36-inch primary mirror, have your friendly neighborhood welder join a strip of ¼- by 1½-inch flat steel to both sides of middle and bottom rungs.

5.3.5 Drill Holes Before Welding

After you get home with the rungs and side rails, all you have to do is drill and tap a bunch of holes. Mark off the hole locations as accurately as you can because once they have been drilled, you can't put the metal back on. (You learn stuff like this in dental school, but about teeth.)

Boring the holes without a drill press is tough, but don't run out and buy one for this job. Borrow an evening on a friend's drill press or use one at the welding shop. Every high school in the country has at least a modest metal shop. Borrow some time in their shop after school. If you bring your own drill bits and safety glasses most folks are agreeable. If they won't let you use the equipment for liability reasons give the instructor a few bucks and have him or her do it.

Think ahead: *Drill all the holes before welding the frame.* If you do the welding first, you won't be able to get the frame under the chuck to drill the collimation holes in the middle rung nor the mounting holes in the side rails. This seems pretty elementary, but you just might amaze yourself at how dumb you can be when you're in a frenzy of excitement while building a telescope. This suggestion is based on personal experience.

The locations of the mounting holes in the side rails and the top rung are not critical, but the locations of the collimation bolt holes and the off-center pin bolt holes are. Take your time and lay out these holes accurately. It's a good idea to lay out the tailgate side rails and rungs on a table top to see how it will look after welding. Place each of the tailgate members according to the dimensions you have worked out. From heavy paper or cardboard, cut out a "mirror" and center it on top of the rungs. Here is what to look for:

- The two split bolts that hold the sling must fit between the side of the mirror and the side rail of the tailgate. Keep the split bolts just far enough from the side rails so you can fit a nut and a wrench on them later. Allow ⅛ inch for the welding bead at the junction of the rung and the side rail. Leave at least ½ inch between the split bolts and the side of the mirror so there will be room for wrapping the sling.

- The two side pins in the bottom rung must be placed to allow ¼ inch between the side of the mirror and side of the pin. Make sure the hole locations in the rung will keep the mirror from sliding around, but also make sure that the mirror stops against the side pins and never touches the split bolts. Plastic and wood are friendly to glass; the steel split bolts are not.

Determine the size and thread you will use for the collimation bolts and drill appropriate body holes for that size thread. If you plan to use levelers as your collimation bolts, then you want a ⅜-16 thread, and the correct tap-drill size for this thread is 5⁄16 inch. Tapping the holes for the collimation bolts requires threading both the top and the bottom sides of the steel rung. Be sure you run the tap all the way through in one pass. Do not thread one side, stop, turn the rung over, and then thread the other side. The threads won't mesh. Another suggestion based on personal experience.

Save some money and forgo buying a whole set of taps just for threading these three holes. When you get the 5⁄16-inch collimation holes drilled in the middle and bottom rungs, take them down to the hardware store and have the clerk tap them for a couple of bucks.

5.3.6 Weld Using a Jig

After all the holes are made, place the five members of the tailgate frame on a welding jig. A welding jig is nothing more than a piece of plywood that holds the frame members in position so that a welder can spot-weld (or "tack") each junction. After the welder has tacked the frame together, he or she can remove it from the jig and complete welding it together.

A good way to position and hold the steel pieces of the tailgate in the jig is to screw short lengths of 2 by 4 to the plywood so that the parts of the tailgate frame fit snugly between them. Alternatively, you can use a whole mess of "C" clamps to hold the members in alignment. The advantage of using 2 by 4's is that you can reuse the jig and make identical tailgates for your friends.

Above all else, keep the jig absolutely square. If the jig is crooked, you will face hassles when it comes time to fit the tailgate into the mirror box. When a plywood mirror box and a steel tailgate have a contest to see which is stronger, the tailgate always wins. An out-of-square tailgate will split the corners of the mirror box or warp the mirror box out of shape, and you'll have to live with the vexatious sound the mirror box makes when it scrapes the inside of the rocker as you are observing. (But, if despite your best efforts, the frame gets welded out of square, all is not lost. Set the less-than-90°-corner on the basement floor and whack the opposite corner with a heavy hammer a couple of times.)

After the welder is done, clean up the frame members and the joints and paint the frame. Flat black primer from a spray can does a nice job. The primer looks attractive and you don't need to apply a finish coat.

Gentle reminder: Do not forget to make the inside dimensions of the mirror

Fig. 5.16 *The welding jig for a tailgate is a piece of plywood with blocks to hold the side rails and rungs accurately in position for welding. Here, Ed Kriege shows how wooden blocks hold the frame members in place.*

box ⅟₃₂ inch bigger than the size of the tailgate. Shimming a loose tailgate frame is a cinch. Trying to squeeze an oversize tailgate into an undersize mirror box is a nightmare.

5.4 Mount the Tailgate in the Mirror Box

The tailgate is mounted flush with the bottom of the mirror box. In a typical 20-inch telescope, the top rung is attached to the mirror box with two ⁵⁄₁₆ by 2-inch carriage bolts, with the nuts on the inside (that is, the frame side) of the panels that make up the box. Four ⁵⁄₁₆- by 1¼-inch flat head machine bolts secure the side rails (see **Figure E.1**).

Countersink the heads of the bolts in the mirror box so that the tops of the heads of the bolts are flush with the plywood surface. If you countersink the heads of the bolts, it is not necessary to leave more than a ³⁄₃₂-inch clearance between the mirror box and the inside of the rocker box.

Ordinary carriage bolts also work, but the heads of carriage bolts are round and normally not countersunk, so if you use carriage bolts, be sure to allow adequate clearance between the bolt heads and the inside of the rocker.

When you install the tailgate, drill the two holes for the top rung first. Install the frame and tighten these two bolts. Only then should you drill the side rail mounting holes, insert the bolts, and tighten them. If you drill the holes for the side

Fig. 5.17 *A good weld is stronger than the metal it holds together. If you don't have welding equipment, there are lots of shops where you can have small welding jobs done for a very reasonable price.*

Fig. 5.18 *To make the split bolts for adjusting the sling, you make a saw cut 2 to 2½ inches down the length of a machine bolt. The split segment holds the sling material so that it does not slip when you take up slack under the mirror.*

rails first, when you tighten the upper bolts they will pull the frame up a bit and the holes for the side rails won't line up. Drill the mounting holes one or two drill sizes larger than the bolt diameter. It makes inserting the bolts and installing the tailgate much easier. There—another headache prevented.

5.4.1 Collimation Bolts

If you want to make building your telescope costly and complicated, machine special collimation bolts and knobs. Fortunately, there is an easier way. The ordinary leveling screws you find under file cabinets, refrigerators, and washing machines are ideal for collimating telescope mirrors. (Just another example of how an ordinary hardware-store item solves a telescope problem.) If you use machine bolts, then when you go observing, you have to remember to bring a wrench to collimate the primary. Levelers have a built-in knob, so they eliminate the wrench altogether. You can even turn levelers with a mitten on.

For telescopes in the 20-inch size range, look for levelers that have a metal disc factory welded to a ⅜-16 threaded portion at least 2½ inches long. (For mirrors over 30 inches aperture, look for a those with a ½-inch threaded section.) The manufacturer may fit a gray plastic cap over the welded disc to prevent the metal from scratching the floor. The disc and cap make a tough collimation knob that is easy to grip. What's more, since levelers are designed to support heavy loads, you can lift the mirror box out of the rocker and set it right on the ground. Even the biggest Dobsonian in the world won't squash the collimation knobs. For a dollar each, they can't be beat.

If you can't find something similar in a local hardware store, you can always custom fabricate your own. Simply ask Mr. Welder to weld a large flat washer to the head of a ⅜- by 2½-inch hex-head machine bolt. It won't look as attractive as a leveler, but it will do the job nicely. You can make it look nicer by placing a rubber floor cup—the kind used under piano and bed legs—over the washer. Be sure to select a washer size that fits the cup before you have the washer welded to the bolt head.

Standard bolts do an adequate job—but fast mirrors (that is, those with focal ratios of *f*/6 and less) need to be collimated precisely. If the optical axis is off by as little as ¹⁄₁₀ inch, the star images you see at the focus will show noticeable coma. This corresponds to turning a ⅜-inch collimation bolt with National Coarse (NC) threads only ⅛ of a turn. (Note that ⅜- and ½-inch NC bolts have 16 and 13 threads per inch respectively. National Fine (NF) bolts have 24 and 20 threads per inch, respectively.)

To make collimation easier, look for levelers or bolts with National Fine threads. Most hardware stores stock them. Ask for ⅜-24 or ½-20. The larger number of threads per inch makes collimating your primary mirror silky smooth. And while you're in the hardware store, don't forget to purchase a matching tap for the holes in the tailgate—or save a few bucks and have the hardware store clerk tap them for you. Thread the three collimation bolts into the tailgate. Adjust the bolts to leave about ½-inch of thread between their heads and the bottom of the rungs.

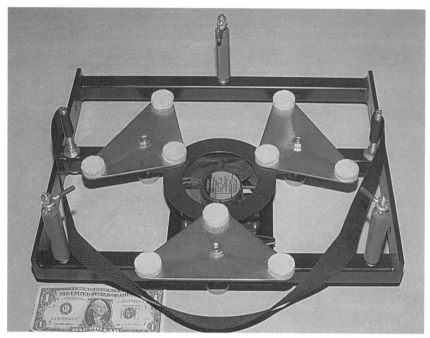

Fig. 5.19 *An efficient mirror cell is a thing of beauty—tidy, symmetric, and every part has a specific function. The results are also pleasing from the eyepiece end! This 9-point cell is perfect for 12″ to 18″ mirrors.*

5.4.2 Flotation System Hardware

The next "things" you'll need from the hardware store are all-metal lock nuts for use under the three balance bars in the flotation system. Lock nuts have a spherical surface on top that serves as a pivot for the bars. The lock nuts are tough to turn but once they're in place they won't work loose. That's important when you collimate the primary: the bolt and nut must turn together when they drive the bar up or down. Put a little Vaseline on the bolt before installing the nut. Don't use lock-nuts that have nylon inserts; they will slip.

Run the nut down to about ¼ inch above the rung. This leaves about ¾ inch of bolt travel which is more than enough to collimate even the most badly misaligned telescope. All that remains now is to assemble the flotation system onto the collimation bolts and install the split sling bolts and side pins.

The three bars of the 18-point flotation system pivot on the top beveled surface of the lock nuts. Set the bar with its duet of triangles onto the all-metal lock nut. To prevent the bar from popping off the collimation bolt when you hit a pothole in the road, cut an all-metal lock nut in two (so it's not as long) and install it at the very end of the bolt. (If you don't cut it in half it will be too tall and might contact the glass.) The top of the cut lock nut should lie flush with the end of the bolt but not touch the bar. If it touches, the bar will not be free to float. Leaving a

Fig. 5.20 Slip the fabric of the sling through the slot in the split bolt and then turn the bolt to take up the slack. The split-bolt arrangement shown here allows you to adjust the mirror sling from behind the primary mirror.

small gap under the locknut leaves the bar free to pivot.

The triangles must be aligned or the bars will rotate out of position. Cut a retaining ring from Kydex or Mylar. If the ring is 2 inches wide, it will nicely cover the apices of the triangles. For a typical 20-inch telescope, the outside diameter of this ring should be 14 inches and the inside diameter 10 inches. Attach the ring to the triangles with double-sided foam tape. Kydex is rigid enough to prevent the triangles from rotating, yet flexible enough to allow them to pivot slightly. If you don't trust tape, six tiny screws and nuts work just as well (see **Figure 5.13**).

5.4.3 The Sling and Split Bolts

Dobsonians suspend the primary mirror in a sling. As the tube assembly is rotated down, more and more of the primary is supported by the sling. The best material for the sling is 1½-inch-wide black nylon webbing. You can find it (or a similar material) at most hardware or sporting goods stores. If you can't find nylon webbing, visit an auto scrap yard and get a seat belt from an old car. You'll need about three feet of webbing for a 20-inch mirror.

Two split bolts hold the sling. By turning the bolts, you wind the strap around them and adjust the tension on the sling. The nice thing is that you make this adjustment from the back of the telescope. (Trying to adjust the sling from the front of the mirror box is difficult and frightening. When you bend over and reach in, sweat will drip from your face and fall onto the optical surface of the mirror. Take it from us: split bolts that you can adjust from behind the mirror are great!)

Split bolts are made by partially hacksawing two ⅜-16 by 4-inch-long fully threaded hex-head machine bolts. (For 30- to 40-inch mirrors, use 5-inch-long ½-

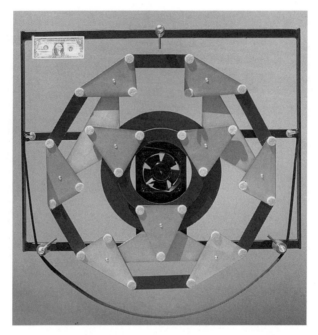

Fig. 5.21 *Although this 27-point cell for a 36-inch mirror is enormous, the open framework allows air to circulate around the mirror. Note that two rings of Kydex plastic keep the nine small triangles properly oriented.*

20 bolts.) If you can't locate fully threaded hex-head machine bolts, weld a nut to one end of a 4-inch length of threaded rod.

Split the bolts the long way by hacksawing the bolts lengthwise 2 to 2½ inches deep. Insert the bolts through the middle rung of the tailgate on opposite sides of the primary. Spin an ordinary nut down over the split and tighten it against the tailgate rung. Place the ends of the sling webbing into the splits.

Later, when you install the mirror, you will drape the sling loosely around the side of the glass and attach the bottom of the sling to the mirror with a piece of double-sided tape. The tape keeps the sling from slipping off the side of the mirror when the mirror box is vertical. After you set up the scope, point it toward the horizon, loosen one of the split bolts with a wrench, and from the back side of the tailgate turn the bolt until the slack is drawn up from the sling. As you turn the bolt, the mirror rises effortlessly until it almost touches the upper side pin. With a second wrench, reach through the tailgate and tighten the opposing nut to lock the bolt in place—a tough job completed in 30 seconds. Remember this is all done from behind the mirror cell of the fully assembled telescope—something you might not appreciate until you have tried to adjust a sling from the front.

5.4.4 Side Pins

To prevent the primary mirror from sliding around the mirror box during trans-

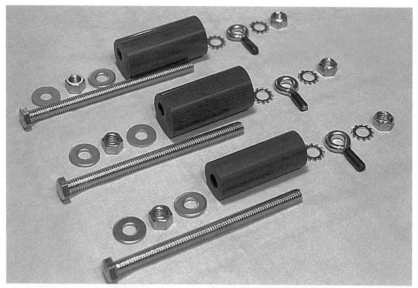

Fig. 5.22 *Side pins protect your mirror from the inevitable jolts and bumps you'll drive over on the road to that fabulous deep-sky site. An off-center hole through the two bottom pins allow you to set the size of the gap between the mirror and the pin.*

Fig. 5.23 *Unless you hit a huge bump, the mirror clips on the side-pins' bolts never touch the mirror. The clip is an "eye" from a eyebolt that fits over the machine bolt and is secured with a lock nut (which in this picture has a box-end wrench slipped over it).*

Fig 5.24 *Install the fan assembly on the back of the tailgate before installing the tailgate in the mirror cell. The fan helps the large mirror to come to equilibrium with the air around it.*

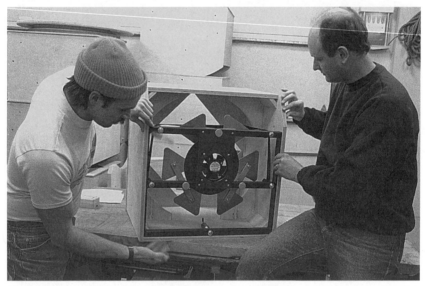

Fig. 5.25 *Be sure to leave ¹⁄₃₂ inch clearance between the tailgate and the mirror box. It is easy to shim a slightly undersize tailgate to fit its mirror box, but it is acutely embarrassing to find that the tailgate is slightly too large and will not fit.*

port, install three side-pin assemblies 120° apart around the mirror. Each pin is a short cylinder of wood or plastic that the glass of the mirror can bang against without damage. The two side pins on the bottom rung of the tailgate are 1-inch diameter by 3-inch-long dowels with a ⅜-inch hole drilled ⅛-inch off-center down the long axis. Because the hole is eccentric, you can rotate it closer or farther from the mirror.

The pin mounted in the center of the top rung is identical to the other two except that the dowel is drilled exactly down the center. All three pin assemblies are held firmly to the tailgate with ⅜ by 5½-inch fully threaded hex-head machine bolts or 5½-inch lengths of ⅜-inch threaded rod. For telescopes with apertures of 30 inches or more, substitute ½-inch mounting bolts and 1½-inch plastic pins for the bottom rung, and 1¼-inch diameter pins for the top rung.

Make three mirror clips from ordinary eyebolts. The "eye" should have a ⅜-inch or ½-inch opening to match the bolts you use. Place a length of shrink wrap tubing over the threads of the eyebolt. When it is warmed over a flame it will shrink to fit the threads, making a cute, low-tech mirror clip. It can be difficult to find eyebolts with the right opening unless you know where they are kept. Go to the shelf where the turnbuckles are sold. Turnbuckles use eyebolts that have just the right hole size. Buy two, unscrew and throw away the middle part, and keep those right-size eyebolts.

Install each bolt in the frame, run a nut down its length, and tighten the nut securely. Slide a lock washer down the bolt, then slip the side pin on the bolt, then another lock washer, another nut, then a mirror clip, another lock washer, and finally an all-metal lock nut on top. The lock washers prevent the pin from turning.

At this stage of completion, considerable time may pass before you do anything more with the mirror cell. If you build the cell early in the construction of your telescope, as we recommend, you may be engaged making other parts for weeks or even months. If the mirror cell is among the last items you expect to complete, install the optics as described in **Section 11.2**.

Chapter 6
The Secondary Cage

If constructing the upper end of your telescope before the rest seems counterintuitive, consider this: The whole purpose of a telescope is to let you aim some big optics at the stars. Logically, then, your telescope begins with the optics and you build the rest of the telescope around them. The function of the secondary cage is to provide a rigid supporting framework for the secondary mirror and eyepiece. After making the mirror cell, constructing the secondary cage is the next logical step.

It is extremely important to keep the secondary cage as lightweight as possible, consistent with being reasonably strong and very rigid. A lightweight upper end is the secret to keeping the rest of scope from becoming too heavy because, in an *f*/5 telescope, *every extra pound you add to the secondary cage adds up to six pounds to the total weight of the telescope*. One pound goes on the secondary cage and five pounds go behind the mirror to balance it.

The secondary cage consists of two lightweight plywood rings separated by four short aluminum struts and a focuser board. The secondary is carried on a spider, and the rings are lined with thin, tough Kydex plastic to block stray light. The secondary cage for a 20-inch telescope weighs in at about 15 pounds, complete.

In telescopes of 25 inches and under aperture, you can attach the secondary cage directly to the truss poles. When you assemble the telescope, you climb the ladder carrying the secondary cage and place it atop the poles, and clamp the poles into the tube sockets. To make telescopes larger than 25 inches aperture easy to set up, before you set the secondary cage atop the telescope, attach a connecting ring to the top of the truss. The truss ring is light and easy to carry and attach to the poles. You then lower the telescope and attach the secondary cage to the truss ring. This extra step makes assembling "The Big Ones" no more difficult than assembling the smaller sizes.

If you build the secondary cage as we recommend in this chapter, your scope will have a low-profile mirror box and rocker. This is desirable because the shallower the mirror box and rocker are, the more rigid they get. When large telescopes are turned in azimuth, the mirror box and rocker are the sources of most of the flexure. The low-profile mirror box and rocker may look strange, but they

Fig. 6.1 *The "business end" of a big telescope is the secondary cage, the place where the starlight that strikes the primary mirror comes to focus, and where you'll find the eyepiece. Sheryl Johnson uses her 20-inch telescope to teach astronomy. Photo by John Sanford.*

work great. When you follow an object, the telescope responds, and when you let go, the object remains in the center of the field. There is no rebound. These benefits follow as direct consequences of the lightweight secondary cage.

6.1 Dimensioning the Secondary Cage

The secondary mirror sits in the middle of the converging cone of light from the primary and reflects it to the eyepiece on the side of the telescope. The main purpose of the secondary cage is to hold the secondary mirror and eyepiece in their proper positions. (The secondary cage is also a handy place to put a Telrad finder and a digital setting-circle readout, since these are often used near the eyepiece.)

The distance from the secondary to the focal plane is the sum of one-half the inside diameter of the upper tube assembly, the thickness of the board that supports the focuser, and the height of the focuser. The inside diameter of the secondary cage is set by the optics: the secondary cage must not block any incoming light. You need to find a properly baffled focuser, and the focuser board should be as thin as possible.

6.1.1 The Truss-Tube Advantage

In the standard Newtonian telescope with a solid tube, it is usually considered desirable to allow several inches clearance between the inside of the tube and the incoming cylinder (actually, a truncated cone) of light from the stars. If the tube is too close to the light path, so the thinking goes, air currents from the warm tube wall degrade the image.

Because the truss-tube design is open, tube currents do not exist. Air flows

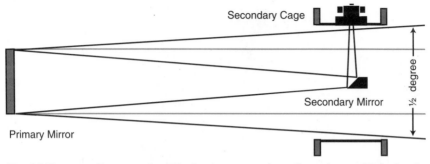

Fig. 6.2 *The secondary cage should be just large enough to allow a beam of light that deviates ½-degree from the optical axis to pass through without obstruction. The angle shown in the drawing has been exaggerated for clarity.*

through the tube, or if you have a cloth shroud over the truss, the shroud and tube components quickly reach air temperature and hence generate no air currents.

In the traditional cardboard-tube design, a 20-inch mirror needs a tube at least 24 inches inside diameter, and will almost certainly be subject to tube currents. However, in a truss-tube telescope the inside diameter of the secondary cage can just barely clear the path of the incoming light. Half an inch on either side is plenty. A secondary cage with an inside diameter of 21 inches is sufficient for a fast 20-inch mirror, and because there is no tube, there are no tube currents.

6.1.2 The Inside Diameter of the Secondary Cage

In a little while, we will argue (and hopefully convince you) that the field of view for a large-aperture telescope at low power is about ½°. This means that the incoming light paths from individual stars are cylinders that can deviate by ¼° from the optical axis of the telescope, and that the diameter of the incoming beam at the position of the secondary cage equals the diameter of the primary plus the sine of ¼° times the distance between the primary mirror and the end of the telescope.

Suppose that you have just one star centered in the eyepiece. The light entering the telescope is a cylinder with the same diameter as the mirror. The light from a star at the edge of a ½° field of view is angled at ¼° to the optical axis, so that the off-axis beam diverges from the on-axis cylinder by ¼°. When it reaches the secondary cage, a distance just a little less than the focal length of the mirror, it will deviate from the on-axis beam by the distance times the sine of ¼°. Since we need to account for the ½°-distance across the whole field, we'll fudge a bit and say the divergence of the beam is the focal length times the sine of ½°.

The inside diameter of the secondary cage is thus the sum of the diameter of the primary mirror plus the divergence, or:

$$T_{sc} \approx D + \sin(\phi)F$$

in which:

D = diameter of the primary mirror

Fig. 6.3 *Even though the secondary cage looks substantial, it's really very light. Consider what's left after you take away the Kydex—just two rings, four struts, and a focuser board.*

F = focal length of primary mirror

ϕ = angular field of view.

The sine of ½° is 0.00873. We recommend leaving a bit more clearance around the light path, so use a value of 0.01. The equation is thus:

$$T_{sc} = D + \frac{F}{100}.$$

Therefore, for any open truss telescope the rule is that the inside diameter of the secondary cage is the diameter of the mirror plus the focal length of the mirror divided by 100. For a 20-inch *f*/5 scope, the rule works out as follows:

$$T_{sc} = 20 + \frac{100}{100}$$

$$T_{sc} = 21 \text{ inches.}$$

This rule works because it is dead simple to remember and it adds a decent fudge factor automatically.

Sample problem: You are a traditionalist who has always wanted to build a 24-inch *f*/8 telescope. Forgetting momentarily about the massive lead counterweights you would need in the bottom of the mirror box to balance such a long telescope, determine the inside diameter of the secondary cage.

The answer:

$$T_{sc} = 24 \text{ inches} + \frac{192}{100}$$

$$T_{sc} = 25.92 \text{ inches}.$$

You would probably round this up to an even 26 inches.

6.2 Constructing the Secondary Cage

The secondary cage is the second major component you will build. The primary reason for building it so early is to get an accurate weight so that when you dimension the mirror box and truss tube, the whole thing will balance. It is not necessary that you complete every step in the construction, but you absolutely must finish everything that influences its weight.

In the foregoing sections, you have determined:

• The inside diameter of the secondary cage,

• The minor axis of the secondary mirror, and

• The (optional) secondary mirror offset.

Before you can weigh the completed or semi-completed secondary cage, you need the following:

• the low-profile focuser,

• the spider assembly,

• the secondary mirror holder,

• the secondary cage finder (we recommend a Telrad), and

• one eyepiece of the type you plan to use most often.

Order these parts early so you won't have to stop working if delivery takes a while.

Before purchasing wood, reread **Chapter 3** and at the very least, browse **Appendix A** on plywood. We trust that you are convinced of the value of solid hardwood multi-ply plywood for the rings in the secondary cage. Plywoods like ApplePly and Baltic Birch really are worth the extra money. They have twice as many layers and virtually no "holes" or voids. These plywoods provide more strength per unit volume than softwood plywood, so you can make narrower, and therefore lighter, rings. If you must use softwood plywood, make the rings wider and thicker to compensate for the weakness of the material.

For apertures from 15½ to 20 inches, the rings should be fabricated from ⅝-inch hardwood veneer hardwood core (HVHC) plywood. For 22- to 40-inchers, use ¾-inch HVHC plywood.

The rings should be 1½ to 3 inches wide depending on the size of your scope

Table 6.1
Secondary Cage Dimensions

Aperture	*f*/ratio	Cage I.D.*	Ring Thickness	Ring Width	Strut O.D.**	Strut Length
14.5	4.5	15¼	½	1½	1	9
15	4.5	15¾	½	1½	1	9
16	4.5	16⅞	⅝	1½	1	10
17.5	4.5	18⅜	⅝	1⅝	1¼	10½
18	4.5	18⅞	⅝	1⅝	1¼	11
18	5.5	19	⅝	1⅝	1¼	11
20	4.0	20⅞	⅝	1¾	1¼	12
20	5.0	21	⅝	1¾	1¼	12
22	4.5	23	¾	1⅞	1¼	13¼
24	4.5	25⅛	¾	2	1½	14½
25	4.0	26	¾	2	1½	15
25	5.0	26¼	¾	2	1½	15
Use connecting ring on 30-inch and above.						
30	4.5	31⅜	¾	2¼	1½	18
30	5.0	31½	¾	2¼	1½	18
32	4.5	33½	¾	2⅜	1½	20
32	5.0	33⅝	¾	2⅜	1½	20
36	4.5	37⅝	¾	2½	2	22
36	5.0	38	¾	2½	2	22
40	4.5	41⅞	¾	3	2	24
40	5.0	42	¾	3	2	24

All dimensions in inches.

* Minimum diameter rounded up to the next ⅛ inch.

**Use four struts on 14½- to 30-inch scopes; use seven struts on scopes 32 to 40 inches.

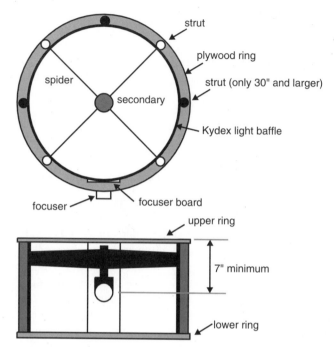

Fig. 6.4 *The secondary cage consists of two plywood rings, four or seven (30" and larger) struts, the spider and secondary mirror, the focuser and focuser board, and a Kydex light baffle. The structure must be strong and yet as light as possible.*

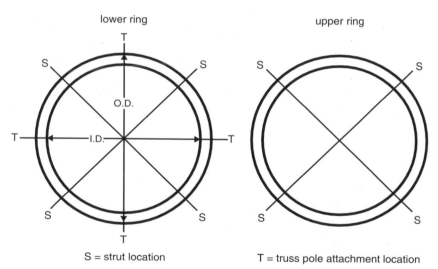

S = strut location   T = truss pole attachment location

Fig. 6.5 *Mark the four strut locations on the upper and lower rings, and equidistant between them mark the locations of the truss pole attachments (T). Dobsonians of 30-inch aperture and larger require three additional struts for greater strength.*

and the type of plywood you use. For a 20-inch telescope, we recommend making the rings 1¾ inches wide. This leaves ¼ inch of wood on either side of the 1¼-inch aluminum strut tubes that run between the rings. Earlier, we saw that the inside diameter of the secondary cage of a 20-inch *f/*5 telescope should be 21 inches. Using rings that are 1¾ inches wide means the outside diameter of the rings will be 24½ inches. Since hardwood plywood is expensive, decide how you're going to place these big rings on the sheet. Make a drawing of the plywood sheet at 1:8 or 1:10 scale and move the circles around to use the sheet effectively. Sometimes it's heartbreaking to see how much wood you'll waste.

6.2.1 Order the Spider, Secondary Holder, and Focuser

See **Appendix E** for suppliers of spiders and secondary holders. For a 20-inch telescope with rings that measure 21 inches across the inside diameter, you would order the 21-inch custom extra-heavy-duty spider. "Extra-heavy-duty" means the spider vanes are wider and have two set of mounting studs on each vane.

If you plan to mount the secondary mirror centered in the tube, order a standard secondary holder sized to fit the secondary mirror you have ordered. If you plan to offset the secondary mirror, ask for a secondary holder with a built-in offset. (You will need to specify the offset for your telescope.) Since an offset secondary holder is a custom order, expect to pay a premium price. Alternatively, you can order an undrilled secondary holder and then drill and tap an offset hole in the secondary holder yourself, or you can order a standard holder and make a small adapter that displaces the secondary mirror holder by the correct offset distance.

As we discussed in **Section 4.4**, the height of the focuser influences the size of the secondary mirror and the need for baffling. "Ultra-low-profile" focusers (focusers less than 2 inches tall) may allow stray light to enter the eyepiece, so if you order one of these, be prepared to add some extra baffling. "Standard" focuser units that stand 4 or more inches tall can cause vignetting at the edge of the field of view, especially with fast mirrors, so we recommend that you avoid these. The best compromise is often a low-profile focuser that stands between 2 and 3 inches high when it is fully racked in.

6.2.2 Cut the Rings

To mark circles on the plywood, make a "stick compass." At one end of an approximately ½- by ½-inch stick of the proper length, drill a hole just big enough to hold a pencil snugly. Measure from the pencil point to the desired radius, that is, at half the diameter, and there drill a pilot hole for a small finishing nail. The nail should fit tightly.

Locate the center of the hole where your scale drawing indicated you would use the sheet most efficiently. Tap the nail lightly into the plywood so it can't wander as you swing the compass. Draw circles for both the inside and outside diameters. Next—and this is important—use a straightedge and protractor to mark eight points. Four of these are for the struts and four for the truss-pole attachments.

Fig. 6.6 *Peter Welbourne takes three or four passes with a router, making each pass a bit deeper, to cut the outside radius of the secondary cage rings. Whenever you use a tool that throws off chips and dust, remember to wear proper eye protection.*

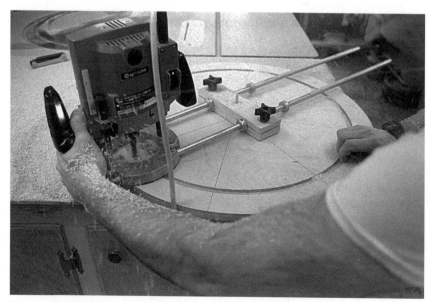

Fig. 6.7 *The inner disk is screwed to the bench top so that it cannot wander as you rout the inner radius of the secondary cage ring. If you are not familiar with the safe use of power tools, ask a knowledgeable friend to help you.*

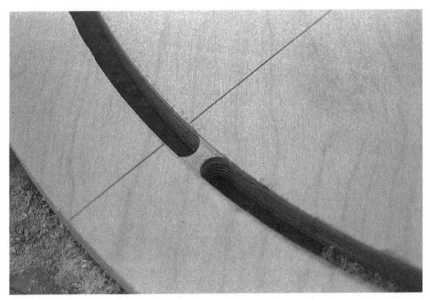

Fig. 6.8 *The secret of cutting a perfect secondary cage ring is to leave two small "lands" connecting the outer ring to the inner disk. These hold the ring in place until the rest of the cut is finished. They can then be removed with a slow, careful cut.*

Check to be sure these eight points are equidistant, that is, 45° apart. If they are not, the spider vanes will be slightly out of line and you'll have stars with eight-spike diffraction patterns instead of four spikes. Mark these points before the rings are cut because it's difficult to mark them later, when the center of the circle is gone. Overlooking a detail like this can force you to waste an afternoon salvaging what might otherwise have been an easy task.

If you have a saber saw, cut along the outside diameter and remove the disk of wood. Although it isn't strictly necessary, removing the excess makes cutting precise circles easier work for the router. As you cut, stay close to the outside of the line without touching it.

To rout a ring from plywood, attach the plywood disk to the bench top with eight small wood screws. Insert four of the screws where the four struts will later attach so that you do not have to make extra holes in the rings. Countersink the screw heads so they do not interfere with the motion of the router. Screw the inner disk to the bench top with the other four screws—but you need not worry about the screw locations because the inner disc is scrap. Screwing the disc to the bench top is better than clamping it because the clamps get in the way of the router and you can't clamp the inner disc anyway. With both the ring and the inner scrap disc firmly screwed down to the bench top you can confidently cut out the ring.

If you have a circle-cutting attachment for your router, attach the pivot point to a bolt through the center of the disc. Most routers have some fairly convenient places to attach such accessories. If you don't have one of these attachments, mount the router on one end of a piece of ¼-inch scrap plywood. At the other end of the plywood, drill a hole at the desired radius for the bolt. Remember that for the inside cut, you must measure from the outside of the router bit, and for the outside cut, you must measure from the inside of the router bit. Secure this radius-cutting jig with a ¼-20x2-inch carriage bolt through the center of the disc. The idea is that the router should not be able to wander, so drill a hole slightly undersize and press the bolt through from the underside. Drill a matching ¼-inch hole in the router attachment— be sure you have a snug fit on the bolt so there won't be any wobble.

The distance between the bolt hole in the plywood router attachment and the router bit should be the radius you desire for the cut. Check the radius on a scrap piece of plywood. If the radius it too large or too small, drill another ¼-inch hole in the plywood attachment at the correct radius. It may take several tries before you get it right. Using this method, the upper and lower secondary cage rings are guaranteed to be exactly the same dimensions.

Use a good quality carbide router bit to make the cuts. Ordinary high-speed-steel bits may be cooked black after the first ring is cut, so don't cut corners when it comes to buying a bit. The glue layers in hardwood ply are very hard, and they will dull the tool quickly. Sharp tools are safer than dull tools, and cheap tools usually get dull quickly. You may as well spend the money—you'll need this router bit again and again, when it comes time to cut the ground board, the side bearings, and the rocker side arcs.

If you have never used a router before, get someone who knows how to show you the ropes. *Go slow and be careful. Wear eye protection and keep your fingers away from the bit.* You will be delighted at what a fine job the router does, but when the router bit is cutting, chips fly in all directions and the cutter is, well, a cutter, and fingers are softer than plywood. If you have a phobia about using high-powered tools, get someone who is comfortable with them to do the job for you. Promise the moon and stars—after all, you can deliver!

Cut the outer diameter in several passes. Make the first pass a very shallow cut through the top veneer or two. This severs the wood fibers and prevents the face veneer from lifting during subsequent passes. If you are cutting high-quality hardwood plywood, take three or four trips around the circle, each time lowering the cutter a little bit, to cut all the way through.

After the outside cut is made, move the router in and rout the inside diameter. As before, cut the ring in several passes. After you've made the cuts, unscrew the ring and check your work. The cuts should be clean, sharp-edged, and smooth. It is very satisfying to produce a couple of nearly perfect rings in a fine plywood.

Gently round the inside edges with sandpaper to prevent the grain from catching your knife blade when you trim the Kydex light baffle later on. Radius all the outer diameter edges of the rings with a ⅛-inch radius router bit. This will leave them rounded and smooth to the touch. The outside edges are visible, and during set-up and observing they are handled extensively. Spend some extra time to finish the outer end grain edges so they look smooth and attractive. Fill and sand any end-grain voids.

The plywood ring cut-outs are scrap. They are too small for a ground board, so forget about that right now. Maybe you can give them to someone for making a smaller Dobsonian. Or make a Lazy Susan for the dinner table, and have rotating mashed potatoes.

6.2.3 Make Three Rings for "The Big Ones"

Secondary cage rings for telescopes in the larger-than-25-inch class are just the same as those for smaller telescopes, except that you should use seven struts. Four of these struts support the spider and the other three provide extra support for the cage rings. The secret behind the construction of secondary cages is that the thin plastic light baffle, wood rings, and struts all weigh very little. However, even the lightest cage can become unwieldy to carry up a ladder and attach to a set of truss poles on apertures larger than 25 inches. "The Big Ones" require an auxiliary *connecting ring* to hold the eight truss poles in alignment. Why is this? In smaller telescopes, the observer scampers up the ladder with the secondary cage and leans away from the ladder, and deposits the secondary cage atop the truss poles. It works, and it's quick and easy.

"The Big Ones" call for a different approach. Reaching out with a secondary cage the size of a bass drum and setting it atop the truss poles is a tall order. For example the secondary cage for a 30-inch telescope weighs around 25 pounds and is an ungainly 3 feet across. There is considerable danger that the observer and the secondary cage could plummet into the mirror box.

The safe solution is to assemble the telescope horizontally. With both feet securely on the ground, you simply walk over with the secondary cage and connect it to the connecting ring. **Figures 11.17** through **11.20** show how this works. The connecting rings hold the tips of the poles together and also supports the secondary cage.

6.2.4 Drill the Strut Seats

Four struts join the two plywood rings. The struts themselves are aluminum tubing, the same stuff you will later use for the truss tube poles. Recessed holes cut to the diameter of the struts locate and help the struts to seat properly, and they also prevent the struts from pulling inward under tension of the spider vanes. (More about struts below.)

Align the two plywood rings and clamp them together. Because the outer edges of the rings are rounded slightly, it looks better if you place the strut mounting holes about ⅛ inch inward from the center of the ring. Drill ⅛-inch locating holes through both rings at the four strut locations. Before you unclamp the rings, mark their orientation so you know which pairs of holes match up perfectly. Then, using a spade bit or Forstner bit, countersink the four 1¼-inch strut-mounting holes to a depth of ¹⁄₁₆ inch into one side of each ring at the four points you marked with location holes.

On the lower ring, the remaining four marks between the strut locations

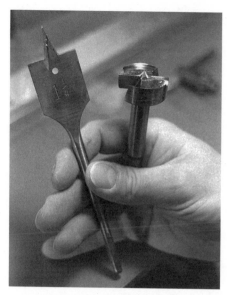

Fig. 6.9 Drill the strut seats with a 1¼-inch spade bit (left) or a 1¼-inch Forstner bit (right). These tools allow you to make clean, shallow holes that locate and hold the aluminum struts that join the secondary cage rings.

Fig. 6.10 After drilling the shallow holes that serve as seats for the struts, enlarge the original pilot holes to provide clearance for the ¼-20 bolts that secure the struts to the rings. Threaded inserts in the struts make this aluminum-to-wood joint extremely solid.

show where to place the truss pole attachments. If you already have the truss pole attachments, drill pilot holes for the screws that will later hold them in place. Stain, seal and varnish both rings. Congratulations! The rings are finished.

6.2.5 Make the Struts

The length you choose for the struts determines the length of the secondary cage. To prevent stray light from reaching the secondary or entering the eyepiece on a 20-inch telescope, the secondary cage should be a minimum of 12 inches long, and proportionately longer on bigger scopes. Here is another rule of thumb:

Cut the struts to 60% of the primary mirror diameter.

A 25-inch Dobsonian would, for example, need struts 0.60 times 25, or 15 inches long. This rule of thumb does not provide complete baffling. Purists recommend that baffling extend *above and below* the secondary mirror at least one full diameter of the primary, which on a 20-inch scope results in a secondary cage 40 inches long. Think about it: To assemble the telescope, you will have to carry the secondary cage up a ladder at night by yourself. The length we recommend is easy to handle and provides baffling that is entirely adequate for reasonably dark sites.

The main cause of image fogging is stray light entering through the top of the telescope opposite the focuser. Light sneaking in below the Kydex baffle will usually be blocked by the full-length cloth shield over the truss tube. If the Kydex above the focuser extends ⅓ the primary's diameter above the center of the focuser, sky light cannot reach the eyepiece. On a 20-inch telescope, position the spider and secondary so that the focuser is at least seven inches from the upper end of the secondary cage. On very large telescopes, you can place an auxiliary light baffle made from a short length of Kydex opposite the focuser at the top of the tube. Attach this extra baffle with Velcro so it can be placed and removed easily. See **Figure 12.11.**

The struts separating the two plywood rings are made from the same tubing used for the truss. We recommend aluminum tube 1¼-inch O.D. for telescopes 22-inches and under, 1½-inch O.D. for 24- through 30-inchers, and 2-inch O.D. for 32-, 36-, and 40-inchers. The 1¼-inch tubing is available at most hardware stores or lumberyards, and is sold in six-foot lengths. The wall thickness should be about 0.055 inch.

Cutting thin-wall aluminum tube is a snap with a tool called a tube cutter. You can get one at the hardware store for around $20. It will cut tubing to length more precisely than you can with a hacksaw. You might as well buy one now because you'll need it later for cutting the truss poles to length.

The location of the holes for the spider will depend on the dimensions of the spider you order. Normally, the holes turn out to be 1 to 2 inches from the upper ends of the struts. This means the upper end of the spider and secondary holder lie about 1½ inches inside the secondary cage. After you have cut the four struts to length, drill the mounting holes for the spider.

Fig. 6.11 A great way to insure that your struts have their holes in the same location is to construct a strut jig—a block of hardwood with holes drilled in the correct positions. On a drill press table, insert each strut and run a drill through the holes in the jig.

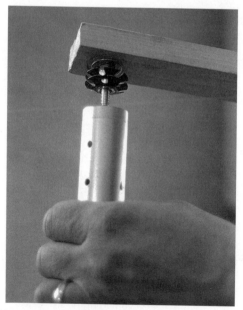

Fig. 6.12 Threaded inserts look like stiff metal flowers—and they work like the proverbial umbrella in a chimney. Attach a threaded insert to a bolt run through a small block of wood then press it into the tube. Once in place, the insert cannot be pulled out. Also see **Figure 8.20**.

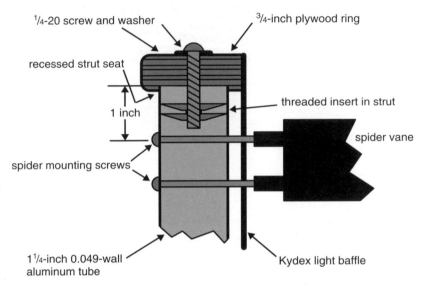

¹/₄-20 screw and washer

³/₄-inch plywood ring

recessed strut seat

1 inch

threaded insert in strut

spider vane

spider mounting screws

1¹/₄-inch 0.049-wall
aluminum tube

Kydex light baffle

Fig. 6.13 *Construction of the upper end of the secondary cage is shown in this cross-sectional view. The secondary cage rings support four struts, and the struts support the vanes of the spider which carry the secondary mirror. Double-sided foam tape secures the Kydex light baffle to the rings.*

6.2.6 Attach the Struts to the Rings

Threaded inserts are used to attach the aluminum struts to the plywood rings on the secondary cage. There is no need to epoxy the tubing to the wood, and no need to use machine threaded plugs and press-fit them into the tubes. Threaded inserts are off-the-shelf metal "flowers" that you force into the tube and—*presto*—eliminate the hassles.

Simply press a threaded insert into each end of each strut. Run a short bolt through a hole drilled into a scrap of wood, screw an insert on that bolt and with a hammer tap the inserts into place—once inserted they never come out. An even niftier way is to put a machine screw into a drill press (obviously the drill press should not be running) and spin the insert on up to the chuck. Stand the tube on the drill press table and press the insert into the tube by lowering the chuck with the handle of the drill press.

You can now attach the rings to the ends of the cage struts using ordinary ¼-20 by 1¼-inch machine screws. Best of all, threaded inserts allow for future disassembly if you need to rebuild that part of your scope. Threaded inserts are cheap, only about 55¢ apiece. They weigh next to nothing and come in assorted sizes and threads.

After you've inserted one into each end of the struts, assemble the secondary cage by passing the proper size machine screw through the ring and into the threaded insert of the strut. Congratulations! You have just created a strong, lightweight "cage" for the secondary mirror and focuser.

Fig. 6.14 *The components of a secondary cage await assembly. Converting a set of prepared parts to a complete the secondary cage takes only about 30 minutes. The toughest part is sticking the Kydex light shield in place.*

Fig. 6.15 *The secondary cage seems to appear from nowhere as Pete inserts bolts into the struts—suddenly two wooden rings become the upper end of a big telescope. With nothing more than the rings and struts in place, the secondary cage already feels strong.*

Fig. 6.16 *Before you tighten the bolts that hold the struts, orient the holes in the struts so they point diametrically across the cage to the holes on the opposite side. Here Pete Welbourne insures that the screws that hold the legs of the spider will insert easily in the secondary cage.*

6.2.7 The Focuser Board

The focuser board is a ¼-inch thick piece of Baltic Birch or similar high-performance plywood with a reinforcing strip of wood on each side of the board. Don't get lazy and use a thick piece of plywood: we're trying to make the secondary cage as light as possible.

Trim the focuser board to fit between the two rings of the secondary cage. Drill the hole for your focuser with a hole cutter. Be sure that the focuser hole will end up directly opposite the secondary holder. To be sure the alignment is accurate, mount the secondary holder so that there is an extra ¼ inch between it and the spider hub. That way, if you miss slightly with the hole in the focuser board, you can reposition the secondary holder to compensate.

Position the focuser board exactly between two of the struts. Secure it with some wood glue and a couple of drywall screws on each end.

6.2.8 Install the Kydex Light Baffle

You can use any type of plastic or metal inside the secondary cage, but we strongly recommend Kydex. Kydex is inexpensive. It is waterproof. It is easy to cut and drill. It is dull on the inside (to absorb light) and shiny on the outside (so it looks good). It is black through and through, so if it gets scratched, it stays black. It accepts double-sided tape. It maintains its shape. It comes in long sheets so you can wrap a single piece around the whole secondary cage and hide the seam under the focuser board.

Kydex is sold in 4- by 8-foot sheets in nine thicknesses from 0.028 inches on up. You want the 0.028 in black for the secondary cage light baffle; this stuff retails for $38 per sheet. It is primarily sold as a commercial wall covering. It won't dent like aluminum, yet it's nearly as light. If you feel creative, Kydex is available in a rainbow of colors and five surface patterns. And, as you will soon appreciate, it can be trimmed with a sharp knife and drilled for the spider screws.

But beware, Kydex deforms at 180° Fahrenheit. If you wrap your scope in clear plastic under the hot, daytime sun at a star party, the greenhouse effect will warp your Kydex. Always cover your scope loosely with one of those ubiquitous *blue* plastic tarps.

Elsewhere in your telescope, you may have already used Kydex to retain the flotation cell triangles. Cut into a 2-inch-wide ring and attached with double-sided tape to the apices of the triangles, the Kydex prevents the triangles from rotating, yet still allows them to pivot.

Double-sided tape is used to attach the Kydex to the secondary cage. Double-sided tapes are versatile. Cloth carpet tape, 3M weatherstripping tape, and automotive trim mounting tape are a few that work well in this application. Buy it in ½-inch width, which is just right for holding Kydex inside the secondary cage.

Cut a piece to length using the formula for the circumference of a circle: $C = \pi \times ID_{sc}$. For a 20-inch telescope with the 21-inch rings, 21π equals 66 inches. Cut a piece this length and a couple of inches wider than the length of the secondary cage.

Kydex has a mind of its own—it expands and contracts with temperature. Wrapping it into the secondary cage on a hot day or in a hot room is asking for trouble. It will contract enough on cold nights to separate itself from the inner surface of the rings. When you press it down in one place, it pops up somewhere else. It's maddening. Be clever. Roll up the Kydex and put it into your freezer. Or wait until night or for a cold day to wrap the Kydex. When it warms, you'll have a nice tight fit.

Wrapping the Kydex inside the secondary cage is *very* tricky. If you want to reinvent the wheel, feel free to do so, but if you want to save time and materials, pay close attention to the following.

Apply a thin coat of contact cement to the inside edges of the plywood rings and let it dry. Contact cement yields a better bond with adhesive tapes than uncoated wood. Apply a layer of double-sided tape to the inside of both rings and remove the protective backing. On the inside of the focuser board, mark in pencil a starting line that is perpendicular to the plywood rings. Apply a length of double-sided tape adjacent to the line.

Lay the secondary cage on its side on the work bench with the focuser board down. Roll the length of Kydex tightly into a roll a few inches across. *Remember: dull side in, shiny side out.* Starting behind the focuser board, align the edge of the Kydex with the pencil line on the focuser board. Press the end of the roll down onto the plywood rings evenly.

Unroll the Kydex gradually, proceeding around the secondary cage by roll-

Fig. 6.17 *To install the light baffle, roll up the Kydex with the shiny side facing out. Position the roll inside the secondary cage taking care to align the end of the roll exactly perpendicular to the rings, and then slowly unroll the Kydex and press it firmly against double-sided tape on the rings.*

Fig. 6.18 *Trim the loose end of the Kydex roll so that it butts tightly against the start of the roll. If the Kydex has formed a slight helix, don't worry as long as the roll is wide enough to cover the double-sided tape on both rings.*

Fig. 6.19 Trim the ends of the light shield with a sharp utility knife, taking care to avoid cutting into the wood of the ring. If you apply a steady, even pressure, the Kydex will part smoothly and you'll have a neat-looking job.

ing it on the bench as you go. After you're all the way around, trim away the excess along the top and bottom of the secondary cage with a very sharp utility knife. Try not to mar the wooden rings with the blade. If you beveled the inside edges a bit, the blade won't catch as easily.

Here is another trick. Although Kydex is quite resilient, when it is creased or hit hard, it shows a dull, whitish mark. Warm the area—gently, gently, gently—with a heat gun or blow torch and the mark magically goes away. If you heat the Kydex too strongly, you'll *melt* a hole in it. Go easy when you apply the heat.

6.3 Connecting Ring for "The Big Ones"

The connecting ring is a single plywood ring exactly the same dimensions as the two rings of the secondary cage. The connecting ring has the eight truss pole fasteners permanently mounted to it on one side and on the other side, simple snap latches that mate with catches on the secondary cage. (Snap latches can be found at most hardware stores, marine supply outfits, and Reid Tool Supply. They are commonly used on tool boxes to keep the lid closed tight.)

The connecting ring is light—just five pounds for a 36-incher— so you can carry it up a tall ladder with one hand, set it on top of pole tips, and clamp the poles in place. See **Figures 11.17** and **11.18**. Once the truss is locked together with the connecting ring, you can safely lower the tube to a horizontal position. There is no need to worry about pole failure or splitting a pole block, because the eight poles share the load.

Fig. 6.20 *Secondary cages for "The Big Ones" are enormous—too big to simply carry up the ladder and attach. The solution is to attach a connecting ring to the top of the truss, and then with the telescope horizontal, attach the secondary cage to the connecting ring (see also* **Figure 11.20**).

Attached to the connecting ring is a short length of rope with a safety hook on the end, the same kind used on dog leashes. After pulling the telescope horizontal, you secure the connecting ring to a heavy object on the ground. A cement block, the bottom of your ladder, a car bumper, or anything else substantial can hold the telescope.

The secondary cage has a pair of hooks on the bottom ring, so you carry the secondary cage over and hang it on the connecting ring by the hooks. With your hands free, you engage the latches. After the secondary cage is latched onto the connecting ring, the telescope is balanced. You can undo the rope hook and tilt the telescope upwards to observe.

Take down is the reverse. The most important thing to remember is this: Secure the rope to the connecting ring to a weight on the ground before you remove the secondary cage. As soon as you detach the secondary cage, the telescope becomes quite unbalanced. You could get hurt and the telescope could be damaged as the poles swing to the upright position.

As a safeguard, we strongly recommend that you attach a second rope from the connecting ring to the secondary cage. If you forget to secure the telescope to the ground, the unbalanced truss assembly will rise as you try to lift the secondary cage, and the rope will prevent you from removing it completely.

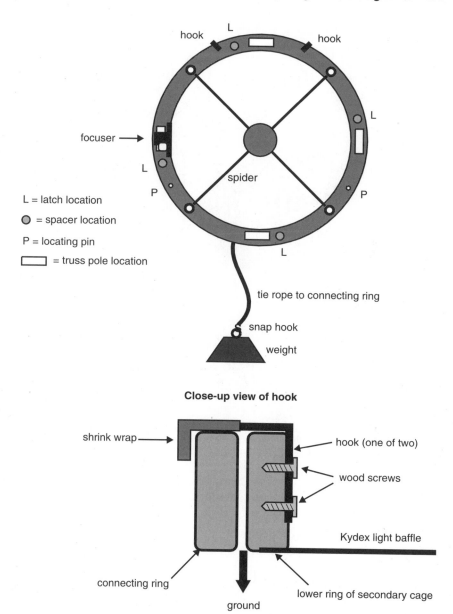

L = latch location

⬤ = spacer location

P = locating pin

▭ = truss pole location

Fig. 6.21 *Attaching the secondary cage of a large telescope is a one-person operation with the help of hooks and locating pins. Use a tie rope and weight to hold the telescope horizontal. Lift the secondary cage into position and drop the hooks over the connecting ring. When the locating pins enter the locating holes, you know that the secondary cage is in the correct position. Latches hold the secondary cage securely on the connecting ring. See also* **Figures 6.22–24** *and* **11.17–19***.*

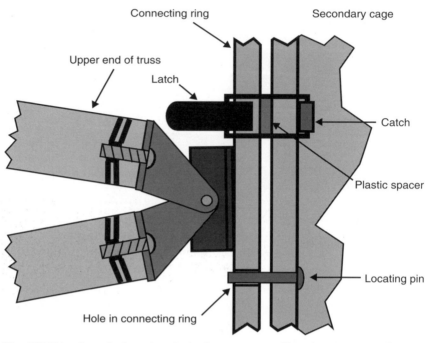

Fig. 6.22 *This schematic shows how the latches, spacers, and locating pins secure the secondary cage on one of "The Big Ones." The plastic spacers are chair glides and the locating pins are carriage bolts. The latches and spacers should be as close to the truss tube connectors as possible.*

You need to attach the secondary cage on the connecting ring precisely the same way each time you set up the telescope. The simplest way to do this is with a couple of locating pins. The pins are nothing more than a pair of ¼-20 carriage bolts located on the bottom ring of the secondary cage, spaced 180° apart and protruding far enough to pass through matching holes in the connecting ring. Glue them permanently in place. When you set up the scope, align the two bolts to the holes in the connecting ring and perfect collimation is assured.

To prevent the connecting ring from scuffing the bottom ring of the secondary cage, put four ordinary plastic chair glides equally spaced around the bottom of the secondary cage. Chair glides are about ¼ inch thick, and they act as spacers to provide clearance for the screw heads that connect to the struts between the secondary cage and the connecting ring.

Locate the spacers as close to the latches and truss pole tips as possible, preferably directly underneath. If you don't place them close enough, the rings will bend and warp towards one another as the latches are drawn tight. If the top of the secondary cage is 12 o'clock, mount the two hooks that it hangs on at 11 and 1 o'clock. (The focuser is at 9 o'clock). Attach the tie-down rope at 6 o'clock.

Fig. 6.23 It looks difficult, but it's not. To attach the upper end of this enormous telescope, you lift the secondary cage into position and the hooks hold it. You can then let go and secure it atop the truss tube by setting the latches.

6.4 Install the Hardware

Although your secondary cage is nearly complete, you may want to postpone taking the final steps—adding the secondary mirror and focuser—until you are ready to determine the length of the truss tubes. The reason for waiting is that nothing bad can happen to the secondary mirror or the focuser while they are sitting safely in their boxes—so let them sit. Only when you are ready to put them into service should you expose them to the dangers of fingerprints, curious children, and accidental falls to the floor.

6.4.1 Install the Focuser

Since you have already drilled a hole in the focuser board for the focuser, it's a snap to install. Put it in place, put bolts through the holes in the base, and it's done. Easy—in fact, too easy. To check that it is in the correct place, install the spider. Drill holes through the Kydex baffle, then push machine screws through each of the struts to engage the studs on the spider. Run in the machine screw tight enough that the spider makes a nice "twang" when you pluck the vanes.

Make a dummy secondary mirror from a sheet of stiff paper. You can photocopy the special ellipse template in this book (**Figure 11.2**) and cut it out with scissors. The template is a 45° ellipse with a minor axis equal to the minor axis of the secondary mirror, and it has lines down the major and minor axes. If you plan to offset the secondary, make a mark at the offset center, some small distance down the major axis of the ellipse. Install the dummy in the secondary holder with the major axis oriented so that the offset center is closer to the focuser than the true center.

Fig. 6.24 This close-up shows how the latches hold the secondary cage of a 36-inch Dobsonian in place. Even in a monster telescope, the bulky secondary cage weighs relatively little, so the latches that hold it can be ordinary light-duty hardware-store latches.

Install the secondary holder in the spider hub. Be sure to leave an extra ¼ inch between the spider hub and the holder base. Place a coarse alignment sighting tube in the focuser. The cross hairs in the sighting tube should coincide with the intersection of the major and minor axes if you do not plan to offset the secondary, or with the offset center if you are offsetting the secondary. **Figure 11.10** shows how this will look to you.

If you do not see perfect alignment, not to worry. This is why you left some leeway for adjustments. Move the secondary up or down the tube until the cross hairs coincide with the true center or the offset center of the dummy secondary. If the focuser board is not strictly straight, you may be able to achieve alignment by shimming the focuser. If you figured the distances correctly, between shimming the focuser and shifting the secondary along the optical axis, you should see the center of the secondary neatly on the cross hairs of the coarse alignment tube.

If the alignment error along the optical axis is greater than ½ inch or so, you figured something wrong, and you may need to change the position of the secondary mirror by moving the spider. Better that you found it now rather than later, but it's no fun to drill another set of holes for the spider vanes.

Chapter 7
Building the Mirror Box

7.1 Holding It All Together

The mirror box is one of most important components of your telescope. It carries the tailgate and mirror cell, provides a foundation for the altitude bearings, serves as the base for the truss pole assembly, and provides a secure storage place for the most expensive and fragile part of the telescope, the primary mirror. Despite its importance, the construction of the mirror box itself is fairly easy. The toughest parts to make are the altitude bearings and truss pole attachments, components that attach to the mirror box proper. As always, attention to design, materials and good carpentry determine the final level of performance.

7.1.1 Why the Mirror Box is Stubby

The low profile of the mirror box results from and contributes to a whole group of design features that greatly reduce the total weight of the modern-style Dobsonian. The light secondary cage and truss tube shift the center of gravity of the tube toward the mirror, allowing the mirror box to become shorter, lighter, and more open. And, because the mirror box is short, the rocker can be lower, lighter, and more stable.

You may think that these "little details" do not much matter, but they do. Consider one small detail that shaves a couple inches and a few pounds off the telescope: the lower corner of the mirror box is rounded off so that the mirror clears the bottom of the rocker by about an inch. This allows the rocker box to be about two inches shorter. The rounded-off corner is always hidden by the sides, so it doesn't even look funny.

But each time you climb the observing ladder, you benefit from this little detail; that two inches of height you saved adds up to less fatigue in your legs and more starlight in your eyes. The telescope is also smaller, lighter, and easier to transport.

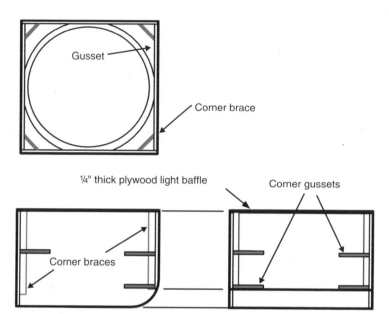

Fig. 7.1 *The mirror box is a square wooden structure reinforced with interior braces and gussets for greater strength, and a thin light baffle across the top. The lower corner is cut away so that the telescope can swing in a very low rocker.*

7.2 Mirror Box Materials

Use the best grade of plywood you can afford. Good plywood makes for stronger corners and reduces the potential for warping. For telescopes up to 25 inches aperture, go with ⅝-inch hardwood-veneer hardwood-core (HVHC) plywood; for bigger telescopes, go with ¾-inch HVHC. If top-notch HVHC plywood, such as ApplePly, Baltic Birch or Finn Ply, proves hard to find, you can fall back on any of the multi-ply marine, aircraft, or cabinet-grade plywoods.

The very best plywood for a large portable Dobsonian is ApplePly. Produced in the United States, the face veneers are a gorgeous species of maple and the core laminations are made from ¹⁄₁₆-inch veneers of Western Red Alder. This wood is a strong, lightweight hardwood that weighs one-third less than birch, so a telescope built with ApplePly weighs less than one built with Baltic Birch or Finn Ply. With ApplePly you get an unusually high strength-to-weight ratio, just what you want for a big plywood telescope that you haul around the countryside. And unlike European plywoods which are sold in 5 by 5-foot sheets, ApplePly is available in the standard 4 by 8-foot size.

In purchasing top-grade HVHC plywood, your goal is not prettiness but strength. The appearance of the face veneer is not important—even though these plywoods usually look great. What matters is inside: how many plies and what they are made of. Ordinary birch veneer plywood has half as many layers, and it's full of voids, defects, and cheap softwood fillers; in short, it's junk wood with a

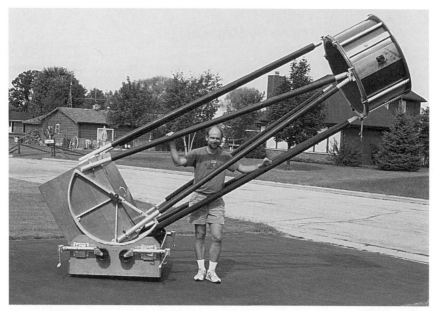

Fig. 7.2 *The mirror box is the unsung hero in the Dobsonian: the central component that joins the mirror, the mounting, and truss tube. By keeping the mirror box short and open at both ends, the mirror cools rapidly and you see sharper, clearer images when you observe.*

pretty face. Use HVHC. You get what you pay for.

7.3 Designing the Mirror Box

The dimensions for the mirror box depend on the size of the primary mirror and the balance point of the telescope (see **Table 7.1**). The size of the primary mirror defines the width of the mirror box, and the balance point determines its depth. Since the balance point depends on a variety of other factors, we will first look at the width and depth of the mirror box.

7.3.1 Height and Width of the Mirror Box

Because of the free flow of air through the mirror box, it is not necessary to leave a large gap between the mirror and the sides of the mirror box; experience shows that a gap of ½ to 1½ inches is sufficient. This gap allows enough clearance for installing and removing the mirror. For a typical 20-inch telescope, a box with inside dimensions of 22¼ by 22¼ inches is sufficient. We recommend that you construct the box ¹⁄₃₂ inch wider than nominal. If the steel tailgate cell turns out a bit larger or slightly out of square, it will still fit into the mirror box. It is easier to shim any open space between the box and the mirror cell than to split open a side of the mirror box because the tailgate fits too tight.

These dimensions assume that you follow our recommendation later in this

Table 7.1
Recommended Mirror Box Dimensions

Aperture	ƒ/ratio	Box Inside Dimensions	Minimum Depth	Thickness of Plywood
15	4.5	16½	15	½
18	4.5	20¼	16	⅝
20	5.0	22¼	19	⅝
25	5.0	27½	22	⅝
30	5.0	32¾	27	¾
36	5.0	39	32	¾
40	5.0	44	38	¾
All dimensions in inches.				

chapter and attach the truss poles to the outside of the mirror box. We recommend this because it makes for a stiffer truss and a lighter, more compact telescope. Since the inner diameter of the mirror box generally turns out to be about 5% larger than the inner diameter of the secondary cage, the truss poles will have to be angled a degree or two inward. However, you want the truss poles angled slightly because the tube assembly is then as compact as possible without blocking any incoming light and the cloth light shroud can be pulled down to a snug fit around the truss poles.

7.3.2 How Balance Affects the Mirror Box

Finding the depth of the mirror box is a Catch-22 situation: you cannot determine the precise balance point until the telescope is built but you cannot build the telescope until you know the balance point. The solution of this dilemma is to *calculate* where the tube will balance and then build the telescope. Let's now look into how this is done.

The depth of the box depends on the balance point of the completed telescope tube. The tube consists of the mirror box, truss poles, and secondary cage. By design, the balance point of a modern-style Dobsonian is at the top edge of the mirror box, exactly.

Balance is achieved by adjusting the depth of the mirror box and the length of the truss poles. If you follow the plans in this book, your telescope will either roughly balance, or end up just a bit bottom-heavy. However, if you want to build a telescope with a different size primary, focal ratio, or mirror thickness, then you will have to determine how deep to make the mirror box so that the tube assembly balances at the top of the mirror box.

There is just one rule for balancing:

Make the mirror box deeper than the calculations call for.

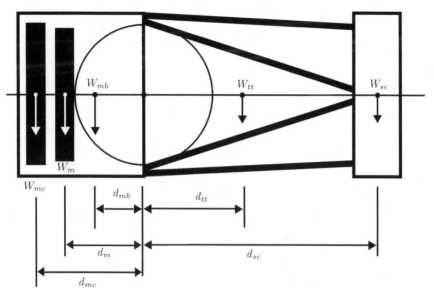

Fig. 7.3 *The key to a balanced Dobsonian is calculating the height of the mirror box. This crucial dimension affects not only the torque contributed by the mirror box, but also the torques generated by the mirror and its cell. By design, the balance point of the telescope occurs precisely at the top of the mirror box. W = weight and d = depth of components.*

If the mirror box turns out to be too shallow, the telescope will come out top-heavy. With a top-heavy telescope, you have to add a lot of dead weights to achieve balance. If an *f*/5 telescope turns out three pounds top-heavy, then you will have to add 15 pounds to the tailgate. With a bottom-heavy telescope, you might have to add a few ounces to the secondary cage to fine tune the balance.

By making the mirror box an inch or two deeper than your calculations, you give yourself some leeway. If you build the telescope and find out that the mirror box is too deep, don't add weights. Simply move the tailgate up, cut the excess off the mirror box, and reattach the tailgate. The problem is neatly solved.

Important note: When you *shorten* the mirror box, you must compensate by *adding* to the truss poles. To avoid having to purchase a whole new set of pole tubes, read **Section 8.1.3** on determining the truss pole length. The idea is to figure out what length the poles should be and then cut them two inches too long. After everything is working right, you trim the poles to their final length.

7.3.3 Calculating the Balance Point

The balance point for a telescope is calculated using torque equations which are basically eighth grade math. It's fun to play "what if" games with the design. But before you get down to determining the depth of the mirror box, you need to know two things:

- the weight of the secondary cage, and

- the distance between the primary and secondary mirrors.

You will also need to make pretty good guesses about some other weights and dimensions, then proceed toward the final dimensions through a series of approximations.

The total weight of the secondary cage includes everything you will *ever* attach to the secondary cage. The most accurate method is to weigh the completed secondary cage with the secondary mirror installed, the focuser in place, the truss pole connectors attached, and a Telrad, finder telescope, or digital setting circle computer installed—plus the heaviest eyepiece you own in the focuser.

In the normal process of building a telescope, you will have completed the secondary cage before the mirror box, so you can weigh it. If you have not yet completed the secondary cage, you will need to make an educated guess. In a typical 20-inch telescope, a fully-loaded secondary cage weighs about 15 pounds.

The distance between the center of the secondary mirror and the primary equals the focal length of the primary mirror minus the distance from the secondary to the focal plane. The only variables that you have any control over are the height of the focuser and the diameter of the secondary cage. Measure the inner radius of the secondary cage and add the thickness of the focuser board and the height of the fully racked-in, low-profile focuser.

In a typical 20-inch telescope, the secondary cage is 21 inches inside diameter, so the radius is 10½ inches. The focuser board is ¼ inch thick. With the draw tube fully racked in, a typical commercial low-profile focuser is 2½ inches high. To that, add another ¼-inch for focuser travel.

The distance from the secondary mirror to the focal plane at the top edge of the racked-in focuser is thus:

$$10½ + ¼ + 2½ + ¼ = 13½ \text{ inches.}$$

If the focal length of the primary is 100 inches, then 13½ inches is used up between focus and the diagonal, leaving 86½ inches from the center of the diagonal mirror to the vertex of the primary mirror.

However, the focal length of the primary mirror may depart by an inch or two from the nominal figure. That's okay. For now you only need to know approximately how far the secondary cage will be from the surface of the primary mirror so you can get a rough idea of the location of the balance point of the tube assembly and from that calculate the depth of the mirror box. In this example, the rounded-off distance between the primary and secondary mirrors will be close to 88 inches. You can adjust everything later.

For starters, pick an arbitrary mirror box depth: just declare that it's 20 inches deep. Now, the surface of a 2-inch thick mirror sitting on a 2-inch thick cell assembly will lie about four inches from the bottom of the mirror box, so the surface of the primary mirror will be about 16 inches below the top of the mirror box. We already determined that the distance between the mirrors would be around 86½ inches, so subtracting the 16 inches that the primary lies from the top of the mirror box, you find that the top edge of the mirror box lies about 70½ inches from the

Fig. 7.4 *The mirror box holds the different functional units of the Dobsonian telescope to-gether. The mirror cell, the truss tubes, and the side bearings attach directly to the mirror box; remove the mirror box and you no longer have a telescope.*

center of the secondary cage.

Now figure out the weight of the truss tubing. You can get an approximate weight from the pounds per foot column in **Table 8.1**. The truss poles will be about 70 inches long. In that table you see that 1¼-outside diameter tubing with an 0.049-inch wall weighs 0.217 pounds per foot. Eight poles each 70 inches long totals 560 inches, or 47 feet. Multiply this by the weight in pounds per foot:

$$47 \text{ feet} \times 0.217 \text{ pounds/foot} = 10.2 \text{ pounds.}$$

Call this 11 pounds, since you will want to wrap the poles with foam pipe insulation. It looks nice and prevents frozen fingers and a banged head.

The idea behind torque equations is this: to balance the telescope, the total torque trying the rotate the telescope clockwise must equal the total torque trying to rotate it counterclockwise. When the opposing forces are equal, the telescope is balanced. The torque from any component of the telescope is just the weight times the distance from the axis of rotation.

The center of rotation is the top edge of the mirror box, a "given" designed into the truss-tube Dobsonian. You know the approximate weight of the secondary cage (15 pounds) and the truss poles (11 pounds), and that's everything above the center of rotation. You also know the approximate distance of the secondary cage (70 inches) and the center of gravity of the truss poles (half their length, or 35 inches), and that's all you need. Of course, the figures for your telescope will probably be somewhat different.

The torque these components exert on the tube is their weight times the distance from the axis of rotation. So, if you can add and multiply, you can do torque equations.

Secondary cage torque: 15 pounds x 70 inches = 1050 inch-pounds

Truss pole torque: 11 pounds x 35 inches = 385 inch-pounds

Total torque: 1435 inch-pounds

There—that's the total torque needed to balance the telescope below the fulcrum. But this is not what it *really* will turn out to be. You'll soon be adding a few gadgets and then a few more gadgets, and so the torque at the upper end will rise. External light baffles, little star atlases, red reading lights, coma correctors, filters, Barlows, tiny tape recorders, digital setting circles, bags of Fritos, and on and on and on. The upper end of the tube will never be lighter than the day you finish building it.

So be aware of this and build the mirror box deep enough to provide sufficient torque to overcome these inevitable add-ons. Don't think you're immune. Rational thinking will fail as you contemplate the ads for various trinkets.

Unless you plan to add lots of heavy weights to the bottom of the mirror box, plan for at least another pound of gear hanging off the secondary cage. Doing the math:

1 pound x 70 inches = 70 inch-pounds.

And for good measure, add a bit more for a new wide-angle eyepiece for a total of 100 inch-pounds. With this addition, the total torque above the center of rotation is 1535 inch-pounds. To balance the telescope, you'll need at least 1535 inch-pounds below the top of the mirror box.

7.3.4 The Depth of the Mirror Box

Even though you have not yet built the mirror box, you can figure out the amount of torque it will have. You can do this knowing the weights of the mirror cell, the primary mirror, and the weight of the plywood you will use for the mirror box.

If you haven't fabricated the mirror cell, just pile all the parts together on a scale to get their total weight. If you are waiting for the primary mirror to arrive, phone the optical company and get the exact weight. The weight of the mirror box is calculated from the weight of the plywood you intend to use. For ⅝-inch Baltic Birch, figure two pounds per square foot. For other types, weigh a sheet of plywood on the bathroom scale and divide the total weight by the area in square feet.

Now, determine the total area of the mirror box (see **Table 7.1**). You already figured out that it will be 22¼ inches square inside. If it matches your guess of 20 inches deep, then:

> 23½ inches x 20 inches deep x 4 sides = 1880 sq-inches
> Divide by 144 sq-inches/sq-ft = 13 sq-feet
> 13 sq-feet x 2 pounds/sq-ft = 26 pounds.

With corner bracing and gussets, you should anticipate adding a couple more pounds, so a 20-inch deep Baltic Birch mirror box will weigh around 28 pounds. The optical shop says that your 2-inch thick 20-inch primary weighs 50 pounds, and the pile of mirror cell parts weighs in at 21 pounds. You have everything you need to calculate the torque.

For each component, the distance used to calculate the torque is measured from the axis of rotation at the top of mirror box to the center of gravity of the component. The mirror cell is roughly 2 inches thick so its center of gravity is one inch in from the bottom of the mirror box, or 19 inches from the top of the box. So figure the torque as:

> 21 pounds x 19 inches = 399 inch-pounds.

The center of gravity for the primary mirror lies about 3 inches from the bottom of the mirror box. You get that because the 2-inch thick mirror sits on the 2-inch high mirror cell. The center of gravity of the mirror is in its center, 3 inches from the bottom of the mirror box and 17 inches from the top of the mirror box. Figure the torque:

> 50 pounds x 17 inches = 850 inch-pounds.

Notice one significant advantage of using a two-inch thick mirror: it really helps balance a low-profile telescope. As an exercise, calculate the balance point for a 1½-inch-thick mirror that weighs 38 pounds.

The center of gravity of the mirror box lies 10 inches from the top of the box:

> 28 pounds x 10 inches = 280 inch-pounds.

In theory, we should add the torque contributed by the side bearings, but their contribution is small and hard to calculate, so we'll ignore them for now. Adding the torques of the various components:

> Mirror cell: 399 inch-pounds
> Mirror: 850 inch-pounds
> Mirror box: 280 inch-pounds
> Total torque: 1529 inch-pounds.

It certainly looks like we cooked the numbers! The upper end exerts a torque of 1535 inch-pounds and the lower end exerts a torque (in the opposite direction) of 1529 inch-pounds. The telescope is slightly top heavy (by 6 inch-pounds).

Try some paper experiments. Suppose you make the mirror box 21 inches deep instead of 20 inches. What happens? Well, here's how it works out:

Fig. 7.5 *Although it appears to be no more than a big crude box, the mirror box is a precise component of your telescope. It is crucial that you make every effort to construct the mirror box as close to square as possible. The top surface must be flat, and the entire structure must be strong.*

> Mirror cell: 21 pounds x 20 inches = 420 inch-pounds
> Mirror: 50 pounds x 18 inches = 900 inch-pounds
> Mirror box: 30 pounds x 10.5 inches = 315 inch-pounds
> Total torque: 1635 inch-pounds.

Now the telescope is bottom-heavy by 100 inch-pounds. This shows how useful it can be to add an extra inch or two when you cut out the panels for the mirror box. A small difference in the depth of the mirror box makes a large difference in the balance.

If you're into such things, set up a spreadsheet to calculate the balance. As you complete components such as the secondary cage, or when components such as the primary mirror arrive, weigh them and then run the spreadsheet again. As the telescope nears completion, you will have tight control over its balance.

7.4 Balancing the Telescope

When all is said and done, every telescope gets built and then gets fixed. Even if your torque numbers are close, a bit of fine tuning always remains. Eyepieces such as the 35 mm Panoptic are extremely heavy; when you switch to your trusty 6 mm orthoscopic, the telescope drifts up. It can be tricky balancing the scope for such a wide range of eyepiece weights. Don't give up—get it right. A properly balanced smooth moving telescope is a joy to use.

The remedy for a bottom-heavy telescope is to add some expensive gadgets to the top. (If you are a cheapskate, just add weights.) If you are really close to balanced and are just a teeny bit bottom-heavy, try putting an extra set of batteries inside the Telrad. Whatever goodies you add, secure them well so they don't fall onto the primary.

As a general rule, it is better to have a telescope that is slightly bottom-heavy. You can correct the balance by adding a few ounces to the top end. It's not so easy with a a top-heavy telescope.

The remedy for a top-heavy telescope is to add weight to the tailgate. You will need the excess weight at the top times the focal ratio of the mirror to counterbalance the tube assembly. If you want to install a jumbo-size finder on the secondary cage, or if you often observe with very heavy eyepiece and Barlow combinations, you will have to add *lots* of weight.

In a typical 20-inch *f*/5 telescope, you must add *5 pounds* of steel flat stock to the tailgate for every *1 pound* you add to the top end. You can obtain strips of flat steel stock from your local welding shop. Ask for 1½-inch wide by ¼-inch thick. Have them cut pieces 20 inches long for a 20-inch telescope and 25 inches long for a 25-inch scope. In other words, make them the same length as the primary aperture. Drill two ⅜-inch holes in the flat stock to match the locations of the two carriage bolts that hold the tailgate to the mirror box.

Replace the carriage bolts on the mirror box with 3-inch carriage bolts. This will let you put as many strips as you need to balance the scope. Attach the strips of steel one at a time to the upper rung of the tailgate while the scope is fully assembled.

An alternative to adding steel weights is to balance a telescope with chain. See **Figure 11.21**. At the hardware store, purchase a couple of S-hooks, a threaded eyebolt that will accept an S-hook, and a 6-foot length of chain that weighs about 10 pounds. If you decide to add a small finder telescope to the secondary cage, then buy four S-hooks, two eyebolts, and 20 pounds of chain.

With the scope fully assembled, drill a small hole through the top rung of the tailgate and install the eyebolt. Make sure it clears the rocker when the scope is rotated towards the zenith. When you go observing take the chain. Attach one end of the chain to the eyebolt in the top rung of the tailgate using the S-hook. Leave the rest hanging free. As the scope is turned upward, the chain will pile up on the ground.

If you need more weight, hang both ends of the chain on the S-hook. By moving the free end of the chain up or down, or by changing where you hook the chain on the S-hook, you fine-tune the balance of your telescope. Pretty soon you'll become an expert.

The chain method is dynamic. The chain applies more weight when the telescope is aimed toward the horizon and less weight when the scope is aimed higher. This happens automatically because the chain piles up on the ground as the scope is aimed upward. The chain may clank a bit when you slew in azimuth, but what do you care? You're among the galaxies . . .

Another way to balance a top-heavy scope is with a bag full of lead shot. Sew a strip of Velcro to the side of a cylindrical cloth bag. Put a zipper or drawstring

Fig. 7.6 Internal braces and gussets add strength to the mirror box. Note that the corner brace ends 2 inches from the bottom of the mirror box to allow room for tailgate installation.

on one end. Insert a Ziploc plastic bag and fill it with shot (shot won't escape from the Ziploc bag). Glue a mating strip of Velcro the length of the mirror box. The Velcro lets you stick the bag of shot to the mirror box. You can adjust the amount of shot in the bag or change the position of the bag on the strip of Velcro on the mirror box to achieve balance.

The bag of shot is a very forgiving counterweight. If you drop a 25-pound bag on your toes, it won't hurt. Drop a solid steel counterweight on your toes and you're off to the emergency room. Spoils your whole night!

Balancing telescopes is a never-ending process. One warm, dry night you will observe using the telescope with the light baffle off, and on a later night, several pounds of dew will weigh it down. You may want to observe with a big Barlow and a heavy Nagler, or you could get tired of star-hopping and attach digital setting circles to the secondary cage. Whatever happens, keep the telescope balanced and you will be a happy observer.

7.5 Split Blocks for the Mirror Box

Split blocks connect the truss poles to the mirror box. The design and fabrication of the split blocks is covered in **Chapter 8**, but in this section we consider *where* the split blocks (or any of the alternative truss pole attachments you may decide to use) should be mounted on the mirror box. This seemingly trivial point has a major effect on the dimensions of the telescope.

For a given size mirror box, mounting the truss poles outside the mirror box allows it to hold a primary mirror approximately two inches larger than the same mirror box with the poles mounted inboard. Why would anyone build a mirror box

Table 7.2
Recommended Side Bearing Dimensions

Telescope Aperture	Side Bearing Diameter	Side Bearing Thickness
15	22	1⅛
18	24	1⅜
20	26	1½
25	32	1⅝
30	37	1¾
32	39	2
36	44	2⅜
40	48	2½
All dimensions in inches.		

that *could* hold a 25-inch mirror and instead install a 23-inch mirror?

Another advantage is that you can install a plywood light baffle at the top of the mirror box. The light baffle serves as a shear web, that is, a flat membrane that keeps the top end of the mirror box absolutely square. The top of the mirror box carries an enormous load: all the torque from both ends of the scope meets there. Of all the structures you can think of, a square box with an open top is just about the weakest, but a square box with a shear web across it is among the strongest. Mounting the split blocks on the outside of the box also makes the truss tube assembly significantly stiffer.

If you are not yet convinced, here are a dozen advantages that you gain by mounting the truss pole attachments on the outside of the mirror box.

1. The mirror box and rocker are smaller.

2. The mirror box and rocker are easier to transport and store.

3. You can get your observing ladder closer so you don't have to lean as far to the eyepiece.

4. The light baffle greatly increases the strength of the mirror box.

5. You get a broader, more stable base for truss poles.

6. Construction is easier.

7. It is easier to the insert the poles in the dark.

8. There's less chance of dropping a pole on the primary during set up or take down.

9. The truss assembly is slightly conical so the shroud fits better and tighter.

10. The light shroud is farther away from the light path; this matters on nights with heavy dew.

11. Dirt and moisture from the truss poles stays on the outside of the box away from the mirror.

12. You can hook the bungee cords from the shroud to the clamping knobs of the split blocks.

Now you know.

7.6 The Side Bearings

Good looking side bearings have become a Kriege trademark. They are the most frequently copied component of his design though most amateur telescope makers and professional telescope makers don't understand *why* they work so well. In a nutshell, it's all because of the way friction works in Teflon bearings.

Dobsonian bearings rely on friction. When Formica slides over Teflon, there is an optimum velocity where the bearing friction increases when the speed increases. When this happens, the bearing acts like a speed brake, resisting more as the telescope goes faster. With small bearings, the velocity is too low for the speed-brake effect to occur, and the bearing tends to bump and squeak along.

There is a practical consideration, too. If the side bearings are tiny, they sit entirely inside the truss poles. As the bearings get bigger, they eventually begin to interfere with the truss poles. When the bearings become really huge, however, they are wider than the truss poles, so all eight split blocks can be mounted on the outside of the mirror box.

Here are some advantages of large side bearings.

1. Formica on Teflon reaches optimum velocity for ease of guiding.
2. The bigger the bearings, the shorter and stiffer the rocker sides can be.
3. The lower the overall height of the mirror box and rocker assembly, the more compact and easier it is to transport.
4. Big bearings reinforce the mirror box.
5. Big bearings transfer the weight of the tube assembly over a broader more stable base.
6. Big bearings match the azimuth bearing diameter to equalize the force needed to move the scope in both axes.

7.6.1 Making the Mirror Box

Most folks look at a Dobsonian and come away with the impression that anyone could make the mirror box. The truth is anyone can make a mediocre mirror box, but it takes knowledge and skill to make a good one. The criteria for a good mirror box are:

- *The sides of the box must be square.* All four corners must have 90° angles or the box will scuff against the inside of the rocker.

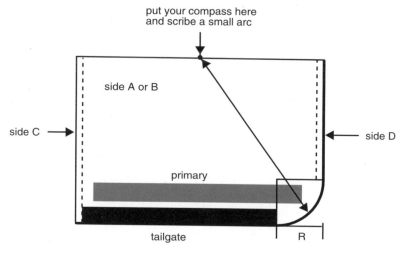

put your compass here and scribe a small arc

side A or B

side C →

← side D

primary

tailgate

R

Fig. 7.7 The curve of the cut-off corner on the mirror box begins at the end of the side rails on the tailgate and ends on the bottom of the D side of the box. The cut can be any shape you like—even a straight cut—but a smooth curve looks best.

Or worse, the tailgate won't fit into the box if the box is out of square.

- *The top edges of your mirror box must lie in a plane.* Hold a cardboard shoe box in your hands. The top four edges are square to the sides. With a hand on each end, twist the box and notice how the top edges no longer lie in the same plane.

- *The mirror box must be strong.* Obviously, structural failure would be a catastrophe: the box is home for your precious primary mirror. But just as important, a weak mirror box flexes. Cheap plywood and poor carpentry produce a box that is too elastic. Excessive elasticity shows up at high power when you are trying to aim the scope—the scope rebounds and bounces when you try to move it.

7.6.2 Cutting the Sides

Meeting the criteria for a good mirror box is not that difficult. The key is to cut the four plywood sides as accurately as you can. A good table saw is best. If you don't have a table saw, a hand-held circular saw will work but it's a lot tougher to make the cuts accurate. Be sure the saw blade is sharp and perpendicular to the fence. Have someone help you pass the big sheets through the saw so the blade does not wander and produce an uneven cut. And, of course, measure accurately and always cut on the same side of the pencil line.

If possible, lay out all four sides next to each other on one sheet of plywood and rip them with one pass. This insures that the sides all have the same height.

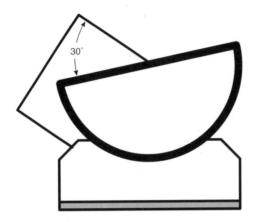

Fig. 7.8 *The center of rotation of the side bearing is located at the top edge and center of the mirror box. The diameter span should be aligned 30° away from the top edge. This allows the telescope to be rotated from horizon to the zenith.*

After each setup, check your work. Stack the four sides together and set them on edge on a table. The top edges should be level and even.

Check all four corners of each side with a steel framing square. If any panel is not perfectly square, or does not match the others, or is not cut properly, then get rid of it and cut out a new one. The box is only as good as its sides.

Cut out the lower corner of the sides of the box (sides A or B in **Figure 7.7**). It may not seem like a big deal, but cutting off the corners lets the mirror box nest inside the rocker about two inches lower, so that the rocker sides don't have to be as tall. The arc of the cut should begin along the bottom edge of sides where the tailgate ends. Set the tailgate frame along the bottom edge of the sides and allow for the thickness of the plywood on the opposite side of the box (side C in **Figure 7.7**). Scribe an arc with a stick compass and cut off the arc with a saber saw or band saw.

If you have already cut out the side bearings, or if you are using commercial side bearings, drill a couple of reference holes through both the side bearings and sides of the mirror box. Doing so at this stage insures that the bearings will be concentric after the mirror box is assembled.

Stack the two bearings together and align the outer edges. Drill a pair of mounting holes through the bearings while they are stacked. Stack the opposite sides of the mirror box together and align the edges so that they are even. Place one of the side bearings on top of them. Align the bearing so that the center of rotation of the bearing is at the top edge and at exact center of the mirror box sides. Align the diameter of the bearing so that it forms a 30° angle with the top edge of the sides. Using the holes in the bearing as guides, drill ⅛-inch diameter pilot holes through both sides (see **Figure 7.8**). You can enlarge the pilot holes later for the mounting bolts. If you make a mistake, it is easier to fill or hide a small hole than a big one.

We describe the alignment for side bearings in **Section 7.6.9**.

Before assembly, make four gussets and four corner braces. Gussets are tri-angular or curved webs of plywood that you attach to the middle of each corner. A corner brace is a narrow strip of wood that runs along the length of each corner. The gussets and braces contribute to the strength of the mirror box.

Cut the gussets from scrap plywood left over from cutting the outside of the rings for the secondary cage. The inner side of each gusset will be curved and will have the same radius as the outside diameter of your secondary cage rings. Trim the outer edges of each gusset until there is an inch of clearance between the light path and the gusset. This clearance leaves a bit of room for your hands and facili-tates mirror insertion. Be sure to trim the two outer sides of each gusset perfectly square. Check them with a framing square. Make all four gussets the same dimen-sion. Compared to plain triangular gussets (which function just as well), curved gussets look more attractive. Besides, you were going to throw away the second-ary cage scrap, so why not put it to good use?

Corner braces strengthen the corners dramatically, tripling the surface area of the joints. Cut the braces 2 to 3 inches wide; the mating edges of each brace make a 45° angle with the hypotenuse. Each corner brace should extend from the top of the mirror box to the top of the gusset and then from the underside of the gusset down to the top of the tailgate (see **Figure 7.6**). Be sure to leave sufficient room at the bottom for the tailgate! It's very discouraging to find that the braces are too long after they have been glued into the box.

7.6.3 The Light Baffle

A thin plywood light baffle will be glued to the top of the mirror box. Use a good quality ¼-inch plywood for this. Rout the hole for the light path before you glue it to the mirror box so if you botch cutting it, you don't have to chisel the baffle off the box. The hole is made as described in **Section 6.2.2** where you cut out the cage rings. Cut the hole ½ inch smaller than the inside dimensions of the mirror box; that is, for a mirror box that measures 23½ inches across the inside, cut a hole 23 inches in diameter. This will leave just enough material to cover the end grain of the mirror box at the middle of each side. Use the cutout to reinforce the dust cover (see **Section 7.8**).

7.6.4 Assembling the Mirror Box

To construct the mirror box, you will need:

- the four plywood sides,
- three to six bar clamps or pipe clamps,
- waterproof aliphatic wood glue,
- a hammer,
- ¾- and 1½-inch finishing nails,
- a damp rag, and

Fig. 7.9 *The two side bearings must be perfectly matched. After clamping the two plywood semicircles together, sand the circumference and diameter faces so the two bearings are identical in size and shape.*

- a large, flat surface, such as a table top, to work on.

Assemble the box top-side-down on your work surface. Working with three of the sides (and therefore, two corners), spread glue along the mating surfaces. Tack each corner with three or four finishing nails: these serve to keep the edges aligned while you put the clamps in place. Without nails, the edges slide on the slippery glue and make clamping a nightmare.

Some folks prefer wood screws to nails, but we don't. Screws can split the end grain of the plywood and they leave big holes to fill over later. Since it is glue that joins the wood and not the nails or screws, take the easy way out: use slender finishing nails.

Glue and nail the remaining two corners.

Wipe excess glue from the outside of the box with a damp rag after nailing and again after clamping. If you don't wipe off the excess glue, wood stain cannot penetrate the wood and you will see every ugly splotch of glue. If you plan to paint your scope, then don't worry.

Do not worry about how excess glue will look inside the box: the inside corners are hidden beneath the corner braces, but wipe the inside so the braces and gussets will fit tightly against the sides of the box.

So far it's been easy. Remembering that the box is upside down, check to see that all four of the top edges evenly contact the flat table top. If they do not, then the top edges of the box are not in the same plane. Push or pull the box into shape. Next, square the corners with the aid of the framing square. When all four corners are square and all four top edges contact the table top evenly, apply the clamps.

Do not even *think* about clamping the box until it is square and all four top

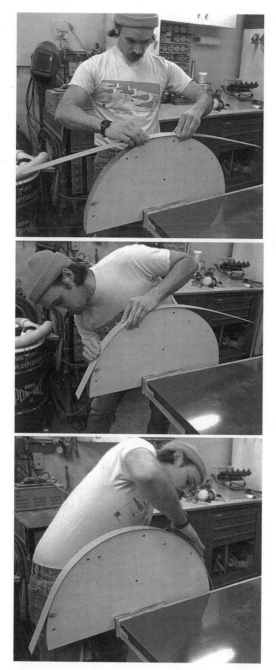

Fig. 7.10 *Pete demonstrates how to bond a strip of laminate to the bearing. Coat each surface with contact adhesive and allow the surfaces to dry until they are no longer tacky. Place the middle of the strip on the center of the bearing, then press down one side and then the other. The contact cement immediately forms a nearly unbreakable bond.*

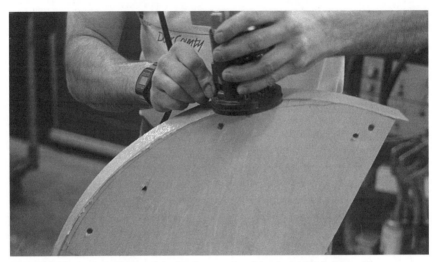

Fig. 7.11 *Clean up the edges of the laminate strip with a hand-held router. Wear eye protection and be careful as you do this: fine dust and sharp chips of laminate fly in all directions. Dull the sharp edges of the laminate with a fine file or with sandpaper.*

edges are in contact with the flat table top. After the bar clamps are on, check the corners again with the framing square. If the box is out of square, remove the clamps, squeeze the box into shape with gentle tapping of the hammer on the corners, and reclamp it. Wipe away all excess glue. Allow at least 24 hours for the glue to set. Don't poke, prod, or play with the box. Leave it alone while the glue sets.

After 24 hours, it is okay to remove the clamps. Install the four corner gussets with glue and nails. Use a couple of scrap sticks of wood to force opposing gussets firmly into the corners while the glue sets. After the gussets are glued in place, install the corner bracing with glue and nails. Attach the ¼-inch plywood light baffle to the top of the box with wood glue and a few ¾-inch finishing nails. Place heavy weights along the edges while the glue sets.

Sink all the outside nails slightly and fill the holes with wood filler. With a router, radius the outside of all four corners, the top edges where the light baffle joins the sides, and the four bottom outside edges. Use a ¼-inch rounding-over bit for this. Do not use the router on the inside edge of the light baffle. Instead, break the inside top and bottom edges of the light baffle by hand with sand paper.

7.6.5 Side Bearing Dimensions

The side bearings on a Kriege-style Dobsonian are large semicircles. Before the Kriege design, the bearings on Dobsonians were full circles, but the top half of the circle is never used—so why lug around the extra weight? The bearings are so large that the two lower truss split blocks on each side lie inside the bearing. In a typical 20-inch *f*/5 telescope, the mirror box is 23½ inches wide and carries side

Fig. 7.12 *One way to insure that the bearings are mounted exactly the same way on both sides of the mirror box is to make a template from ⅛-inch hardboard. The template shows the outlines of both parts and has pilot holes drilled in it. Place the template against each part and use the pilot holes to position the holes in the bearings and mirror box.*

bearings 26 inches in diameter. However, because the bearing is a half-circle mounted at a 30° angle, the side bearings extend beyond only one side of the mirror box.

The thickness of the side bearings depends on the thickness of the sides of the rocker box and the clearance between the mirror box and the inside of the rocker. Because the face of the bearing should fit flush with the outside of the rocker, the thickness of the bearing equals the thickness of the rocker side plus a clearance of between ³⁄₁₆- and ¼-inch. Use the dimensions found in **Table 7.2.**

Fig. 7.13. *Wood glue and lots of pressure are the simple secrets of bonding plywood panels. Apply a thin layer of glue to both surfaces, tack the pieces together with finishing nails, and apply the weight of an automobile overnight.*

7.6.6 Bonding Flat Panels of Plywood

In the next section, you will need to bond two or more layers of plywood. Here's how to do it.

Spread a thin layer of glue over the entire mating surface on one of the components. A big spatula or a flat scrap of wood helps to spread it evenly. Set the other panel on top and align the edges. After they are perfectly aligned secure with four small finishing nails in the corners. These nails are very important as they prevent the two panels from sliding around on the slippery glue causing the edges to misalign. Now find a flat place on your driveway or garage floor. Drive your car so as to position one of the front wheels directly over this place. Set the emergency brake and chock the other three wheels so the car cannot move. Now jack up the wheel with a car jack. Under the wheel first place a scrap piece of plywood that is larger than the components you intend to bond. Next, place a layer of plastic food wrap (like Saran Wrap) over the plywood. Now place your workpiece on top of the plastic followed by two layers of scrap plywood that are approximately the same size as your work piece. Lower the car onto the stack and watch the glue ooze out all around the edges. Wipe away the excess glue with a damp sponge. Let your car stand on this pile for at least 12 hours.

7.6.7 Constructing the Side Bearings

After you have determined the diameter and thickness of the side bearings (see **Table 7.2**), laminate a couple of layers of plywood together with wood glue to achieve the desired thickness. For a 20-inch telescope, bond two layers of ¾-inch plywood to get the desired 1½-inch thickness. Ordinary softwood SVSC plywood is okay if you need to conserve cash, but the more expensive hardwood HVHC grades of plywood are better. Not only are they stronger, but they have fewer voids that, in the perverse nature of things, always seem to turn up at critical stress points.

With your stick compass, mark two disks 26 inches in diameter on a sheet of plywood. Be sure to mark the centers clearly so you'll know where to cut the full disk into halves. Cut the circles with a router and then bond them with wood glue, in that order and not the reverse. If you bond two layers of ¾-inch plywood and then try to cut a circle, you'll find it's too thick for the router bit.

Draw a pencil line through the center point and across the bonded circle, then cut it precisely in two. Clamp the semicircles together and edge-sand the circumference and diameter smooth so you end up with exact duplicates. Mounting them concentric on the mirror box depends on making the two halves identical. (If they are not identical, you will have trouble mounting the semicircles on the mirror box.)

Cut a strip of the laminate wider than the thickness of the bearing. Mark the middle of the strip and the middle of the circumference of the disk with a pencil and test fit it in place. You don't want to find out you've got too much laminate at one end and not enough at the other. Apply contact cement to the circumference of the bearing and to the inside face of the laminate. Let the disk and strip of laminate air

Fig. 7.14 If you want to get really fancy, you can have aluminum side bearings cast at a foundry. For a few hundred dollars, you'll have rock-solid side bearings that look really great. If you're trying to stretch your dollar, however, stick with plywood side bearings.

dry until the cement is no longer tacky to the touch. Starting in the middle of the strip and working out to the ends, press on the laminate with firm hand pressure.

Trim off the overhanging laminate with a router. You must use a special laminate-trimming bit to do this, so buy one. The easiest way to rout off the excess laminate is to clamp the bearing upside down in a vise and run the router around the circumference. **Caution:** Wear safety goggles. Chips of laminate and wood will pepper your face like a sandstorm. It would be an ironic tragedy to suffer a permanent eye injury while building an instrument to help you see the universe. If you are trimming glassboard, wear a respirator mask—you do not want glass fibers in your lungs! So that they won't cut your fingers, sand or file the razor-sharp edges of the laminate smooth and round.

7.6.8 Placing the Side Bearings

Mounting all the split blocks on the outside of the mirror box is a great idea. So are jumbo side bearings. The mirror box holds a bigger mirror, the telescope is more compact and moves like a dream, the light shroud fits better, and the tube assembly looks more attractive. The problem is that jumbo side bearings interfere with two of the split blocks.

To attach the truss poles to the outside of the mirror box with split blocks in the classic Kriege style, it is necessary to cut a section from the side bearings. This allows clearance for the two lower split blocks on each side. Do not actually cut the bearings until they are ready for permanent installation on the mirror box.

Place the bearing on the side of the mirror box so that its center lies exactly

Fig. 7.15 *The solid wood side bearing creates a problem: where to place the front split blocks? Unfortunately, it's tricky to mount the split block on top of the bearing because it has to be angled in steeply, and aesthetically, having the block outside the bearing looks terrible.*

at the top edge and center of the mirror box. With a protractor, angle the bearing so that the flat upper face is tilted 30° with respect to the top of the mirror box (see **Figure 7.8**). This orientation allows the telescope to point from the zenith to the horizon. (It works like this: swinging from the zenith to the horizon requires 90°. The altitude pads are 70° apart. The sum is 160° which is less than the 180° the side bearing provides.)

Set a split block (or a paper template of a split block if you haven't made them yet) on the bearing approximately where it will be mounted later. Turn it to approximate the angle it will have with a truss pole in it and then trace a pencil line around it. This is the minimum amount of bearing material to be removed to clear the block and permit it to be mounted on the outside of the mirror box.

Give yourself a little extra room especially along the circumference. If you use the plans and the bearing dimensions recommended in this book, you will find that the circumference adjacent to the split block may get rather narrow. Try to leave at least 1-inch thickness of plywood at the narrowest point.

If the bearing turns out to be too thin, move the split block inward a little to give this weak area more strength, or figure on making these two split blocks a bit narrower than the others. Mark the "footprints" of the blocks carefully on the side bearings.

Reminder: Do not cut the marked areas from the bearings until you are ready to attach the bearings permanently to the sides of the mirror box. At that time, you will cut the openings and reinforce the disk on the inside with a wedge of plywood.

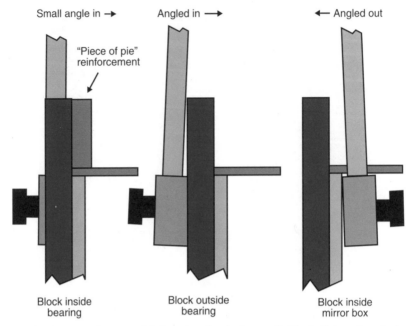

Small angle in →   Angled in →   ← Angled out

"Piece of pie" reinforcement

Block inside bearing   Block outside bearing   Block inside mirror box

Fig. 7.16 *There are three possible locations for the front split block: 1) inset into the bearing, 2) outside the bearing, and 3) inside the mirror box.*

7.6.9 Mounting the Side Bearings

The two semicircular bearings must be mounted on the mirror box so that they are precisely concentric. When you do this right, the two laminate surfaces will be sections of a single imaginary cylinder. The following steps insure a good result.

Step 1. Clamp the two completed bearings together and align the edges as accurately as you can.

Step 2. With the bearings still clamped together, drill three mounting holes through them. Use a drill press so the holes are precisely perpendicular to the surfaces of the bearings.

Step 3. For a thin plywood or hardboard panel, cut a template to the exact dimensions of the sides of the mirror box. Verify that the template matches both sides of the mirror box.

Step 4. Clamp one of the side bearings against the template with the center of the bearing at the center of side of the template that corresponds to the top side of the mirror box, and extend the mounting holes into the template.

Step 5. Place the template against the mirror box and continue the mounting holes into the side of the mirror box.

Step 6. Without turning it around, shift the template to the other side of the mirror box. Continue the mounting holes into the side of

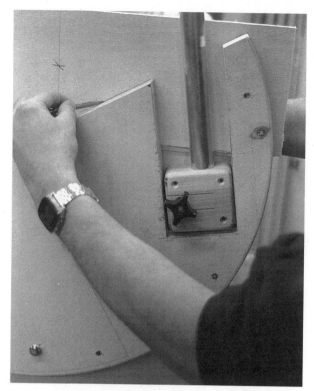

Fig. 7.17 *The inset-block solution introduces another problem: When you inset the split block into the side bearing, if the outer edge of the bearing is not at least an inch thick, the bearing may act "bouncy" at low elevations. Fortunately, you can easily reinforce the side bearing from behind.*

the mirror box.

Step 7. In the bearings, cut the openings for the split blocks that you had marked earlier. Be careful not to crack the narrow part of the rim.

Step 8. Reinforce the isolated portion of each side bearing with a pie-shaped piece of high-quality plywood. Glue a wedge to the inside face of each bearing. Radius the outer edge of the wedge to match the radius of the bearing.

Using the template insures that the mounting holes on opposite sides of the mirror box are located in exactly corresponding positions, and also that they match the holes on the side bearings.

7.7 To Shortcut or Not To Shortcut?

By now you may have realized that building a top-notch telescope is a lot of work.

Fig. 7.18 Here is the solution that saves the day. A pie-shaped piece of plywood reinforces the section of the side bearing that making the cutout has weakened. As an added benefit, the pie-shaped piece makes the side bearing look nice because it fills in the half-circle silhouette of the bearing.

Perhaps you could cut some corners by mounting six of the split blocks outside the box and mounting the two lower blocks inside. On his first Dobsonian, Dave did just that but he found that it was actually more work. The two inside blocks must be angled outward the same amount the other six are angled inward. In addition, you must provide holes where the poles pass through the light baffle. This was more work than making the side-bearing cutouts described above.

A simple alternative is to mount the lower split blocks on the outside of the bearings. You'll need to angle the blocks so that the truss poles mounted on them line up with the secondary cage.

A second alternative is to make the bearings hollow "arcs." Made from cast aluminum, such bearings work very well indeed on telescopes under 25 inches aperture. Although aluminum castings tend to be expensive, almost every city has an aluminum foundry so it's easy to get the work done. Make a pattern from wood and have the foundry cast them in metal; with good eyepieces costing hundreds of dollars, investing a few hundred bucks on aluminum castings may not seem so bad. The resulting telescope will be the envy of every amateur who sees it. Or purchase them from one of the suppliers listed in **Appendix E**.

If aluminum is too expensive for you, you can make arc bearings from plywood if you use Baltic Birch and keep the arc at least two inches wide. Reinforce the unsupported end with a "piece-of-pie" described above. If you don't reinforce the arc, your telescope will act springy when you are viewing near the horizon.

7.8 The Dust Cover

The dust cover for the top of the mirror box must be larger than the hole, so you cannot use the nice big cutout from the light baffle to make it. Get more plywood and cut out a circle ½ inch smaller than the outer width of the mirror box. Reinforce the back of the dust cover with the cutout from the light baffle to prevent the thin plywood from curling. Use cupboard clips to attach it to the mirror box.

Attach the male portion of the clip on the reinforcement because it is thicker and stronger than the thin plywood of the dust cover. Install the handle or knob of your choice. Put a dab of glue on the screws that hold the clips and handle—you don't want a screw working loose and dropping on the mirror.

And while you're thinking about things that might fall on the mirror, we have a question for all the telescope makers who have built mirror boxes with square dust covers and no light baffle. Have you ever wondered why manhole covers are round? Think about it.

Chapter 8
The Truss Tube

Unless you've been out of astronomy for the last ten years, you realize that your large-aperture dream telescope will have an open framework tube. The framework, or truss, consists of eight aluminum poles that connect the secondary cage to the mirror box. This type of tube is stiff and light compared to a solid tube, it is easy to take down and reassemble, and it all but eliminates troublesome tube currents that plague large telescopes with closed tubes. More than any other single factor, the light, rigid truss makes large, portable telescopes possible.

The truss tubes that amateur astronomers build are often called "Serrurier trusses," but technically they are not. A true Serrurier truss functions like the tube that a young engineer named Mark Serrurier designed for the 200-inch Hale telescope on Palomar Mountain. Because of the size of the instrument, flexure was unavoidable, so Serrurier designed a two-truss tube that flexes by an equal amount at each end. Because the truss flexure at the mirror cell and the focus cage is equal, the optics stay aligned. In amateur telescopes, even in very large amateur telescopes, the truss is simply an excellent way to build an extremely light, rigid tube.

At first sight, the truss seems like a lot of extra work—until you consider the alternative. For a typical 25-inch aperture telescope, a solid tube has to be 28 to 30 inches in diameter and perhaps 10 feet in length. Such an instrument is difficult to transport without a lot of helpers. For many years, the Sidewalk Astronomers *did* haul such a beast around in an old school bus, but setting it up for a night of observing was a real circus act. Of course, the hubbub attracted a great deal of attention, and that's exactly what John Dobson and his group needed to attract viewers to their public star nights.

The open truss eliminates the fuss by replacing about 8 feet of solid tube with a knock-down aluminum skeleton. The truss solves the handling problems that occur when you try to observe with a telescope that is substantially larger and heavier than your own body. When it is not needed, you take the truss apart and have only to store and handle a relatively compact bundle of aluminum poles.

However, open tubes are not a panacea. For small instruments, the classical closed tube makes perfect sense. Between the authors, we have designed and built telescopes from tiny 2.4-inchers to 36-inchers that are two stories tall, and every

Fig. 8.1 In the days before truss tubes, the Dobsonian and its derivatives were large and heavy. This highly innovative telescope and mounting by Steve Dodson housed a 22-inch f/7 mirror in an unwieldy 16-foot long spiral-wound paper tube.

size in between. Each time we plan a new telescope, we ask, "What is the most effective way to support the optics so that I can get starlight into my eyeball?" The answer depends on the size of the telescope.

For telescopes up to 12.5 inches aperture and up to 72 inches in length, a one-piece solid-wall cardboard tube is great. Most people can easily handle a tube 4 to 6 feet long by themselves. A tube 4 feet long easily fits into the back of a car, and a 6-foot tube can ride in the family minivan.

Telescopes that have 14-inch to 16-inch mirrors are in the transition area. A solid tube is no problem providing that the builder is young, strong, and owns a large vehicle. Problems arise not so much from the weight of the tube as from its awkwardness. Wrestling an 18-inch-diameter seven-foot-long tube is no picnic. You can barely get your arms around it, let alone maneuver it delicately into or out of your car—but it's not impossible. With such telescopes, the choice is yours whether you go with a solid tube or an open truss.

Telescopes 17.5 inches and larger more-or-less *have to be* built with a truss tube. If it weren't for pioneering telescope makers like Ivar Hamburg, the large-aperture revolution might have stalled out. Instead of a monster tube the size of highway culvert, Ivar realized that a jumbo telescope tube could collapse into a compact mirror box, a compact secondary cage, and a bundle of lightweight aluminum poles.

Skeptics initially proclaimed that portable telescopes with truss tubes would be so unstable they would have to be recollimated every time you observed. With a decade of experience to the contrary, we know that's not true. Truss tubes can be assembled precisely over and over again. The secret of reliable and repeatable assembly is to make the truss poles all the same length. The secondary cage then mounts exactly the same way every time you set up the scope. It can be no other way.

8.1 How to Build a Serious Truss Tube

Muskies are large freshwater predator fish native to Canada and the northern United States. Catching a 50-pound trophy fish requires proper equipment. Tales of broken poles and "the one that got away" are ubiquitous. That's why serious big-game fisherman use bigger diameter fishing poles. Likewise, serious big-game astronomers need to use bigger truss poles.

Tube selection for the truss poles is a major factor in truss-tube telescope performance. Undersized poles flex. You soon learn to recognize undersized poles by their feel: you aim the telescope and it feels "willowy." Undersized poles are a serious problem with large, heavy telescopes. When you get that willowy feel, pointing and tracking performance simply aren't up to snuff.

Let's get back to fishing poles. Try this experiment: seat a fishing pole firmly in a bucket and then rotate the bucket by pressing on the tip of the pole—because that's what you do with a telescope, you press on the secondary cage to move the mirror box and rocker. What happens? The pole bends. At some point the bucket rotates, of course, but pushing on the whippy tip of the pole gives you poor control.

So get a bigger pole. Try the same experiment with a broomstick in the bucket. When you push, the bucket moves.

Now put yourself back on the top of the observing ladder watching the Blinking Planetary at 300 power. You press on the secondary cage, exerting a torque. If that torque is faithfully transmitted to the mirror box and rocker, you get the right kind of motion. The Blinking Planetary moves smoothly to center of the field of view. No fuss, no muss—just smooth response.

Elastic deformation in the truss poles is fairly common. Go to any big star party and scout out some telescopes with long, slender truss tube poles. Return that night and observe the effect: the image in the eyepiece bounces around wildly each time you aim the scope. You push it and it pushes back. You center the Blinking Planetary with a light push, and as soon as you let go of the telescope, it bounces back out of view. Real annoying at higher powers.

Aiming one of these "spaghetti" scopes is like lifting a heavy fish out of the water with a skinny fishing pole. The fishing pole bends and bends until finally the elastic deformation is all used up and the fish rises out of the water. Aiming a big Dobsonian like that, you pull and pull on the secondary cage until the elastic

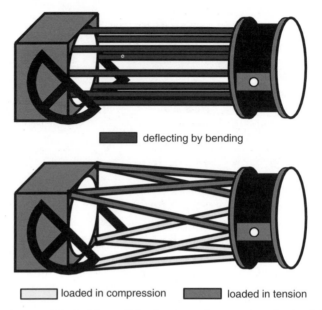

deflecting by bending

loaded in compression   loaded in tension

Fig. 8.2 *A tube constructed with parallel poles droops because the poles bend. In contrast, the truss tube is composed of triangles that can deform only by lengthening a pole under tension or shortening it under compression. When it come to stiffness, there is no comparison: the truss tube is far superior.*

deformation in the poles is used up, then the scope moves forward. Smart fisherman use bigger diameter poles to catch the big ones. So should you.

8.1.1 Flexure in Truss Poles

Understanding the geometry of the truss and how poles behave under stress is more satisfying than simply cloning a telescope from blueprints. Suppose for a moment that you build a big Dobsonian that has its secondary mirror and eyepiece mounted on one single pole. When you try to move the telescope, the pole bends. Engineers understand exactly how this happens:

$$\text{Deformation is proportional to } \frac{\text{Force} \times \text{Length}^3}{OD^4 - ID^4}.$$

In this equation, deformation is the amount of bending at the end of the pole, the force is the weight of the secondary cage, OD is the outside diameter of the pole, and ID is its inside diameter. If you know the wall thickness, the inside diameter is the outside diameter minus two times the wall thickness.

You can see that even though this telescope would be quite limber, increasing the OD of the pole pays off handsomely. If you keep the wall thickness constant, doubling the diameter cuts the deformation by factor of about 8. (Of course, the ultimate extension of this principle is to make the pole large enough to put the

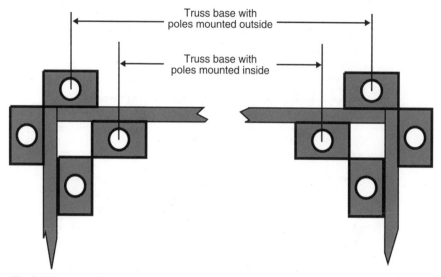

Fig. 8.3 *Mounting the poles outside the mirror box instead of inside the box leads to a stiffer truss because the base of the truss triangle is longer. When you're trying to coax maximum performance from your telescope, every bit helps.*

optics inside—then you have a conventional telescope!)

Suppose that you decide to beef up the design using eight poles instead of one. You arrange them parallel to each other. (Important: parallel poles are not a truss.) Since the poles now share the force, the deformation is one-eighth that a single pole. Use eight big poles and you have a pretty stiff telescope. Look at pictures of the 100-inch Hooker telescope on Mount Wilson and you'll see the basic parallel pole design. But notice how much reinforcement they had to add to it.

Each of these designs asks the poles to do something that they are not good at doing, that is, staying straight under a bending load. If there were just some way to put the poles into tension and compression, they would be very strong indeed!

Well, that's exactly what the truss does: it loads the poles under compression and tension. Of course, there will always be some bending force present, but the beauty of the truss concept is that it ties the tips of each pair of triangles together at the mirror box and at the secondary cage. The upper pole in each triangle is under tension, and the lower pole is under compression. The triangles transmit the weight of the secondary cage and any aiming forces along the axis of the individual poles, thereby loading them under compression and tension. Bending becomes a secondary consideration.

If you don't believe all this, there's a simple experiment you can do. Buy a box of spaghetti and a small bottle of white glue. Carefully select two groups of eight strands and cut them all to the same length. From stiff cardboard, cut four circles about four inches diameter.

On one pair of cardboard circles, mark eight spots on a three-inch diameter

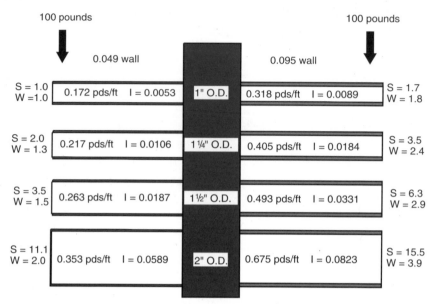

Fig. 8.4 *The stiffness (S) and weight (W) of a pole depend on its diameter and wall thickness. Taking the 1-inch diameter 0.049-inch-wall tube as the standard for stiffness and weight, you can see that increasing the diameter of the tube rather than its wall thickness is the most effective way to construct a truss that must be both stiff and light in weight.*

circle. With an awl, punch eight equally spaced holes the same diameter as the spaghetti strands in each cardboard circle. Glue eight of the spaghetti strands parallel to each other between the two circles and secure each strand at each end with a spot of glue. Set this unlikely contraption aside until the glue has dried fully.

Repeat the experiment, but this time mark four pairs of holes on each circle. The holes should be as close together as it is practical to punch. Arrange the spaghetti strands in a truss configuration and secure them at each end with a spot of white glue. Allow a day or so for the glue to dry fully.

Compare these model tubes. The tube with the parallel strands is weak and wobbly. If you stand it on end and pile coins on top, you will see the spaghetti bend. It might even collapse. The truss is quite a different story. Although the spaghetti strands are weak, the truss assembly is strong. When you apply a light load with your fingertips, the spaghetti strands do not bend until the load is great enough that they buckle. The forces in the truss are compression and tension. The reason you need stiff poles is so they won't bend enough to throw your system out of alignment.

Once again, however, you can see how important the diameter of the poles is. The spaghetti truss fails when a pole buckles and collapses, that is, when compression suddenly causes bending and permanent deformation. If you repeat the whole experiment and substitute soda straws for spaghetti strands, you will see that the relatively large diameter straws resist bending and buckling. A tube model made with eight soda straws can easily support several pounds.

Without going into the details, the truss itself behaves somewhat like a hollow cylinder. The longer the truss, the more prone it is to flexure; and the larger its overall diameter, the stiffer it is. What flexure you see results largely from the bending or buckling of the individual poles. To make a truly stiff truss tube, you need to use large-diameter thin-wall tubing in a truss configuration having the largest possible "diameter." This means you should attach the truss poles to the outside of the mirror box, to maximize the diameter of the truss.

8.1.2 The Diameter of the Truss Poles

You can purchase aluminum tubing in many sizes if you look in a big city Yellow Pages. **Table 8.1** shows the relative stiffness for all types of commercially available tubing, so you don't even have to do the math. "Second moment of area" is an engineering term that describes the stiffness of the particular size and shape tube.

For big Dobsonians, we recommend tubing with 0.049-inch wall thickness. It is both light and strong. For big instruments, avoid the 0.035-inch wall tubing: it is just too thin. It might crush at the pole attachments, or worse, a pole might buckle in the middle. From the table it is clear that increasing the wall thickness has an effect on stiffness, but that the outside diameter is far more important. Be that as it may, the tube must have sufficient wall thickness to resist crushing and buckling.

Look at the numbers. A 1½-inch O.D. tube with a wall thickness of 0.049 inches is twice as stiff as the same thickness tube in a 1¼-inch O.D., and over three times as stiff as 1-inch O.D. tube. For a 50% increase in weight, you get a 300% increase in stiffness. That's good design.

Increasing the wall thickness is relatively unimportant for a given tube O.D.; doubling the wall thickness does not even double the stiffness, but it does double the weight. For a 100% increase in weight, you don't even gain 100% in strength. That's not too bright.

We strongly recommend 1½-inch by 0.049-inch tubing on a 25-inch telescope. It is nearly twice as stiff as 1¼-inch tubing with the same wall thickness but a 7½-foot length is only six ounces heavier. Sure, it's a whole lot less expensive to buy small-diameter tubing for your truss assembly—that's why there are so many Dobs using 1-inch diameter pole. For the same wall thickness, a truss tube made with 1-inch pole is three times flimsier than a truss tube built with 1½-inch pole. The difference shows up when you observe because there is three times more flexure in the truss assembly when guiding and aiming those scopes.

As you saw earlier, the length of the poles is a very important factor in their stiffness. On a typical 20-inch *f*/5 Dobsonian, the poles turn out to be about 70 inches long. On a 20-inch *f*/6 Dobsonian, they would need to be 90 inches long. That extra 20 inches results in a truss assembly with twice the flexure! How so? Remember that the deformation of the pole is proportional to the cube of its length. Well, 70 cubed is 343,000 and 90 cubed is 729,000. In going from a 20-inch *f*/5 mirror to a 20-inch *f*/6 mirror, it would be prudent to jump from 1¼-inch tubing to 1½-inch tubing to regain the lost stiffness. Well, you can spend a lot of time doing this type of math, or you can follow a simple rule of thumb. It works as follows:

$$\text{Diameter of truss pole} = \frac{\text{focal length}}{80}.$$

Divide the focal length of that big mirror you bought by 80 and the result is the recommended truss pole diameter.

Flexure in the truss shows up when you are guiding the telescope at high powers. Think of big truss poles as insurance, and spend a few dollars to get them right. When you've just spent several thousand dollars on a fine set of optics, why pinch pennies on a hundred dollars' worth of aluminum tubing? If you want proof, attend star parties until you get a chance to observe with a 25-inch or 30-inch telescope built with 1-inch diameter truss poles. The experience will convince you of the importance of large-diameter poles.

8.1.3 The Length of the Truss Poles

By now you are probably wondering how to determine the pole lengths. When it comes time to do this, you'll need the finished secondary cage with the secondary mirror in place and the mirror box with flotation cell and the primary mirror in place. (By the time you're ready to determine the pole lengths, these parts should be complete.) You will also need to have made the pole sockets and the pole seats, which are described later in this chapter, in **Sections 8.2** and **8.4**.

Carry all of these parts outside and set the mirror box on its side on a large picnic table or a temporary table made by placing a sheet of plywood on two heavy-duty sawhorses (see **Figure 8.5**). You need to have a clear view of a *very* distant object—something several miles away. If you do not have a reasonable horizon, prop up one end of the picnic table or plywood sheet so you can pick off a star low in the sky. (There are plenty of telescopic stars, so you really don't have to aim at one.) The idea is to get a focused image of any star, or a miles-distant radio tower, mountain, or tree line. Place the secondary cage in front of the mirror box at roughly the right distance and align the optics. You don't need to do a precise alignment—just good enough that the mirror forms reasonably sharp images.

Objects that are closer than a few miles away increase the distance from the mirror to the focal point. You can figure out whether an object at some distance will push out the focus too much using the following equation:

$$\frac{1}{f} = \frac{1}{i} + \frac{1}{o}.$$

The distance i is how far the image will form from the mirror, f is the infinity focal length (the focal length of your mirror), and o is the distance to the object you are viewing. Make sure you keep the same units throughout. Pull out your

Table 8.1
Properties of Thin-Wall Aluminum Tubing*

O.D.	Wall	I.D.	Pounds/foot	Relative Second Moment of Area**	Primary focal length
¾	0.035	0.680	0.093	0.10	60
¾	**0049**	**0.652**	**0.127**	**0.14**	**12.5-inch *f*/5**
¾	0.058	0.634	0.148	0.15	60
¾	0.065	0.620	0.164	0.17	60
1	0.035	0.930	0.125	0.25	80
1	**0.049**	**0.902**	**0.172**	**0.34**	**14.5- to 17.5-inch *f*/4.5**
1	0.058	0.884	0.202	0.39	80
1	0.065	0.870	0.225	0.43	80
1⅛	0.035	1.055	0.141	0.36	90
1⅛	0.049	1.027	0.195	0.49	90
1⅛	0.058	1.009	0.229	0.56	90
1⅛	0.065	0.995	0.255	0.62	90
1¼	0.035	1.180	0.157	0.50	100
1¼	**0.049**	**1.152**	**0.217**	**0.68**	**20-inch *f*/5**
1¼	0.058	1.134	0.255	0.79	100
1¼	0.065	1.120	0.285	0.87	100
1⅜	0.035	1.305	0.173	0.67	110
1⅜	0.049	1.277	0.240	0.92	110
1⅜	0.058	1.259	0.282	1.06	110
1⅜	0.065	1.245	0.315	1.17	110
1½	0.035	1.430	0.189	0.88	120
1½	**0.049**	**1.402**	**0.263**	**1.203**	**25-inch *f*/5**
1½	0.058	1.384	0.309	1.39	120
1½	0.065	1.370	0.345	1.54	120
1½	0.083	1.334	0.435	1.90	120
1¾	0.035	1.680	0.222	1.41	140
1¾	0.049	1.652	0.308	1.93	140
1¾	0.058	1.634	0.363	2.25	140
1¾	0.065	1.620	0.405	2.49	140
2	0.049	1.902	0.353	2.91	160
2	**0.065**	**1.870**	**0.465**	**3.77**	**32-inch *f*/5**
2	0.083	1.834	0.588	4.69	160
2	0.125	1.750	0.866	6.62	160
2¼	**0.049**	**2.152**	**0.398**	**4.18**	**36-inch *f*/5**
2¼	0.065	2.120	0.525	5.43	180
2¼	0.083	2.084	0.664	6.77	180
2¼	0.125	2.000	0.981	9.63	180
2½	0.049	2.402	0.444	5.77	200
2½	**0.065**	**2.370**	**0.585**	**7.51**	**40-inch *f*/5**
2½	0.083	2.334	0.741	9.39	200
2½	0.125	2.250	1.100	13.43	200
3	0.065	2.870	0.705	13.15	240
3	**0.125**	**2.750**	**1.330**	**23.81**	**50-inch *f*/5**

*Tube listed in bold type recommended for aperture and focal ratio in last column

** Relative Second Moment of Area ($I = D^4 - d^4$). The constant 64 in denominator was removed to aid comparisons.

Fig. 8.5 *The telescope maker's first look through a new telescope occurs when it's time to determine the length of the poles. With the mirror in the cell, the builder focuses on a distant object and measures the separation between the mirror box and the secondary cage.*

pocket calculator and let's do an example. The focal length of the mirror is 80 inches and you want to focus on a radio tower 1 mile away. The equation becomes:

$$\frac{1}{i} = \frac{1}{80} - \frac{1}{12 \times 5280}.$$

That works out to

$$\frac{1}{i} = 0.0125 - 0.0000157, \text{ or } 0.0124843,$$

so the object will focus at 80.1 inches. If a star image forms 80 inches from the mirror, the radio tower will focus ¹⁄₁₀ inch farther out. Conclusion: the radio tower is an excellent focus target.

   Place an eyepiece in the focuser (preferably the one in your set with the most inward travel) and move the secondary cage back and forth along the optical axis

Fig. 8.6 Hack saws are named aptly because they really mess up the end of a thin-walled tube. Instead of a saw, use a tube cutter on thin-walled aluminum tube. The tube cutter makes neat, clean cuts that leave the end perfectly square. Get one at your local hardware store.

of the primary until an image forms. (This is a sort of daytime "first light" for your telescope.) Clamp the secondary cage at that location.

Try all your eyepieces and determine the distance between the mirror box and the secondary cage that allows you to focus all of your eyepieces. Measure the distance from the bottom of the split block bore to the area on the secondary cage where you will eventually attach the pole seats. Write this number down. *Add two inches and cut all the poles to this length.* You can shorten the poles later. Don't try to cut the poles to exact length or you may end up ordering more tubing because you just can't rack the focuser out far enough to focus with your favorite eyepiece.

To cut aluminum tube accurately, use a tube cutter. Do not use a hack saw; it will make a mess. Go out and spend $20 on a good tube cutter at the hardware store. They are simple to use and they make very clean even cuts. Practice on some scrap tube first. The only trick is to *avoid cutting off less* than ¼ inch of tubing at a time. If you do, the rollers on the cutter won't have enough surface to ride on. This is important because you start with the poles too long and trim them back. Even if it takes you a couple tries to find the optimum truss pole length, you need to make sure that the last cut means cutting at least ¼-inch from the poles.

8.2 Attaching Truss Poles to the Mirror Box

The truss poles that support the secondary cage over the mirror box must somehow be attached to both components. In a portable Dobsonian, you need to be able to attach and detach the poles each time you observe. For such a simple function, the requirements are surprisingly tricky. Indeed, it took five years of trial and error development for a really good system of attaching poles to emerge.

The attachment devices must meet the following requirements.

- The connections must be *easy to use*. The telescope truss has 16 attachments that must be made in the dark, without tools, by one person.

- To maintain collimation, the mechanism must be *repeatable*, that is, each time the telescope is used, the poles must connect the same way.

- The optics are valuable, so the mechanism must be extremely *reliable*. Failure of a pole attachment could result in a costly accident.

These are tough requirements. During that five-year period, we watched the evolution of truss tube clamps in amateur telescopes, and investigated numerous commercial fasteners. Some of them met some of the requirements, but none met all. Probably the biggest single breakthrough in developing good attachments was the realization that the methods of attachment to the mirror box and to the secondary cage need not be the same. For the mirror box, we recommend wooden *split-block pole sockets*, and for the attachments at the secondary cage, we suggest *seats and wedges*. While these are not the only pole attachments that work, in our experience they best meet the requirements: they are easy to use, reliable and repeatable.

8.2.1 Overview of Pole Sockets

The bottom of the truss poles can be connected to the mirror box a number of ways. Remember, you are building a portable telescope that you will want to set up and take down quickly and easily. You'll also want the truss assembly go together time after time exactly the same way so that the collimation, or optical alignment, doesn't change. A simple wooden clamping device called a split-block pole socket serves very well. Basically the split block is a hardwood block with a bore to accept a truss pole and a slit. By tightening a knob, the block clamps down on the pole. Eight pole sockets are bolted to the outside corners of the mirror box.

Here are some of the reasons we recommend split blocks.

- They grip the pole tightly, so there is no chance of a pole pulling out, yet they cannot crush the lower ends of the poles.

- The pole enters the socket the same way each time. Since the fit is precise, the telescope assembles the same way every time, and optical collimation stays bang-on.

- It is easy to insert poles into the sockets. It takes no tools and needs no light. There are no nuts and bolts to find, misplace or drop on the primary.

- Because the poles stay upright, one person can insert them all. The poles stick up in the right spots and hardly sway, so attaching the secondary cage is easy, too.

- Split blocks are easy to construct in a shop equipped for basic woodworking. And, since they are low-tech devices, there is little to go wrong with them.

We know there are other ways to attach the poles to the box, and you are free to experiment and try to invent better ways if you want to. In the spirit of helpful advice, we offer the following so that you can learn from the experience of those who have gone before you.

A very simple method of attaching the poles is to crush flat the ends of the tubing and drill a bolt hole through each. Builders usually crush about two inches of the pole's length. The crushed ends are quite flexible, so you lose considerable stiffness at the outset.

To connect the poles, you attach them with nuts and bolts. To get the nuts tight, you need a wrench or nut driver, and it is very easy to drop nuts, bolts, and wrenches in the dark—and you know how difficult it is to find a nut that has fallen into grass. (Of course, if the nut lands on the primary mirror, the distinctive sound of metal on glass tells you *where* it fell, and it would be easy to find if the nut did not invariably bounce off the mirror and roll into grass.)

The major difficulty with crushed tubes becomes apparent the first time you set up the telescope. There is nothing to hold the top ends of the poles where they belong when the time comes to attach the secondary cage. The poles flop around and you earnestly wish that you had been born with four hands because you need two hands to hold the secondary cage and two hands to fiddle with nuts, bolts, and tools. In big telescopes, the truss poles may be 6 to 14 feet long, so imagine yourself up a ladder at night holding the secondary cage in one hand, while holding a flashlight in your teeth, while trying to corral two of the poles.

Even as we write this, however, we salute the innovators in the early stages of the Dobsonian revolution. It is impossible to think these things through when you're dreaming up ideas in a warm, well-lit shop. It takes building the new idea and going out under the stars to learn what works and what doesn't. And, of course, you can get pretty good at assembling your telescope especially if you get a little help from your friends. Since the early innovators were always people who had built the largest telescope within a hundred miles, they had plenty of help in setting up. It was an exciting time to be a telescope maker.

8.2.2 Wooden Split-Block Sockets

The beauty of split-block pole sockets is that they are permanently aligned and bolted to the mirror box so that when you insert the poles, the tips all line up properly and stay put. This allows one person to set the secondary cage atop the poles and lock it in place.

The socket consists of a square block of wood with a hole bored down its length. The pole fits snugly into the hole. "Split block" refers to a saw cut in the side of the block that extends to the bore. A carriage bolt with a plastic knob squeezes the halves of the block together, securely clamping the pole.

One tricky aspect of the basic split-block design is that it's difficult to bore a hole that is big enough to let you insert the poles easily and still small enough to grip them firmly. Split blocks are also prone to crack at the end opposite the split. This is not good when it happens at your dark site.

Cabinet maker Pete Welbourne deserves credit for developing a split block that is free of these problems. Pete makes two saw cuts, dividing the block into three sections. The cuts don't extend all the way through the block. He also hollows out the wood slightly on the back side of the middle third to make it more flexible. Tightening the knob pulls the middle third of the block inward to grip the pole. Not only does this allow the block to have a slightly larger bore for the truss poles, but it also grips them tightly. Best of all, Pete's blocks don't break.

The next three sections give the details for making standard split-block sockets, fancy split-block sockets, and simplified split-block sockets. Pick the one that best matches your carpentry skill and ambitions. Remember to make four "left-hand" and four "right-hand" blocks.

Whichever type you chose, it is important to keep the depth of the pole bore exactly the same in all eight blocks. In addition, you need to keep the bolt hole exactly the same distance from the side and bottom of the pole bore in all eight blocks. If you do these things accurately, the truss pole tips will meet at the right spots for the secondary cage.

8.2.2.1 How to Make Split-Block Sockets

To make standard wooden split-block sockets, you will need the following materials:

- 6 feet of ⁵⁄₄– by 4-inch maple, mahogany, or other close-grained hardwood;
- 8 plastic knobs internally threaded ⁵⁄₁₆-NC;
- 8 ⁵⁄₁₆– by 3-inch carriage bolts;
- 8 ⁵⁄₁₆-inch flat washers, and
- 32 2½-inch drywall screws.

Section the maple into 16 pieces 3¼ by 3½ inches. The grain must run the long dimension of each piece, that is, it must run along the 3½-inch length. Bond two pieces together with carpenter's glue, clamping firmly until you have made eight blocks 1¾ inches thick by 3¼ inches high by 3½ inches wide.

To make attractive split blocks, select the prettiest face and rout a ¼-inch radius around the four sides. If you are skillful with a router, round off the four corners. Careful, don't trim off a finger tip! Do not round the edges on the side that

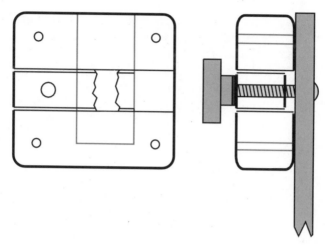

Fig. 8.7 *The split-block pole clamp is made of a hardwood—maple—and captures the pole when you turn the knob and compress the thin middle section. This makes setting up and tearing down your telescope quick, easy and repeatable.*

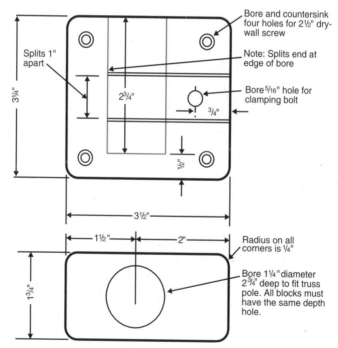

Fig. 8.8 *This plan shows how to make a split-block pole clamp for 1¼-inch aluminum tubing. Note that the depth of the hole must be exactly the same for all eight blocks. Note also that four of the eight blocks you will need for your telescope must be an exact mirror image of the block shown here.*

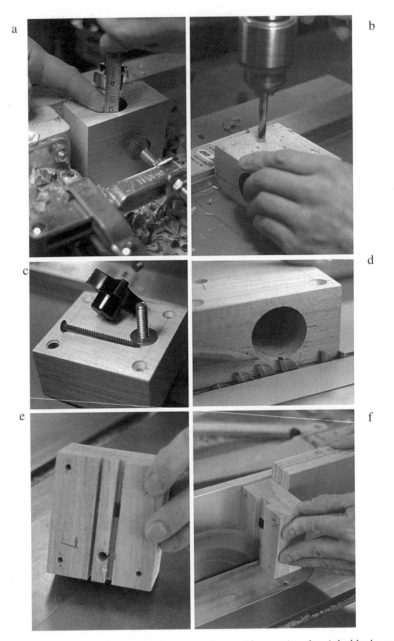

Fig. 8.9 *Watch as we make a split-block pole clamp. After cutting the eight blocks and rounding their out-facing corners, (a) bore a 1¼-inch hole for the pole. Note that the holes in all eight blocks must be exactly the same depth. Next, (b) bore the hole for the clamp bolt and (c) holes in the corners for the drywall screws. On a table saw, (d) set the height of a dado blade so that it is just a bit higher than the bottom edge of the pole hole. After several passes to open the bottom channel (e), check that the pole hole is open. Finally, (f) cut the two splits down two-thirds the length of the block.*

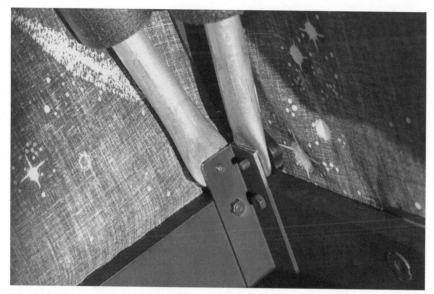

Fig. 8.10 The crushed-tube method is simple and reliable, but it's hard to set up and you risk dropping all sorts of hardware on the primary mirror.

you will attach to the mirror box. Do all of this optional routing *before* you drill the pole bore and cut the splits. If you do it afterward, the guide for the router bit will fall into them and the rounded edge will be bumpy.

The base of the standard truss tube is slightly bigger than the top, so the bores in the split blocks must be angled slightly inward. The angle is small: only one or two degrees. There are two ways to achieve the desired angle: one tricky and one simple.

In the tricky method, you bore the pole hole at a slight angle. Mount each block on a scrap of plywood shimmed up a bit at one end to the desired angle. Set this jig on the drill press table and make the bore with the appropriate size spade bit or Forstner bit (Forstner bits are best, but they are expensive).

The much simpler alternative is to drill the bores square and then when you mount the blocks, shim the bottom end of the block out slightly. The shim will be about the thickness of a popsicle stick. Both methods work well. Whichever you choose, keep the inside faces of the blocks flat and smooth.

For 1¼-inch poles, bore the hole 1¼ inches in diameter. Bore exactly 2¾ inches deep so there is precisely ½ inch of material in the bottom of the block. Drill each of the pole bores to exactly the same depth in all eight blocks. If you make a mistake and drill one of them too deep, then drill them all deeper or make a new block. The bore must be drilled before you channel the back of the block.

Up to this point, the blocks are identical. However, remember that four of them are destined to become left-hand blocks and four are destined to become right-hand blocks. It is easy to get excited and make eight identical ones, so prime

Fig. 8.11 *Split blocks are the best low-tech way to support the truss poles in medium-size Dobsonians. They are easy to make and easy to use. Here you see the in-facing side of left-hand and right-hand blocks. Note that the two types of clamps are not interchangeable: they are mirror images of one another.*

your mind to remember that you are making two groups of four things and not one group of eight.

On the side of the block that mates to the mirror box, cut a channel ¼ inch deep by 1 inch wide. You can cut this with a router or a dado cutter, or make multiple passes with a table saw. Notice that this channel just cuts into the side of the pole bore. Knock out a little of the remaining thin wall of the bore with a screwdriver. Once the splits are cut, this channel makes the middle third of block more flexible. This portion can then freely cam inward, moving the middle third of the hole and gripping the pole firmly.

As shown in the illustrations, cut two splits into the side that will have the clamping bolt hole. Use a hand saw or band saw for best results. (A table saw will also work, but be extra careful. It is difficult to hold work this small.) The splits are 1 inch apart and run along the edges of the channel cut in the back of the block. Be sure not to cut too far. Leave an inch of the block uncut.

Drill the four mounting holes for the drywall screws. Their location is not critical. Countersink the heads. The screws should pass through the holes freely so that when they are tightened, the heads and not the threads will draw the block

Fig. 8.12 The inner surfaces of left-hand and right-hand split blocks are mirror images of one another. When you are making the blocks, be sure to make four left-hand blocks and four right-hand blocks. In the fever pitch of boring holes and cutting dado grooves, it's remarkably easy to get carried away and produce eight identical blocks.

to the mirror box. Drywall screws are black and, when countersunk flush with the face of the block, will look attractive.

Next, drill the ⁵⁄₁₆-inch hole for the clamping bolt. The location of this hole is *very* critical. It must be exactly the same distance from the bottom of the pole bore in every block. The goal is that the tips of all eight poles will end in exactly the same plane. If you do this right, the clamps exert precisely the same pressure on each pole, and the truss will be quick and easy to assemble.

8.2.2.2 Simplified Split-Block Sockets

If you lack the woodworking skills or the equipment needed to make to make standard split-block sockets, you can make simplified split blocks. Skip the stuff about putting a radius on the edges. Make the eight blocks, and drill the pole bores and clamping holes as above. Forget the backside channel. Forget the double splits. Eliminate the two mounting holes adjacent to the clamping bolt. Make one saw cut from the side of the block into the pole bore. Sand off the rough edges and they're done. These clamps look a little crude, but they work very well.

8.2.2.3 Fancy Cam-Action Pole Sockets

If you are really ambitious and have a friend in a machine shop, you may want to try an alternative to the split blocks: cam-action poles. Dave did this on his 25-inch scope. The idea came from the cams used in telescoping golf ball retrievers, adjustable tent

Fig. 8.13 *Simplified split-block clamps are the best option for telescope makers with limited woodworking skills or equipment. The pole hole goes right through the block, and the block is split down the whole side. Metal mending plates serve as the pole stop and also as reinforcement for the clamping side of block.*

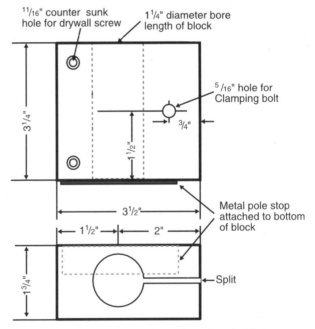

Fig. 8.14 *Plan for traditional simple split block.*

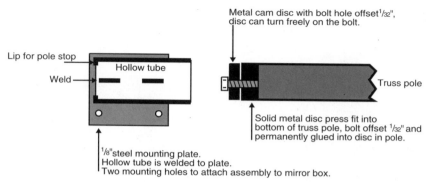

Metal cam disc with bolt hole offset$\frac{1}{32}$",
disc can turn freely on the bolt.

Lip for pole stop

Hollow tube

Weld

Truss pole

Solid metal disc press fit into
bottom of truss pole, bolt offset $\frac{1}{32}$" and
permanently glued into disc in pole.

$\frac{1}{8}$"steel mounting plate.
Hollow tube is welded to plate.
Two mounting holes to attach assembly to mirror box.

Fig. 8.15 The anatomy of cam-action pole sockets. The inside diameter of the hollow steel tube is one-thirty-second of an inch larger than the outside diameter of the truss pole. Note the lip machined into bottom for pole stop. Offset location of bolt and cam disk is greatly exaggerated for clarity.

poles, and the handles on swimming pool cleaners. Put an off-center cam on the end of each truss pole, drop them into a short piece of tubing mounted on the mirror box, and twist. The cams make set up and take down faster and more fun. Although these sound easy, they are actually very tricky to make. We don't recommend making cam-action pole sockets unless you really enjoy a machining challenge.

The cam is a stainless steel disk $\frac{3}{8}$ inch thick and for a 25-inch scope, $1\frac{1}{2}$ inches diameter, the same as the outside diameter of the truss poles. Drill a hole to accept a $\frac{3}{8}$-16 bolt just $\frac{1}{32}$ inch off-center through the cam. Into the ends of each of the eight truss poles, press-fit 1-inch-long stainless steel plugs with an offset threaded hole in them to accept the $\frac{3}{8}$-16 bolt. The offset in the plug is $\frac{1}{32}$ inch. Leave the cam bolt loose so the cam can turn freely on the end of the truss pole.

So far it's all simple lathe and drill press work. Making the 4-inch long stainless or aluminum tubes that retain the poles is the tough part. The inside diameter of these tubes should be exactly $\frac{1}{64}$ inch larger than the outside diameter of the truss poles or the cam won't engage, or even worse, you'll have poles that stick up at all sorts of odd angles making placement of the secondary cage difficult.

The secret of making these is to weld the mounting plate on before boring the inner diameter; if you reverse the order, you'll make the sad discovery that welding warps thin-walled tubes. Machining an inside diameter in such a small tube is difficult because large cutting tools won't fit, and small tools are too weak to reach in more than a couple inches, but a good machinist can do it. To provide a positive pole stop, have a small lip machined into the bottom of each tube. The lip provides friction so the cam holds still when you twist the pole to lock it into position. The lip must be exactly the same distance from the top hole on the mounting plate so all eight tubes can be mounted identically on the mirror box.

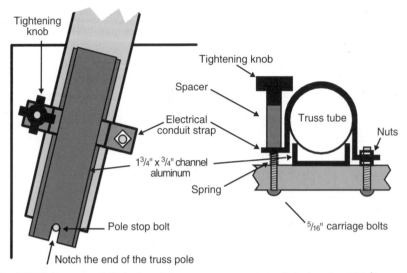

Tightening knob

Tightening knob

Spacer

Electrical conduit strap

Truss tube

Nuts

$1^3/_4$" x $^3/_4$" channel aluminum

Spring

Pole stop bolt

$^5/_{16}$" carriage bolts

Notch the end of the truss pole

Fig. 8.16 *Telescopes of 30-inch and larger aperture require 2-inch poles. This low-tech clamp presses the pole firmly against a ¾-inch by 1¼-inch U-channel with a heavy-duty electrical conduit strap. The notch in the bottom end of the pole orients the pole so that the angle bracket at the upper end of the pole is in the right position.*

8.2.2.4 Strap and Channel Sockets

Truss poles larger than 1½ inches diameter require sturdier mounts than wooden split blocks. You can exert tremendous leverage with a truss pole that is ten feet long, so snapping apart a wood split block is a real possibility on scopes over 30 inches aperture.

What you need is this all-metal clamp that can be made from standard off-the-shelf components. It uses a metal strap to press the pole against a short length of aluminum U-channel. Since the cylindrical pole always rests against the parallel sides of the U-channel in precisely the same way setup is highly repeatable. Select aluminum channel with a width equal to three-fourths of the pole diameter and just deep enough to keep the pole from contacting the base.

An 8-inch length of channel works well. Aluminum channel (or steel if you like things that rust) can be found in most welding shops and some of the better hardware stores. Cut it with a hack saw, or, if you can manage it, with a band saw. If you are stuck with using a hack saw then go ahead and hack off eight pieces each 8 inches long. File the ends and corners smooth.

Aluminum channel gives the pole a solid and precise alignment base so that the tips of all eight poles aren't waving around in the breeze. Attach the 8-inch length of channel to the mirror box with a couple of wood screws at each end. It is not necessary to use bolts because the metal strap will compress the pole down into the channel and press both firmly against the side of the mirror box.

Shim the channel to aim the pole tips inward as needed. Install a small bolt

Fig. 8.17 This heavy-duty pole clamp is made from standard components readily available at any good hardware store, yet it is suitable for telescopes in the over-30-inch class. The pole used on this 36-inch telescope is 2 inches in diameter and has an 0.065-inch wall thickness.

or block of wood at the end of the channel to serve as the pole stop. The pole stops for the eight poles must be exactly the same distance from the top of the mirror box, otherwise the truss pole tips won't fall into the same plane.

The retaining strap can be fabricated from a number of off-the-shelf items never intended for telescope making. For example, the 2-inch-wide steel straps intended for retaining farm equipment bearings work great and cost only $3 apiece. Electrical conduit hanger straps, common U-bolts, and automotive tailpipe and muffler clamps also work well. You decide.

Mount the strap on the mirror box over the pole and channel with a couple of carriage bolts. Place a spring under one side of the strap to force it open when you loosen the tightening knob. Install a spacer and a tightening knob and you've made a tough, cheap, all-metal, no-tools-needed pole clamp.

8.3 How to Align Truss-Pole Sockets

At first prospect, aligning the truss-pole sockets sounds like it might be fussy, tedious and time consuming. It is not. Here is a way to align the sockets that is easy, accurate and quick.

Assume your mirror box is square and that the top edges all lie in the same plane. Place a finished block on the side of any corner on the mirror box. Place it as high and close to the corner as possible. Rotate it to the approximate angle it will have when the pole is in it. Drop a pencil through the bolt hole and make a

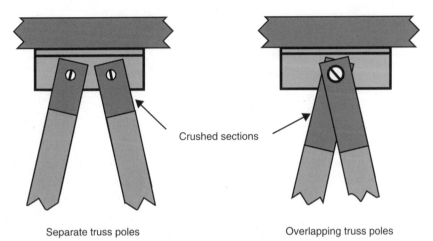

Separate truss poles   Overlapping truss poles

Fig. 8.18 *An easy way to construct the truss is to crush the ends of the poles and bolt them to an angle bracket on the secondary cage. Unfortunately, trusses made this way tend to be difficult to assemble in the field. If the poles are fastened together so that they overlap, the crushed sections must be longer than if the poles are attached separately.*

mark on the box. Remove the block. Measure the distance to the mark down from the top of the box and in from the side of the box.

Using these measurements, put marks for the bolt holes of each of the other seven blocks on all the corners. Do this as accurately as you can. Drill a hole through each mark on the mirror box to accept the clamping bolt. Install all eight blocks with only the clamping bolt and plastic tightening knob. If you've got your poles cut to the correct length plus two inches, you are ready to align the blocks.

Once again: be sure you have left the truss poles two inches longer than the length needed to focus on a star.

Place two of the poles into their sockets on each side and rotate the pair of blocks until the tips of the poles come together at the apex. Tighten the knobs and then drive a couple of dry wall screws through the block into the box. Do the same for each of the other three pairs of poles.

Measure the separation between opposite pairs of poles. Shim the blocks so that the pole tips from the same sides meet and also so that the pairs of pole tips are the correct distance apart to be attached to the secondary cage. Strips of scrap Kydex work great as shims. If you were a perfect craftsman, the shims would be the same thickness for all eight blocks. When the poles meet and have the correct spacing, tighten the screws. After doing the first pair of sides, do the second pair of sides.

If you want to get mathematical about it, the shim thickness is the ratio between the difference in secondary cage and mirror box diameter to the distance between the two. Apply that ratio over the length of the block (or over the length of the flat mounting plate if you are making cams) to get the ideal shim thickness. You can probably attach and align the split blocks without the secondary cage in

Fig. 8.19 Popsicle sticks, scraps of Kydex plastic, or other non-crushable materials are used to shim the split blocks inward so the pole tips meet at the secondary cage. Once the blocks are aligned, you can insert the poles and attach the secondary cage in moments.

about twenty minutes. Leaving the poles two inches too long during alignment insures that they will meet properly in the pole seats.

8.4 Attaching Truss Poles to the Secondary Cage

There are many ways to connect the upper end of the truss poles to the secondary cage. The most important requirement at the upper end is security. Looseness is unacceptable. If the clamp device can fall apart, that's even worse. If a couple of clamps were to fail and the secondary cage come crashing down, there could be severe damage to the scope and observer.

No matter what type of clamp design you select, you still want a telescope that assembles quickly and precisely each time. Keep these principles in mind when you are fabricating truss pole clamps for the secondary cage.

Principle #1: All the truss poles must be exactly the same length. If the poles are different, you will have to recollimate the optics every time you set up in the field, or you will have to number the poles and insert them into the same locations every night. It works, but if you do it, everyone will know that you were careless when you built your truss.

Principle #2: The attachment points on the mirror box and the secondary cage must be consistent. Whatever device you use to hold your poles to the secondary cage, all must be made identical. You can't be sloppy. If one or more of the attachment points at either end of the poles is

Fig. 8.20 Threaded inserts make attaching the poles to the secondary cage easy. Press these into the upper end of your truss poles and you have instant threads for bolting a bracket or clamp to the pole.

different from the rest, you may need to become an expert on nightly recollimation.

8.4.1 The Crushed-Tube Method

The most common and easiest way of attaching poles at the upper end is the "crushed-tube" method (see **Figure 8.10**). Flatten the ends of the truss poles in a vise. Fasten each pair of them to a short length of angle aluminum permanently bolted to the underside of the secondary cage. A couple of small holes are drilled through the angle and through the crushed ends and —you guessed it —they are connected with nuts and bolts. The primary drawback of this method should be obvious: you will drop the bolts on the primary mirror.

It's worse than that: there you are partway up an eight-foot ladder with the scope vertical. How are you going to set the precious secondary cage atop the flat ends of eight poles? You can't do it alone, that's for sure. You need someone else to hold the secondary cage while you manage the nuts, bolts, screwdriver, and wrench.

However, it can be done and it has been done. Many times. The crushed-tube nut-and-bolt design is very secure and repeatable. It is also easy and cheap to fabricate. But it requires tools and two people to set up. If you build your big Dobsonian this way, overlap the pole tips on each side up at the secondary cage. This

Fig. 8.21 Here is yet another way to clamp the poles to the secondary cage. It is simple, reliable and there are no loose parts to fall on the mirror.

eliminates half the fasteners. The disadvantage is that you must flatten more of each truss pole to do it (see **Fig. 8.18**).

With fast mirrors, less overlap is needed. Slower focal ratios mean the triangles formed by each pair of poles get narrower and narrower and more of each pole overlaps its partner. It is quite difficult to crush more than 6 inches of a pole, especially with tubing over 1¼ inches diameter. The poles get weaker and warp out of shape. Sometimes they crack. Interchangeability and repeatable collimation go out the window. If you *must* crush the poles, we suggest that you first heat them with a propane torch. It softens the aluminum and helps prevent splitting and cracking the ends.

8.4.2 The Threaded-Insert Method

A way around this dilemma is to attach angle brackets to the pole ends using threaded inserts. You can then attach the brackets to strips of aluminum angle stock attached to the bottom ring of the secondary cage. Threaded inserts are gadgets that look like little metal umbrellas that you insert into the end of a tube. In the center of each umbrella is a threaded hub. Push the insert in, and it's in for keeps.

This method is great because you do not have to crush the poles. The inserts permanently secure the brackets to the pole end. The brackets then connect to a short piece of aluminum angle stock on the underside of the secondary cage. For 1¼-inch truss tube, tell the clerk you would like a piece of 1¼-inch angle cut up into eight pieces each 1¼ inches long. File the ends smooth and round them slightly to preserve the skin on your fingers.

Through each surface, drill a hole for the bolts. Make sure that each of the brackets is the same, so that the holes mate to the angle aluminum on the secondary cage at exactly the same location for all eight. If you have a friend with a machine shop, have him make you eight pole brackets with one side long enough to overlap each pair. One more thing. You might need to shim the metal angle mounted on the underside of the secondary cage so that the pole tips will lie flush against it. This will be necessary if the width of the mirror box and the diameter of the secondary cage are not the same.

So far so good. You have satisfied the first two basic requirements of clamp design, namely security and repeatability. The crushed-tube or the threaded-insert designs both connect the secondary cage securely to the truss poles precisely the same way every night. The next step is to make the attachments more efficient and eliminate the need for hand tools like wrenches and screwdrivers that make assembly tedious.

Step one is to attach the fastener bolt or machine screw permanently to the short angle on the bottom of the secondary cage. That means fewer parts to drop. Having them welded on is best. (If you use angle aluminum, be sure to use an aluminum bolt so it can be welded on.) An all-metal lock nut will also work if you dog it down tight, but we lean toward welding the bolt to the angle on the secondary cage. Imagine the dilemma you'll be in some night when you are taking down your scope and find that the bolt is slipping as you turn the knob or nut. What do you do up there on the ladder with no tools and only two hands?

Now slip the angle bracket with the hole in it over the bolt and fasten with a wing nut or threaded knob. No tools are needed now for setting up your big Dobsonian. The mechanism is secure, precise and relatively efficient. You can still drop the wing nut or the knob, but at least they are bigger than a nut and much easier to hold.

8.4.3 The Offset-Bracket Method

Getting to the next logical method takes some doing. The threaded-insert method means connecting poles to the angle on the underside of the secondary cage at two separate points. What you want is to attach both poles to the same point, and you encounter the old overlap problem. You could make the brackets really long, but that's a way to lose strength.

Better to make four pairs of "offset" brackets. This type of clamp works well on "The Big Ones"—the 30-inch to 40-inch telescopes. An offset bracket is almost as easy to make as a plain bracket—the only difference is that the hole is offset to one side. This gets it out of the way so the short brackets can overlap and connect to one point (see **Fig. 8.27**).

Because the poles are connected to one another at a single point, forces on the poles cannot twist or warp the lower wooden ring of the secondary cage. Structures made with triangles are inherently the strongest, and that's why it works.

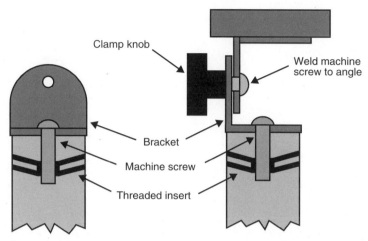

Clamp knob

Weld machine
screw to angle

Bracket

Machine screw

Threaded insert

Fig. 8.22 Angle-bracket pole clamps are both simple and reliable. The threaded insert makes attaching an angle bracket to the pole easy. The receiving angle brackets each hold two poles. Welding the machine bolt to these brackets reduces the danger of dropping them on the primary mirror.

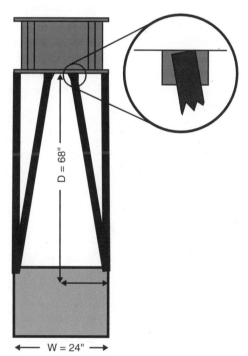

D = 68"

W = 24"

Fig. 8.23 To calculate the angle for the pole seats, divide the offset distance (usually half the width of the mirror box) by the length of the truss tube. This ratio is the tangent of the angle for the pole seats. Above, half the width of the box is 12 inches, and the truss is 68 inches long, so their ratio is 0.176. The arctangent of 0.176 is 10 degrees.

8.4.4 The Seats and Wedges Method

The preceding text describes easy ways to build truss pole clamps that are secure, repeatable, cheap and simple to fabricate, and require no tools to fasten. However, to claim that these clamps can be used *easily* by one person alone in the dark away from home on top of a ladder six feet off the ground is stretching the truth. It is difficult for one person to hold the secondary cage in position over the pole tips, place the tips over the fastener bolts, and one-by-one tighten the knobs. It's a dicey high-ladder act in the dark. The problem is that humans only have two hands and both are needed to hold the secondary cage in place.

Ideally, you would set the secondary cage delicately atop the truss pole tips where it will sit. With your hands free, you then engage clamps to secure the cage in place. Proper split block alignment aids this. The tips of all eight poles must fall under the secondary cage accurately without waving around. The following method of pole seats and wedges comes close to this ideal. Ron Ravneberg, a clever telescope maker from Ohio, invented these little gems in 1987.

8.4.4.1 Making Pole Seats

The pole seats are wood or cast aluminum retainers into which the truss pole tips fit. The tips of the poles are neither crushed nor bracketed. The pole simply slips into a recess that is shaped like an upside-down bucket seat and stays there.

The beauty of this method is that the secondary cage can be placed on top of the pole tips with confidence. The poles are seated. You can let go of the secondary cage and it stays there! Your hands are then free to slip a clamping wedge into place so that the pole cannot leave the seat.

If you want to make your own pole seats from wood, here's how. Start by calculating the tip-back angle of the seat. It is half the angle between the poles. Since you know the width of the mirror box and the distance between the mirror box and the secondary cage, trigonometry can give you the angle. The tangent of the tip-back angle φ is equal to the half the width of the mirror box divided by the distance between the secondary cage and mirror box.

$$\tan\varphi = \frac{\text{Side opposite}}{\text{Side adjacent}}$$

For a 20-inch $f/5$ scope, the side opposite is half the width of the mirror box, or 12 inches. The side adjacent is the distance between mirror box and secondary cage, about 68 inches.

$$\tan\varphi = \frac{12}{68} = 0.176$$

$$\varphi = 10°$$

Select a short length of $5/4$-inch (1 inch thick) clear maple or other close-grained wood about an inch wider than your truss pole diameter. Place the same spade bit or Forstner bit you used to make the holes in the split blocks in a drill

Fig. 8.24 *For telescopes in the 30-inch-plus range, fastening each pair of truss poles to a single point on the secondary cage is structurally superior. Tension and compression forces are not transmitted to the secondary cage as they would be if the poles were connected individually.*

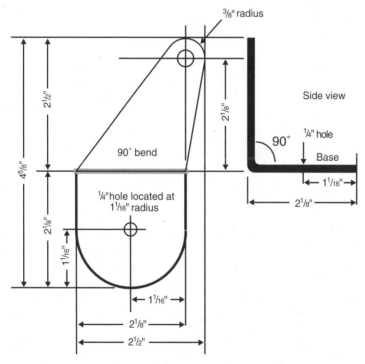

Fig. 8.25 *This plan shows an offset bracket for a 2-inch-diameter pole. Cut the offset brackets from ⅛-inch-thick stainless steel or ⅜-inch aluminum sheet. When you bend them, don't forget to make four left-hand brackets and four right-hand brackets. The offset bracket allows you to attach two truss poles to a common point on the secondary cage.*

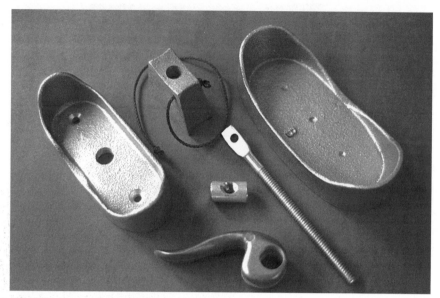

Fig. 8.26 *For small- and medium-sized Dobsonians, nothing beats custom pole seats cast in aluminum, cast aluminum wedges and cam-lever clamps. The pole seats fit the poles precisely and the clamp makes set up a breeze.*

Fig. 8.27 *Offset brackets in action on a 36-inch telescope. No other arrangement offers the security and structural advantage of the offset bracket. On a large telescope, the receiving bracket is attached to a connecting ring rather than the secondary cage to make set up and tear down safer and easier.*

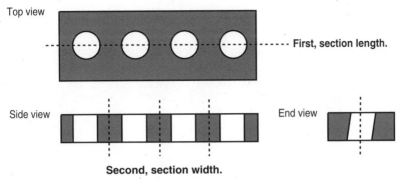

Top view

First, section length.

Side view   End view

Second, section width.

Fig. 8.28 *Cutting the pole seats is easy if you make them all at the same time. Here's the idea: drill four angled pole holes in a block of hardwood. Cut the block the long way to make pairs of seats and then cut the half-blocks again across their length to yield a set of eight pole seats.*

press, and bore eight holes at a 10° angle through the board about an inch apart.

Section the holes in half by ripping through all eight on the table saw. Section the halfholes by cutting between them. This will give you eight little blocks of wood each having a half hole through it at a 10° angle. Space the blocks in pairs about three pole widths apart at four locations around the lower ring of the cage, and attach them with wood glue. Shim the blocks with a slight inward or outward tilt if necessary.

8.4.4.2 Making Pole Wedges

The function of the wedge is to keep the poles seated tightly in the seats. The wedge is nothing more than a chunk of material that is drawn up snugly between a pair of poles with a bolt. The bolt passes through the wedge and lower wood ring of the secondary cage. As the bolt is tightened the wedge is jammed up between the poles more and more, driving the poles securely into the pole seats. The wedge design is very sturdy.

If you combine the wedge with a cam lever on the bolt, the mechanism is very easy to operate. One person can lock it without any tools, and collimation stays right where it belongs. There aren't any parts to drop on the mirror either.

The wedge can be made from wood, cast aluminum, or hard plastic. A block of wood half as thick as your pole diameter and trimmed to the angle of the poles is needed. It does not even need concave mating surfaces to match the curve of the poles. One telescope we saw had short sections of 2-inch PVC pipe between the poles instead of wedges. Almost anything will work as a wedge.

Quick-release cam levers make assembly a breeze. Simply engage the wedge, push the bolt up through the lower ring of the secondary cage and secure with a hitch pin or small nail. Throw the cam lever and you're done. You'll be observing while others are still looking for nuts and bolts in the grass.

Making pole seats and clamps is a bit like making your own focuser. You can

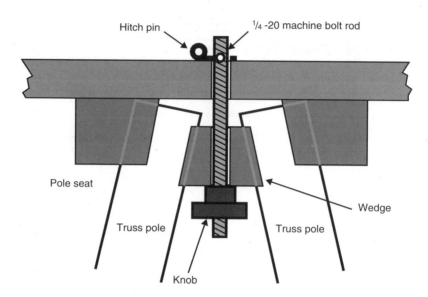

Fig. 8.29 *The wooden pole seat and wedge design makes a great build-at-home truss-tube clamp system for small- to medium-size Dobsonians. Drill a small hole in the bolt to admit a hitch pin. After inserting the poles, tighten the knob until the wedge holds the poles snugly. Alternatively, use a cam-lever for quick and easy clamping.*

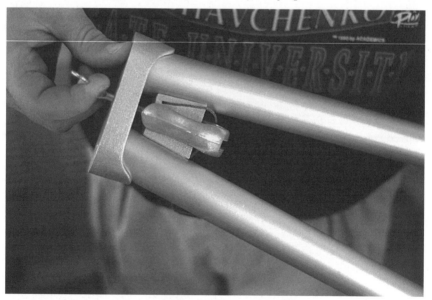

Fig. 8.30 *Pole seat clamps hold the truss-tube poles snugly so your secondary cage is secure. Whatever type of pole clamps you decide on, be sure that there are no loose parts to fall on the primary mirror during set up and tear down.*

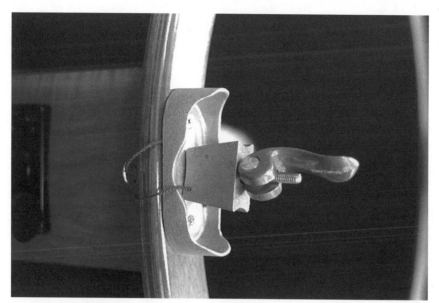

Fig. 8.31 *Pole clamps remain attached to the secondary cage when the telescope is disassembled. Here you see a cast aluminum pole seat and cam-lever clamp on the secondary cage of a 25-inch telescope. See* **Appendix E** *for suppliers.*

Fig. 8.32 *If you expect to set up and take down your telescope infrequently, wooden wedge and pole seats held with a captive hex-head machine bolt make a simple and reliable system. The captive bolt engages a tee-nut threaded into the lower ring of the secondary cage. Note that a short chain prevents the wedge from falling on the primary mirror.*

do it, but unless you have the skill and equipment you can end up with a unit that isn't half as good as commercial ones. Luckily you can buy ready-made truss-pole clamps from a variety of suppliers (see **Appendix E**).

Chapter 9
Bearings

After several chapters full of sawing and pounding, this chapter is an anomaly. There is nothing here for you to build. You have already constructed your telescope's side bearings in **Chapter 7**, and you will not construct the azimuth bearing until you build the rocker in **Chapter 10**. If you follow the recommendations in those chapters, your telescope should work well. Nonetheless, we want you to read this chapter carefully so you understand how important the bearings are to the overall success of your telescope.

Over the years, lots of telescope makers tried to get the smooth and easy motion of a properly made Dobsonian, but they failed. You cannot just copy the materials and "the look" of a design without understanding how it works. Both the materials and the geometry matter. You need to use high-grade Teflon, low-friction laminates, the proper surface coatings, and have a well-balanced scope. This chapter is here to give you the background you need to understand how Dobsonian bearings work.

9.1 How Dobsonian Bearings Work

Generally, fingertip pressure of a pound or two is ideal for guiding a big Dobsonian. If it takes less to move the telescope, then balancing is a headache and the wind may blow it around. If it takes more, the telescope feels stiff and sluggish. The goal is to attain just the right amount of friction—not too little and not too much—so the telescope moves with fingertip ease and stays pointed where you leave it.

Amazing as it seems, for almost 20 years no one knew how to calculate the optimum size for the bearings in a Dobsonian. Most people just guessed. Sometimes the bearings came out too stiff and sometimes they came out too loose. To understand how the bearings in his own 20-inch telescope worked, Richard Berry researched and wrote an article, "How to Control Friction in a Dobsonian Telescope," published in *Telescope Making #8,* that explored how Dobsonian bearings work and what you have to do to make them work *properly*.

The article showed that the telescope's weight, length, bearing materials, bearing placement, and bearing diameter all play roles in determining how much

Fig. 9.1 *The low-tech Dobsonian design eschews the expensive and complex tapered roller bearings and massive polar shafts so prized in telescopes built in the 1960s and 1970s. The elegant simplicity of bearings that rely on sliding friction gives the Dobsonian hard-to-beat stability and smoothness of motion. Here, in the palm of one hand, are the bearing surfaces of a 20-inch Dobsonian telescope.*

force you have to exert to move it. If you want to examine the equations, see **Section 3.2**. Otherwise, keep reading and we'll explain the details.

9.1.1 What is Friction?

Before Dobsonians, most telescopes had to be clamped. The observer loosened the clamp, pointed the telescope at the desired place in the sky, and retightened it. There was a clamp for declination and another clamp for right ascension. Observing could be a pain, especially if the clamps were not easy to reach from the eyepiece. The Dobsonian was the first telescope designed to make use of (rather than fight) friction in the bearings.

We are all familiar with friction. When you put this book on a table it stays. You have to push it to make it move. Friction is what keeps the book in place. If there were no friction, the book would slowly gather speed and slide off the table. If you tried to stand to pick up the book, your feet would slip and you would fall. If you reached the book, it would slip between your fingers.

Friction results from electrical forces between the molecules on the surfaces of objects. These forces vary depending on the composition of objects, so the friction between objects varies. If, for example, you place this book on a table, gravity presses the book against the table. The molecules in the plastic laminate in the

Fig. 9.2 *This "sled" was used to measure the friction of Teflon against Formica laminate. A heavy lead weight exerts a load (or "normal force") on the Teflon pads, and the spring scale measures the force needed to overcome friction and move the sled. Their ratio is the coefficient of friction.*

cover are close enough to the molecules of the varnish coating the table that they form bonds. You can slide the book across the table, but doing so requires breaking the molecular bonds between the book and the table.

When scientists first investigated friction in the sixteenth century, they discovered an interesting relationship: the force you must exert to push an object (like a book) across a surface (like a table) is a constant fraction of the weight of the object. If the book is heavy, you have to push it harder. The force needed to overcome friction and cause the object to move depends on the weight of the object and materials that make up the objects and the surface it slides on. The ratio between the force you push with and the weight on the book is called the coefficient of friction. The size of the book does not matter, only its weight. The equation is:

$$F = f \times N$$

where F is the force you must apply, N is the weight of the sliding object, and f is the coefficient of friction for those specific materials. And that is the law of friction.

In the everyday world, things are a little more complicated. Instead of the object's weight (which operates only straight down), you must use the normal force between the object and the surface. This is the force applied perpendicular to the surface. On a flat surface, the normal force is the weight, but on a tilted surface, the normal force is less. Gravity, springs, or muscle power also can supply the normal force.

If an object is heavy enough to deform the surface, friction is erratic. (A 500-pound book, for example, would splinter the fibers in the table surface.) If the table is contaminated with molecules that bind more or less strongly to the book, friction changes. (Some obvious possibilities are oil and honey.) But the machines we build are special cases: they are *designed* to behave in certain ways. In an engineered object like a Dobsonian bearing, with smooth and uniform materials, the law of friction holds up remarkably well.

In a Dobsonian telescope bearing, a surface made of a plastic resin such as Formica slides on small blocks of Teflon. The weight of the telescope presses the Formica against the Teflon, and an observer pushes the telescope, causing the Formica to slide on the Teflon. The law of friction holds, and the telescope behaves predictably.

You can examine the action of friction if you place a little sled with Teflon runners on a sheet of Formica. On top of the sled you place a weight such as the tube of your telescope. For the sake of argument, the weight and the sled together tip the scales at 100 pounds. The coefficient of friction between Formica and Teflon is 0.10. Question: how much force does it take to move the sled? The answer is 100×0.10, or 10 pounds. If you hook a spring scale to the sled and increase the strength of your pull, when the sled moves along at a steady pace, the spring scale will measure 10 pounds.

9.1.2 Friction in the Altitude Bearings

If there were no friction, imagine what might happen to your telescope. There it would sit balanced on four frictionless pads. If it were one-millionth of an ounce out of balance, the light end will rise and the heavy end would sink. As many machinist telescope makers have found, this is pretty much how telescopes mounts with ball-bearings actually behave.

Now consider a typical Dobsonian. The telescope hangs between two large bearings with Formica surfaces. These sit on Teflon pads. Question: If the telescope weighs 100 pounds, how much force must you exert to move the telescope? Well, you've already solved this problem: the answer is 10 pounds. If you attach the spring scale to *the outer edge of the bearing*, it will read 10 pounds when the telescope is moving at a steady pace.

However—and this is the big however—that's not where you push a Dobsonian. You push the Dob from the upper end, on the secondary cage. The telescope is a big lever. If the radius of the bearing is 10 inches and the distance from the center of the bearing to the secondary cage is 50 inches, then the force you must exert on the upper end is ⅕ what it is at the bearing. To exert a 10-pound force at the bearing, you need only push on the telescope with a 2-pound force.

The force necessary to move something with a lever is equal to the force needed to move it without the lever divided by the ratio of the moment arms. (The ratio is called the mechanical advantage.) That's why it is easier to open a can of paint with a screwdriver than with a penny: the length of the prying section is about a ¼ inch for both, but the moment arm of the penny is only ½ an inch, while that of the screwdriver is 8 inches. The penny gives you a mechanical advantage of ½ divided by ¼, or 2. The screwdriver gives you a mechanical advantage of 8 divided by ¼, or 32.

Fig. 9.3 Hardware and builder's supply stores sell a remarkable variety of laminates that you can evaluate for use in your telescope. Business-card size samples are usually free. For small- and medium-sized telescopes, the best materials are kitchen countertop laminates; for big tele- scopes, a graffiti-proof fiberglass washroom wall covering works exceptionally well. Note Teflon bearing pads at left. Pad with hole is an azimuth pivot pad for larger scopes.

On a telescope, the moment arm is the length of the tube assembly from the pivot point at top of the mirror box to the place where you apply force with your hand. Big telescopes have a long moment arm. On a 20-inch *f*/5 Dob, the moment arm is about 70 inches, and on a 40-incher, it can be 150 inches.

Think about the consequences. The bearing always carries the same weight, so it resists being pushed with the same force regardless of its radius. However, the smaller the radius of the bearing, the greater the mechanical advantage, and the less force is needed to move the telescope. The bigger the bearing, the more the telescope resists moving. If you build a Dobsonian with traditional small-diameter side bearings and a long tube, the telescope will move easily—in fact, it may move too easily! A breath of wind can set it drifting across the sky.

Large side bearings offer more resistance to motion, and they also offer other good things. If the bearings are large enough, they strengthen the mirror box and substantially reduce the height of the rocker sides. These are big advantages. A more subtle advantage is that large bearings also increase the linear velocity of the bearing over the Teflon when you are tracking a star. In his article in *Telescope Making #8*, Berry noted that when "the linear velocity [is] high enough for Te- flon's regulating effect to occur, ... the buttery feel characteristic of Dobsonians [is] attained."

The law of friction assumes that the amount of friction *does not depend* on the speed with which the object is sliding on the surface. From daily experience, we already know this is not completely true. Most objects require a larger force to start than they do to continue moving, that is, the coefficient of friction for a static object is greater than the coefficient of friction for a moving (kinetic) object. Try it with the book on the table: you push harder and harder but once the book starts to move, you can keep it moving with less pressure than you needed to start it. (Sometimes this stickiness is called "stiction.") Most tables of friction give both the static and kinetic coefficients of friction.

Teflon is unusual stuff. The coefficients of static and kinetic friction are almost the same. When a telescope with Teflon/Formica bearings starts to move, it doesn't suddenly jump and then stall, it just starts to slide. However, Teflon's coefficient of friction increases as the sliding speed of the bearing increases. The faster you try to go, the more the telescope resists. Teflon acts like a speed regulator—it is very easy to start and keep moving at low speed, but it slowly becomes harder to push at higher speeds.

When it comes to *engineering* a great telescope, the name of the game is to adjust the altitude bearing to give the right amount of resistance to motion. For the telescope to move in the altitude axis, these factors control how much force the observer must apply:

- the length of the telescope,

- the weight on the bearing,

- the materials in the bearing,

- the radius of the bearing, and

- the separation of the bearing pads.

You have very little direct control over the first two factors because they depend on the telescope. However, by choosing a large-aperture, thin mirror mounted in a truss tube, you have effectively specified a rather long yet light-weight telescope. Indirectly, that means that with low-friction bearing materials, you will eventually find that you need a large-radius bearing. But we're getting ahead of ourselves.

Teflon and Formica make the Dobsonian what it is. You will build your telescope with materials that have about half the coefficient of friction of the classic, smooth, counter-top material. Instead of the classic smooth Formica, we recommend a slightly pebbled plastic laminate or a fiberglass material made for graffiti-proof washroom wall coverings. As for the Teflon, we now know that virgin Teflon is better than the "whatever you can find" Teflon used in telescopes two decades ago. In addition, we know that waxing the laminate surface reduces the coefficient of friction and all but eliminates stiction (see **Section 3.2.7.3**).

The final control over friction is to alter the angular spacing between the Teflon pads. Spacing the pads farther apart increases the normal force the Teflon pads must carry, which increases the force needed to move the telescope. Put them too far apart and the bearing may jam. Spacing the pads closer together makes the bearing move more easily, but if they are too close, the scope may become easy to tip. Experience has shown that placing the Teflon pads 65 to 75 degrees apart works best. This spacing transfers the weight of the telescope to the strongest areas of the rocker. The result is a very stable support with just the right amount of friction to make tracking easy.

The general strategy in engineering a telescope is to build the telescope as light as possible, and to choose bearing materials with low coefficients of friction. This allows you to make the side bearings very large, thereby gaining the advantages of a short, reinforced mirror box, low rocker sides, and the speed-regulating effect of Teflon for exceptionally smooth motion.

9.1.3 Friction in the Azimuth Bearing

The azimuth bearing on a Dobsonian is based on the same principles as the altitude bearing, but the geometry is different. The Formica surface is on the bottom of the rocker box and it slides on three Teflon pads attached to the ground board. You can see right away that the bearing is upside-down, but the reasons for this may not be immediately obvious. There are two: first, the bottom of the rocker is stiff so that the surface does not deform, and second, the upside-down surface collects less dirt. An additional benefit of putting the pads on the bottom is that the ground board does not need to be strong; it serves only to hold the pads in place over the feet. That means less weight for you to lug around.

For the telescope to move around the azimuth axis, these factors control how much force the observer must apply:

- the length of the telescope,
- the altitude angle of the telescope,
- the weight on the bearing,
- the materials in the bearing, and
- the radius of the bearing.

Just as it did in the altitude axis, the telescope tube acts as a lever, giving the observer a mechanical advantage in moving the bearing. There is, however, an important difference. As you raise and lower the telescope, the effective length of the lever changes. By raising the telescope, the point on the telescope that the observer pushes moves closer to the axis of rotation; the mechanical advantage of the telescope-and-bearing-radius lever decreases. The mechanical advantage is the ratio of the distance between the point the observer pushes and the rotational axis to the radius of the bearing. When the telescope is pointed straight up, pushing on the telescope does not move it. Observers call this the "Dobson's hole" and avoid pointing their telescopes straight up.

(Actually, it's not that hard to move a Dobsonian near the zenith. The trick is to push one side of the telescope while you pull the other side. It takes a fair

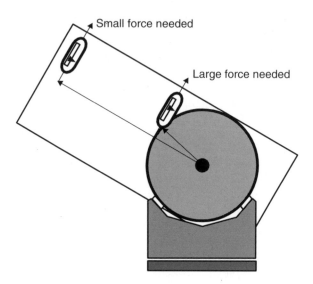

Small force needed

Large force needed

Fig. 9.4 *The bearing and the tube act as a lever; the ratio between them gives an observer up by the eyepiece considerable mechanical advantage in moving the telescope. Here, spring scales represent the force necessary to move the telescope. By altering the ratio of the radius of the bearing to the length of the tube, the builder controls the amount of force the observer must exert.*

amount of effort, but the telescope will turn. Simply pushing or pulling on the telescope will tip it over, but *pushing* one side and *pulling* the other rotates it.)

In designing a Dobsonian, you would like to have the force necessary to move in altitude and azimuth exactly the same. However, because the azimuth force changes with the altitude of the telescope, this worthy goal is impossible to meet. So, the best compromise is to make the forces equal at the most common observing altitude. Except for comet observers, most people observe objects when they are fairly high in the sky. The most pleasing operation occurs when the axes have equal resistance to motion with the scope aimed 60 degrees off the ground.

To change the force needed to turn the azimuth bearing, simply move the Teflon pads towards or away from the pivot bolt. By bringing the pads in close to the pivot bolt, you can make a heavy telescope turn very easily. However, for the greatest stability, the bottom bearing should be made as large as possible, that is, it should be as big in diameter as the rocker bottom is wide.

To reconcile these opposing requirements, you can add an auxiliary Teflon pad around the pivot bolt (see **Figure 9.5**). If you shim this pad so that it carries some of the weight of the telescope, you can place the bearing pads at a large radius for good stability and still get easy motion.

Consider what happens this way: how much weight does the pivot pad carry and what is its effective radius? Suppose you put shims under the pivot pad so that it carries half the weight of the telescope. (As far as the bearing pads are concerned, the telescope has just lost half its weight, so they will take half the force

Fig. 9.5 *The Dobsonian azimuth bearing is dirt simple: the bottom of the rocker box turns on three Teflon pads. For the best stability, the Teflon pads must be mounted directly over the feet. This simple configuration works well in small- and medium-size telescopes, but in the very largest a fourth pad, the pivot pad located in the center, carries some of the telescope's weight. Picture shows top and bottom of two separate ground boards.*

as they did before.) If the pivot pad is 2½ inches in diameter, its effective radius is a bit over 1 inch, so the mechanical advantage in moving the telescope is huge— maybe 30 or 40 even when the telescope is pointed high in the sky.

This little design trick allows you to design a Dobsonian more subtly, because it means you can set the azimuth friction at will. You set as your goal that the sliding speeds are the same in both axes. That dictates that there must be no large difference between the bearing diameters in altitude and azimuth. If the altitude bearing has a diameter roughly equal to the width of the rocker base, both bearings will be the same diameter.

No matter where the telescope is aimed, the sliding speed in the bearings will give that smooth feel. And at 60° above the horizon, the force needed to move the scope in either axis will be equal. The result is a telescope that drives like a car with power steering. You don't consciously think about moving the steering wheel when you drive a car, and you won't think about guiding when you observe with a properly designed telescope.

Chapter 10
The Rocker and Ground Board

Faithfully transmitting guiding nudges from the observer to the telescope bearings is a tall order. The task is accomplished by the rocker, a cradle that supports the telescope tube and carries the bearing surfaces for both axes of rotation.

The foundation of the telescope is the ground board. The ground board simply transmits the weight of the telescope from the Teflon pads through to the ground and provides a center point—the pivot bolt—around which the rocker rotates. Although it is simple, the pads and feet must be aligned so that the rocker rests on a firm foundation.

To transmit small forces from the observer to the bearings and thence to the ground, the rocker must be extremely stiff. If the rocker is stiff, when the observer presses on the secondary cage, the rocker transmits the force and the bearing responds promptly and smoothly. If the rocker is not stiff, the observer gets image rebound—you pull on the upper tube assembly until the flexure in the tube and rocker is used up. The bearing grudgingly jerks forward and the telescopes moves. As soon as you let go, the tube and rocker rebound, and the image bounces back to edge of the field. Elastic rebound occurs to some degree in all structures, but if you want to enjoy observing with your telescope, the motion lost to elastic rebound must be considerably smaller than the field of view of your highest power eyepiece.

The Kriege-style Dobsonian has huge side bearings mounted on the mirror box, and these allow the rocker to be built with very short sides. The sides of the rocker are constructed from multiple layers of plywood, resulting in still less flexure. A Kriege-style rocker—one that has short, thick sides—is easily 30 times stiffer than the tall, slender rockers built for the pioneering generation of Dobsonians. A well-designed rocker shows you its stuff at the eyepiece: the telescope responds smoothly and instantly to the smallest touch of the observer.

10.1 Sizing the Rocker

Building "The Big Ones" (Dobsonians of 30-inch aperture and over) raises other

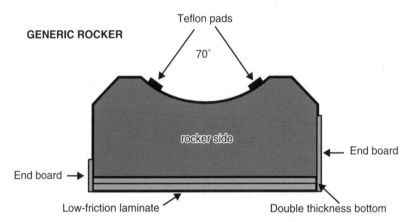

Fig. 10.1 *The rocker is large and simple. It consists of a thick base, two thick sides cut to mate with the side bearings, and two end boards that serve as braces for the sides. The side bearings ride on the Teflon pads mounted atop the rocker sides.*

engineering problems as well. As rocker components become bigger, the overall flexure increases. The rocker sides and bottom are essentially short beams. The greater the distance that a beam has to span between its support points the more flexure it has. It's the old "cube-of-the-length" gremlin again. A rocker built to support a 40-inch Dobsonian must not only support four times the weight as that of a 20-incher, it also must overcome an eight-fold increase in rocker flexure. (The rocker sides and bottom are twice as long, and two cubed is eight.)

To make matters even worse, the magnification of the telescope is also doubled, so that any flexure in the mount is grossly apparent. Observing with a 20-inch Dobsonian doesn't present anywhere near the structural demands as, say, a 36-inch Dobsonian. Care must be taken to design the rocker assembly right.

To design away the concerns described above, you need only increase the thicknesses of the rocker walls and bottom. If you make the azimuth and altitude bearings about the same size as the rocker bottom you will have scored a touchdown. If you adequately size the thickness of the rocker walls and bottom you will have won the game.

However, as much as you might like to, you cannot increase these thicknesses indefinitely. Sooner or later, practical considerations such as the total weight of the mounting catch up with and overtake the need for stiffness. Weight is a problem with telescopes because the weight increases with the cube of its linear dimensions, while the mirror's ability to collect light increases only as the square of the linear dimensions. A fully assembled 20-inch Dobsonian weighs about 150 pounds, but it's difficult to build a 40-incher that doesn't tip the scales at 550 pounds or more. Doubling the aperture often results in a four-fold increase in weight. The point is that as telescopes get bigger they get heavier really fast.

The recommendations in **Table 10.1** reflect a compromise between too much weight and too little stiffness. After you cut out the arcs for the bearings,

Table 10.1
Recommended Rocker Dimensions

Telescope aperture	Rocker side thickness	Rocker bottom thickness	End board thickness	Rocker side clearance
15	1	¾	½	⅛
18	1¼	¾	½	⅛
20	1⅜	1½	⅝	3/16
25	1½	1½	⅝	3/16
30	1⅝	2	⅝	3/16
32	1¾	2	¾	¼
36	2⅛	2¼	¾	¼
40	2¼	2¼	¾	¼
All dimensions in inches.				

glue up layers of thin plywood to achieve the desired thickness. To make 1¼-inch thickness, glue ¾-inch and ½-inch together (or two ⅝-inch). See **Section 7.6.6** to learn how.

From the table, note that the rocker bottom doesn't need to be all that thick even for the very largest apertures. The reason is that the azimuth pads are mounted on the ground board with the same diameter as the width of the rocker bottom so that the pads are almost directly under the stiffest component, the sides.

The top of the rocker sides have an arc cut out to match the radius of the altitude bearings plus an additional ⅛ inch for the thickness of the Teflon pads. For a telescope with 26-inch diameter side bearings, the radius of the rocker arc would be 13⅛ inches. The arc between the pads spans an angle of about 65 to 70 degrees from the center of the bearing.

As you saw in **Section 9.1.2**, friction in the side bearings can be controlled by changing the spacing (the angle μ) between them. The friction is proportional to the secant of the angle μ. If you place them more than 75° apart, the rocker sides are too wide and the telescope moves poorly. If you place them less than 60° apart, the telescope becomes tippy and moves too easily.

10.2 Rocker Construction

The rocker is constructed from a bottom, two sides, two ends and a pivot bolt in the center. The bottom and sides are made from plywood. The dimensions of these parts depend on the size of your mirror box, so the construction of the rocker necessarily follows the construction of the mirror box.

The grade of plywood used to build the rocker assembly is relatively unimportant, especially for the bottom panel. Use a plywood that is free of voids, but

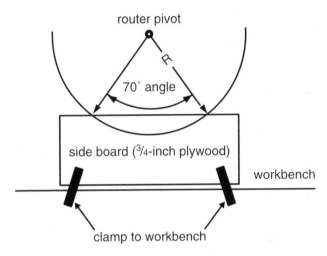

router pivot

70° angle

side board (³/₄-inch plywood)

workbench

clamp to workbench

Fig. 10.2 The radius of the rocker side arcs should be ⅛-inch greater than the radius of the side bearings, and the arcs should subtend a 70-degree angle from the pivot point of the router. Cut one layer of plywood at a time and glue the layers together to reach the side-board thickness that you need. R is radius of side bearing plus thickness of Teflon pads.

other than that proviso, softwood plywoods are acceptable. Because the stiffness of a panel rises with the cube of its thickness, we recommend building up extra-thick side panels and a double-thick bottom. The double thickness on the bottom is not only extra stiff, it also provides twice the surface for joining the sides to the bottom. Consult **Table 10.1** for the optimum thickness.

- *The Bottom Panel:* For a 20-inch telescope the bottom of the rocker should have a width and length ⅜ inch greater than the mirror box. That leaves ³/₁₆ inches of clearance on each side rocker for mirror box rotation. Laminate two or more sheets of plywood using carpenter's glue (see **Section 7.6.6**). Important: alternate the grain of the face veneers before bonding; this greatly enhances resistance to flexure in the final panel. After the sheets are bonded, trim the rocker bottom to final dimension taking care to keep its sides as square as possible. If the cuts tilt inward, the sides may get too close to the mirror box and scuff against it when the scope is moved in altitude.

- *The Sides:* To reach the desired thickness, laminate two or more sheets of plywood using carpenter's glue. The length of the rocker sides should be the same as the rocker bottom.

The height of your rocker box is tough to determine. The sides must be high enough that the exposed edge of the mirror will clear the bottom by a small but safe margin. The height of the sides, therefore, depends on the depth of the mirror box and the location of the back side of the mirror inside the mirror box. Here's

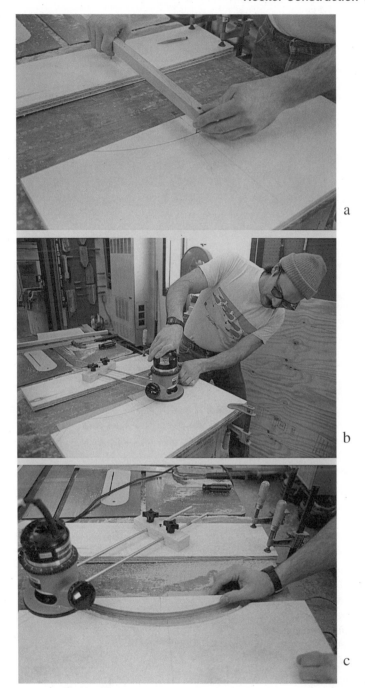

Fig. 10.3 *Here you see the making of a side board: marking the arc with a stick compass (a), then setting up and (b) swinging the router to cut the arc, and (c) there it is! The router allows you to make accurate, repeatable, and smooth cuts in plywood.*

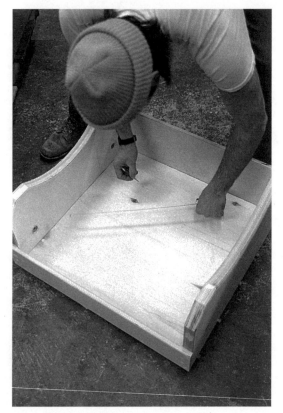

Fig. 10.4 *The rocker in rough: even for a 20-inch telescope, the rocker is pretty large. If you are thinking of building one of "The Big Ones," imagine what it would be like to construct the rocker for a 36-inch telescope. If the idea appalls you, then think about a smaller telescope; if the idea appeals to you, then perhaps a large telescope looms in your future.*

how to do the calculation.

The distance from the fulcrum of the tube assembly (which is the top of the mirror box in the modern-style Dobsonian) to the exposed bottom edge of the primary mirror is the hypotenuse of a right triangle as you can see in **Figure 10.5**. Since you know the lengths of the two sides, the Pythagorean Theorem gives you this distance—the sum of two sides squared equals the hypotenuse squared. The bottom edge of the mirror lies four inches inside the mirror box. For a 20-inch Dobsonian with a 19-inch deep mirror box, therefore, the bottom edge of the glass is 15 inches from the top of the box. The center of the mirror coincides with the center of the mirror box, so its edge lies 10 inches (the radius of the mirror) from the center of the mirror box. Doing the math:

$$C = A^2 + B^2$$
$$C^2 = 100 + 225$$

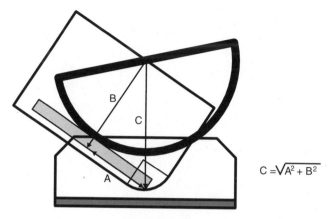

Fig 10.5 *Cutting the lower corner off the mirror box allows you to build a lower rocker box, and that brings the eyepiece down a couple more inches. Use the Pythagorean Theorem to determine the length of C from the radius of the mirror, A, and the distance, B, from the top of the box to the back of the mirror.*

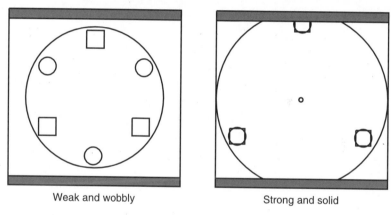

Weak and wobbly   Strong and solid

Fig. 10.6 *A small ground board with offset feet is weak and wobbly. For a strong and solid telescope, the ground board should be as large as the bottom of the rocker and the Teflon pads must be centered over the feet.*

C = 18 inches, approximately.

Eighteen inches is the distance at which the mirror will scrape the bottom of the rocker. Add an inch for clearance to be on the safe side, for a total of 19 inches. This extra inch provides room for adding a shaft encoder for digital setting circles. If you use side bearings with a 13-inch radius, the rocker must support the bottom rim of the bearing 6 inches above the top of the rocker bottom.

Do not forget that the Teflon pads add another ⅛ inch to the height of the rocker sides, and the thickness of the rocker bottom (two layers of ¾-inch plywood) adds another 1½ inches. Therefore, the bottom of the arc must be 7⅝ inches

Fig. 10.7 *When you attach the Teflon pads directly over the feet, the weight of the telescope and the torques you generate when you move it are passed directly to the ground. The result is a mounting that responds instantly and smoothly to fingertip pressure.*

above the bottom edge of the rocker side. This is not a typo: the rocker really is only one-half to one-third as tall as it is wide!

The top of the rocker side is an arc. For best results, make this cut with a router. The pivot point of the router's cut should be 13⅛ inches from the lowest point of the arc, which is 7⅝ inches from the bottom of the rocker side.

Pay particular attention to accuracy when cutting out the arcs. The arcs in the sides of the rocker must be concentric. If they are, the mirror box rests equally on all four Teflon pads.

A good way to approach making these parts with the necessary precision is to cut all four of the ¾-inch plywood layers for the rocker side into identical rectangles. Clamp one of the rectangles firmly to the work bench. Mark the desired radius of the arc on the bench, taking care to center it left to right. Double check by measuring the distance to the pivot from each of the corners of the rectangle.

With a protractor, measure a 70° arc. The pads should be about 70 degrees apart and positioned at the ends of the arc. Cut out the arc with the router's pivot point attached to the radius point.

Before you unclamp the first rectangle, mark its outline so you can place each of the other three rectangles in exactly the same place for the routing. If you cut the rocker bottom and side pieces square, the arcs of the assembled rocker unit will be aligned and concentric. After you have cut the arcs, trim the top of the rocker sides. Glue the plywood pieces together with carpenter's glue.

- *The End Boards:* The other two sides of the rocker are the end boards. They stiffen the rocker bottom substantially. The height of the shorter end board is determined by the clearance necessary for the mirror box rotation. The taller end board at the front of the rocker must be high enough to cover about an inch of the mirror box's short side. In this way, it provides a stop so the telescope cannot rotate too far forward or fall out of the rocker when the entire lower unit is lifted with wheelbarrow handles. A single layer of plywood is adequate for each. Clever digital setting circle owners may wish to install an adjustable stop here. With digital setting circles, initializing the computer requires pointing the tube straight up so the optical axis coincides with the azimuth axis. This must be done each time the computer is turned on. Most people make marks on the side bearing and rocker, but aligning them in the dark isn't easy. With an adjustable stop on the tall end board, all you have to do is raise the scope up until it hits the stop and punch the button to tell the computer the telescope is pointed straight up. It's little things like this that make using a well-designed telescope so much fun! (See **Figure C.7.**)

- *Assembling the Rocker:* Attach the rocker sides to bottom with a generous amount of wood glue and 3-inch drywall screws. The edges must line up precisely—remember, the arcs on the rocker sides must be concentric. After you drive the screws home it is a good idea to firmly clamp the side to the rocker bottom with 3 or 4 pipe clamps until the glue hardens. In the same way, attach the end boards with glue and 2-inch drywall screws. Clean up the joints, sand and finish the wood as you prefer. Paint the inside flat black.

- *Bonding the Laminate to the Rocker Bottom.* Refer to **Table 3.5** and accompanying text to determine which type of antifriction laminate to use on the rocker bottom. Countertop laminates like Wilsonart's Ebony Star and wall coverings like glassboard are bonded using contact cement. Cut out a piece of laminate that is about ¼-inch smaller on each side than the bottom of the rocker. Using a paint roller, spread contact cement over the bottom of the rocker and the mating surface of the laminate. When the adhesive is no longer tacky, align the edges and place the laminate on the rocker bottom. Get it aligned right the first time. Once the two surfaces kiss you won't get a second chance. Hint: if you ever need to remove laminates that are bonded with contact adhesive use an old dish towel and clothing iron. Start in a corner. Double up the dish towel and place it on the laminate. Put a hot iron on top and heat the laminate until it loosens.

Finally, on the inside surface of each rocker side place a felt pad just like the ones used for the mirror cell points, to keep the mirror box centered in the rocker. The felt pads prevent the distracting sound of the mirror box scuffing the inside of the rocker in the dead of the night. If the felt pad rubs against the mirror box, it should not mar the finish.

10.2.1 Ground Board

The lowly Dobsonian ground board supports the entire telescope, retains the Teflon pads of the azimuth bearing, and holds the pivot bolt. It looks so simple. Yet because it looks simple, telescope makers sometimes throw together the ground board without any thought as to why some scopes rotate better than others. Poor ground boards have turned many homemade Dobsonians into telescopes with the stability of Jell-o. Yet there is no magic to getting the ground board right, only basic physics.

Three points define a plane, so we have three Teflon pads to define the azimuth bearing. The pads are spaced 120° apart around a central pivot, usually a large bolt. The rocker unit sits on top these three pads and rotates about the bolt in the plane defined by the pads.

Modern professional scopes like the 10-meter Keck telescope on Mauna Kea use steel wheels for the azimuth bearing. The low-tech Dobsonian uses three Teflon pads. These three pads support the entire weight of the telescope.

However, the ground board *should not support the pads.* For real stability, the pads should transfer the weight of the telescope *through* the ground board directly to feet underneath, and thence to the ground.

If the Teflon pads are placed anywhere but directly over the feet, the telescope will suffer from the "trampoline effect." Remember when you were a kid in gym class? The best spot to bounce on the trampoline was the middle. The worst place—the place with no bounce—was over one of the corners by a leg. The same happens with the ground board: when you put the pads directly over the feet, the support is rock solid, with no bounce.

Some time when you are at a star party, look for a telescope with a square ground board and three pads spaced 120° apart. At very best, only one of the Teflon pads can be placed over a foot. In the modern-style Dobsonian, as long as the pads are directly over the feet, a piece of cardboard would make a satisfactory ground board. All the ground board does is to hold the pads and pivot bolt in place.

In fact, if you want to site your telescope in an observatory, we recommend that you eliminate the ground board altogether. Set a pivot bolt permanently into the floor of your observatory and glue the Teflon pads to the floor with panel adhesive. You can't get more solid than that.

By the same logic, the best location for the three azimuth pads is over the feet and directly under the sides of the rocker. The weight of the scope is carried down

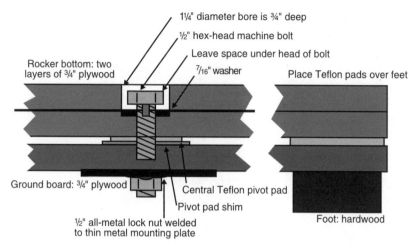

1¼" diameter bore is ¾" deep

½" hex-head machine bolt

Leave space under head of bolt

Rocker bottom: two layers of ¾" plywood

⁷⁄₁₆" washer

Place Teflon pads over feet

Ground board: ¾" plywood

Central Teflon pivot pad

Pivot pad shim

½" all-metal lock nut welded to thin metal mounting plate

Foot: hardwood

Fig. 10.8 *This cross-sectional view shows how to mount the pivot bolt. The 1¼-inch counterbore allows you to reach the head of the bolt with a socket wrench. If you think you might ever wish to install digital setting circles, drill a ¼-inch hole in the head of the bolt.*

from the side bearings to the rocker sides. The sides are much stiffer than the bottom of the rocker. Accordingly, the Teflon pads should be out far enough to be under the solid rocker sides.

If you put the pads very far inward, then they are under the bouncy center of the rocker bottom. The rocker bottom acts like an upside-down trampoline. In the worst-case scenario, you would have three pads on a square ground board at a radius much smaller than the width of the rocker base. The result would be a rubbery rocker and a telescope that would wobble whenever you aimed it in altitude. Sadly, some people have built telescopes just this way.

The old rule of thumb for the ground board was to make it the diameter of the primary mirror. On scopes under 18 inches aperture that guideline works fine. For larger Dobsonians, however, the diameter of the rocker is considerably larger than the primary. Accordingly, for scopes larger than 18 inches, give the ground board a diameter equal to the width of the rocker.

In summary, to design the rocker/ground board combination right, follow these ground board guidelines:

- make the diameter the width of the rocker,
- place the feet near the outer edge, and
- place the Teflon pads over the feet.

10.2.2 Ground Boards for "The Big Ones"

"The Big Ones"—Dobsonians of 30-inches aperture and larger—are harder to build than "ordinary telescopes." Experienced telescope makers already know that the farther out you position the azimuth pads, the tougher it is to rotate a large, heavy telescope. As the bearing size increases relative to the telescope, it takes more pressure to turn it. There are two ways to reduce friction, and you may wish to use both on a large, heavy telescope.

The first is to decrease the coefficient of friction between the Teflon pads and laminate by finding better bearing materials. **Chapter 9** went into this at some length, so you should already be using the best materials available. Without getting into exotic materials, it is difficult to beat a pebbled laminate or glassboard surface.

The second option is quite clever but not very elegant. Because there is virtually no resistance to rotation at the pivot, slip a piece of Teflon around the pivot bolt and shim it to reduce the weight on the three regular Teflon pads. The key is to make the pad thick enough to reduce the weight of the telescope on the pads. You can determine the proper amount by trial and error. Put enough shim under the pivot pad that the scope rotates easily, but not so much that it becomes tippy.

When you do this, the load from the central pivot pad is much greater than it is in a smaller telescope, and the ground board may deform somewhat. In larger telescopes, therefore, we recommend making the ground board 1½ inches thick, that is, made with two bonded layers of ¾-inch plywood. Without greater than normal stiffness, the ground board will not be able to carry the load in the center, and telescope will remain too hard to turn in azimuth.

10.3 Constructing the Ground Board

For a 20-inch telescope, the rocker bottom will be 23⅞ inches across. From ¾-inch plywood, cut a disk of the same diameter. Use a less expensive grade of plywood for the ground board; no one will see it and it's down in the dirt anyway. If you want to get fancy, you can make a triangular ground board as long as you don't change the radius of the pads. Attach three wooden feet to the bottom of the ground board near the edge and spaced 120° apart. To determine the size of the Teflon pads, see **Table 3.5** and **Section 10.3.4.2**. Attach the Teflon pads directly over the feet. Make three feet from scrap plywood or better yet, a tough hardwood, 3 by 3 by 1-inch thick.

10.3.1 The Azimuth Pivot

The universe revolves around the pivot bolt in the bottom of your telescope. Well, not quite. The pivot bolt provides a center of rotation and keeps the rocker from wandering off the ground board. The pivot is a ½-inch diameter bolt-and-nut combination. The bolt and nut also secure the ground board to the rocker, so you can lift the two together when you transport the telescope.

Since the advent of digital setting circles (see **Appendix C**), the pivot bolt has gained an additional role because the azimuth shaft encoder can be attached to it. The encoder tells a small computer how far and in which direction the rocker is rotated. Digital setting circles can really increase your satisfaction as an observer.

Fig. 10.9 *An all-metal lock nut welded to a mounting plate anchors the bottom end of the pivot bolt. The holes in the mounting plate allow you to attach it firmly to the ground board. The mounting plate can be any shape you desire.*

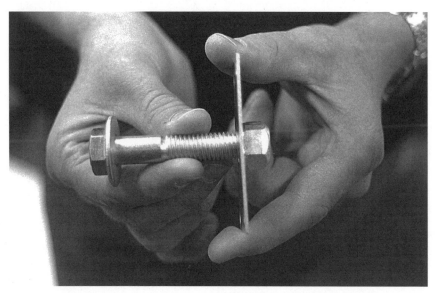

Fig. 10.10 *The pivot bolt is an ordinary ½-inch hex-head machine bolt. Fitting over the bolt, an ordinary ⁷⁄₁₆-inch washer serves as a metal-to-metal bearing. The lock nut in the mounting plate prevents the bolt from turning when the telescope is turned..*

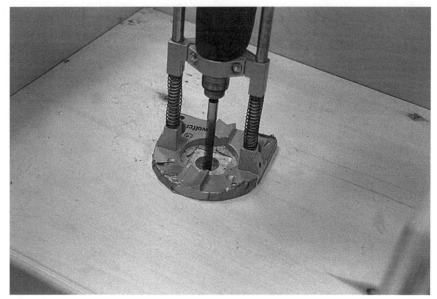

Fig. 10.11 It is important that the center hole for the pivot bolt be perfectly straight, but the bottom board of the rocker is too big to put in a drill press. Instead, use a jig to hold the drill exactly perpendicular to the inside surface of the rocker box.

If you find objects by star-hopping, you spend 90% of the time searching and 10% of the time observing. With digital setting circles it's reversed: you spend 10% of the time searching and 90% of the time observing. You may as well build this option into the pivot in case you decide to try digital setting circles in the future. Both types of astronomy are fun and we observe both ways depending on the mood we're in.

10.3.2 How to Make the Pivot Bolt

The pivot bolt is made from a ½-inch hex-head machine bolt. The bolt defines the rotational axis of the rocker in azimuth, connects the ground board to the rocker, and engages an optional shaft encoder. If you plan to attach an encoder in the future, the length of this bolt should be chosen so that the head of the bolt lies ¼ inch *below* the top face of the rocker bottom. If you plan never to use an encoder, the bolt can be flush with the top of the rocker bottom. Under no circumstances should the head of the bolt protrude above the surface of the rocker bottom: it could catch and damage the tailgate or primary mirror.

The pivot bolt must remain fixed to the ground board. If it were to rotate with the rocker, the shaft encoder would not read the position of the telescope properly. To prevent the bolt from turning with the rocker, have your local welding shop weld a ½-inch all-metal lock nut to a scrap of 1/32-inch-thick steel plate about three inches square. The nut must sit squarely on the plate so that when the pivot bolt

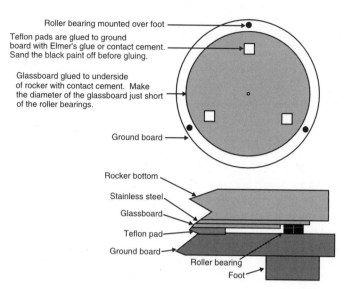

Roller bearing mounted over foot

Teflon pads are glued to ground board with Elmer's glue or contact cement. Sand the black paint off before gluing.

Glassboard glued to underside of rocker with contact cement. Make the diameter of the glassboard just short of the roller bearings.

Ground board

Rocker bottom
Stainless steel
Glassboard
Teflon pad
Ground board
Roller bearing
Foot

Fig. 10.12 If you build a large telescope, you may wish to experiment with a hybrid azimuth bearing. In this design, rollers provide stability and easy motion, and the radius of the Teflon pads is adjusted to give just the right turning resistance. When properly adjusted, the rollers carry most of the telescope's weight. The low-tech alternative, a central Teflon pivot pad, is simpler and works just as well or even better.

engages it, the bolt is precisely perpendicular to the steel plate. In each corner of the steel mounting plate, drill a hole to accept a small wood screw.

Important: **do not** use a nylon-insert lock nut. Nylon inserts do not grip the bolt strongly, and the nylon insert can melt or degrade when the nut is welded to the steel plate.

For digital setting circles, chuck the pivot bolt in a lathe and bore a ¼-inch hole 1 inch deep squarely into its head. It is impossible to bore this hole accurately in the center of the bolt by hand or even on a drill press. Do the job with a lathe: it's a two-minute job and it must be done right.

The completed pivot bolt assembly consists of a ½-inch pivot bolt with a precisely concentric ¼-inch hole drilled in the head and an all-metal lock nut mounted on a small steel plate. You can take the easy way out and order the bolt and nut plate and all encoder mounting hardware from the setting circle supplier.

10.3.3 Installing the Pivot Bolt

For digital setting circles, the hole for the pivot bolt must be centered in the rocker. Carefully measure and mark the location of the center. Drill a ⅛-inch pilot hole through the rocker. Using a 1¼-inch spade bit or Forstner bit, bore a hole halfway from the top surface into the rocker bottom. For a standard 20-inch telescope, this hole will go through one layer of ¾-inch plywood. If you have a drill guide, use it to drill the hole as accurately perpendicular to the rocker bottom as possible. Continue the hole with a ½-inch drill to complete it through the rocker bottom. At the center of the ground board, drill a ½-inch hole all the way through. Keep the drill as square as you can—it is important that the hole be perpendicular to the rocker.

To install the pivot bolt, attach the nut and steel plate to the bottom side of the ground board over the ½-inch hole. Slip the bolt through the hole to insure that the nut and plate are centered on it, drill holes for the wood screws, and insert wood screws to attach the plate to the bottom side of the ground board.

On the rocker, place a steel washer in the bottom of the 1¼-inch hole in the rocker bottom. Washers described as ⁷⁄₁₆-inch have just the right outside diameter to fit snugly in a 1¼-inch hole. The washer has an outer diameter of 1¼ inches and the ½-inch hole in the middle fits perfectly over the pivot bolt. The washer makes a harder and more resistant pivot bearing for the bolt than the wood of the rocker bottom, and it prevents the axis of rotation from wandering. The 1¼-inch hole is big enough to allow a socket wrench in to tighten the head of the bolt. To aid placement, lubricate the threads of the bolt with petroleum jelly. Leave the bolt loose enough that the rocker can turn freely.

10.3.4 Installing Teflon Bearing Pads

Installing the Teflon pads is one of the last procedures you do. The telescope needs to be complete. Read **Sections 3.3.3** through **3.3.4.2** to learn about Teflon and laminates for the bearing surfaces. You will need to determine the size of the Teflon pads that support the altitude bearings and the rocker (azimuth) bearing. Sure you could just arbitrarily cut out a few pieces of Teflon and install them but the scope won't work as well as it could. In order to achieve the best performance Teflon can offer it must be the right size. The Teflon pads must support the proper PSI, Pounds per Square Inch. See **Table 3.5** for advice.

10.3.4.1 Calculating the Size of the Altitude Bearing Pads.

You must know the weight of the tube assembly including the altitude bearings. If the tube assembly is too ungainly to set on your bathroom scale then weigh all the components separately. Don't forget anything. The tube assembly consists of the mirror box, mirror cell, primary mirror, altitude bearings, truss tubes, cage with secondary mirror and eyepiece, finderscope, light shroud. Everything that will be supported by the altitude pads must be included. Add up all the weights of the components. Divide the total weight by four (there are four altitude pads on the rocker). This gives you the pounds per pad. Next divide the pounds per pad by the psi from **Table 3.5** depending on the type of bearing laminate you are using. This gives you the total area in square inches of each pad. The altitude pads are usually rectangular in shape to better fit on the top edges of the rocker sides. Make your pads the same width as the rocker thickness. Since the rocker thickness is already determined all that's left is to divide the area of the pad by the thickness of the rocker side and you've got the dimensions of the altitude pads.

An example will take you through the process step-by-step. Let's say the tube assembly of a 20-inch telescope weighs 140 pounds and the altitude bearings weigh 20 pounds. The total weight of everything that rides on the altitude pads is then 160 pounds. 160 pounds divided by 4 pads equals 40 pounds per pad. The altitude bearing surface is Ebony Star which requires 15 pounds per square inch for maximum slipperiness. 40 pounds per pad divided by 15 pounds per square inch equals 2.66 square inches per pad. The rocker sides were each made from two layers of ¾-inch plywood bonded together so it is 1½ inches thick. Therefore, 2.66 square inches divided by 1½ inches equals 1.77 inches. The four altitude pads for this telescope should thus be cut into rectangles that are 1½ by 1¾ inches.

10.3.4.2 Calculating the Size of the Azimuth Bearing Pads

This is even easier. First you must weigh the rocker. Do not include the ground board. Add the weight of the tube assembly from above to the weight of the rocker. The total is the weight that rides on the three azimuth pads that are attached to the upper surface of the ground board. Divide the total weight by three to get the weight per pad. Then divide the weight per pad by the pounds per square inch depending on what type of bearing laminate you used on the bottom of the rocker. See **Table 3.5** again. The result is the area in square inches of each pad. Since the azimuth pads are squares all that's left to do is to take the square root of the area. The result is the length of the side of each square azimuth pad.

An example will take you through the process step by step. The weight of the rocker is 35 pounds. Added to the 160 tube assembly from above the total weight on the azimuth pads is then 195 pounds. Divide 195 by three pads to get 65 pounds per pad. Since you used glassboard laminate on the bottom of the rocker you see from **Table 3.5** that it requires 12 pounds per square inch. Divide 65 pounds per pad by 12 pounds per square inch to get 5.42 square inches, the recommended surface area of each pad. To get the length of the sides take the square root of 5.42 and the answer is 2.32 inches. The three azimuth pads for this telescope need to be squares that are 2.32 inches on each side. If square roots scare you just press the square root button on any pocket calculator for the answer. After you have determined the size of the pads cut them out of a small sheet of Teflon.

See **Appendix E** for Teflon sources. If you are using etched virgin Teflon you can bond the pads directly to the rocker and ground board with contact adhesive or epoxy glue. However, if you are using standard Teflon there are no adhesives that will stick to it so you have to nail them on with tiny finishing nails. (That's why Teflon works so well in frying pans. Nothing sticks to it.) Be sure to countersink the nail heads so they don't protrude and scratch the bearing surfaces.

10.4 A Handy Option: Bearing Locks

As you know or will soon discover, large-aperture Dobsonians are a magnet for the public. People can't resist touching them. Even seasoned observers want to check out how they move. It's nice that people appreciate these telescopes so much, but you may not feel comfortable with so many eager "drivers" moving your scope.

You can "lock" your telescope by drilling a hole through the rocker bottom and on into the ground board. Drop a long nail through both and—ta-da!—the telescope cannot be moved in azimuth. Point the tube about 25 degrees above the horizon. Drill a hole through one of the rocker sides and into the mirror box. Push a long nail into the hole and your scope is locked in altitude. Once people realize that your telescope does not move, they will leave it alone when you're not around.

Locking the bearings is good policy in windy places. A 20-inch or larger Dobsonian presents a large surface to the wind. In fact, anything more than mild breeze can push the telescope around like a weather vane. If you leave it unattended, lock the telescope to prevent sudden wind-driven motions that could endanger the telescope and anyone standing close by.

10.5 Handles for Portability

Every telescope builder faces a dilemma: setting up the telescope. Even if you have a van or trailer to take your scope to a great dark site, you must still have some way to load it and unload it. Even if you enjoy observing in company, you need to plan for spontaneous observing sessions when you don't have the luxury of inviting a fellow astronomer. Fortunately, Ron Ravneberg (an exceptionally insightful telescope maker) solved the setup problem with wheelbarrow handles.

Ron's inspiration was to temporarily attach wheelbarrow handles *with* pneumatic tires to move the heavy components of big telescopes. With a pair of such handles, one person can easily roll the rocker, ground board, and mirror box with the primary mirror installed into or out of a van or trailer. They make loading and unloading your big telescope a snap.

For telescopes in the 20-inch range, get a pair of wheelbarrow handles five to six feet long. For smaller Dobs, you'll want shorter handles, and for bigger ones, longer. The longer the handles, the more leverage you get. A handy rule of thumb is that the length should be 2¼ times the width of the rocker. Look for solid, knot-free, hardwood handles and stain them to match your scope color. You can find ready-made wheelbarrow handles at most hardware stores and lumber yards.

10.5.1 "It Rolls On Air"

Look for wheels with pneumatic tires. For scopes 20 inches and under, 8-inch diameter by 2-inch wide tires are great. For the 22- to 30-inchers, 10-inch diameter by 2-inch wheels would be perfect, but since most wheels in these diameters are made for use on commercial hand dollies, you may have to accept much wider tires. Do not worry, they are fine. If you shop from an industrial supply catalog, look for the following buzz words: "10-inch wheel with a ½-inch ball-bearing hub,

Fig. 10.13 *The low-profile design of the rocker and ground board in the Kriege-style Dobsonian provides easy roll-out and hassle-free setup as well as great performance at the eyepiece.*

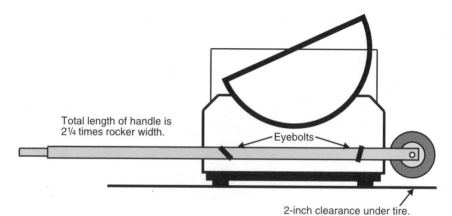

Total length of handle is 2¼ times rocker width.

Eyebolts

2-inch clearance under tire.

Fig. 10.14 *Two 1½- by 1¾-inch hardwood wheelbarrow handles and pneumatic tires make unloading and moving your telescope an easy job. Captured eyebolts hold the handles to the sides of the rocker. Allow ⅜ to ½ inch clearance between each tire and the front board of the rocker.*

2¼-inch offset, 10 by 3.50 air-filled." Obviously, a ball bearing hub is overkill, but since they are off-the-shelf merchandise, order what they have. Quality wheels like these are made to roll around things like cases of beer, so they will probably outlast both you *and* your telescope.

Avoid solid rubber wheels; they do not absorb shocks well. They are fine for level pavement, but you need pneumatic tires because you will be rolling your telescope over bumpy ground. If you don't believe us, try it. With solid tires, the junction of the loading ramp and the vehicle is enough to jar the mirror. Nothing is more uncomfortable than hearing an expensive primary mirror rattling between its cell and the mirror clips. Don't be cheap. Buy air-filled wheels.

Once you have handles and wheels with pneumatic tires, the idea is to place the wheels close to the rocker so that *they* carry most of the weight—this is just the basic principle of leverage at your service. However, since stones occasionally get stuck in the tire treads and can scratch the finish on the front of the rocker, mount the wheels on the handles so that tires clears the rocker by about ½ inch. Attach one wheel to each handle with a fully-threaded carriage bolt that matches the inside diameter of the wheel hub and is long enough to pass through the hub and handle. It usually works out to be a ½- by 5-inch bolt. If you can't get fully-threaded bolts, buy threaded rod. Put the bolt through a hole in the handle and attach it tightly with a nut. Slide the wheel onto the bolt and retain it with a lock nut. Nylon-insert lock nuts are best for this job.

10.5.2 Attaching Handles to the Rocker

To roll out your telescope, attach the wheelbarrow handles to the sides of the rocker, lift, and roll. Two off-the-shelf methods are available: T-nuts and self-tapping threaded inserts. Both of these methods put metal threads into the side of the rock-

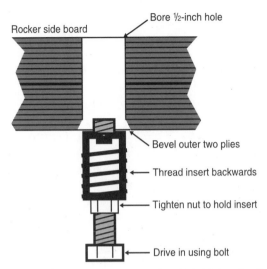

Rocker side board

Bore ½-inch hole

Bevel outer two plies

Thread insert backwards

Tighten nut to hold insert

Drive in using bolt

Fig. 10.15 *To install the self-tapping threaded inserts that hold the wheelbarrow handles, bevel the outer two plies on the side board of the rocker. Thread the insert bottom end first onto a ⅜-16 hex-head bolt and snug it against the stop nut. For an ultra-strong insertion, spread a thin layer of epoxy inside the hole. Use a wrench to tap the insert into the wood, loosen the nut, and then remove the bolt. Once the epoxy sets, the insert will never come out.*

Fig. 10.16 *It takes nothing more than a threaded insert and an ordinary bolt and nut to install metal threads into a plywood panel. Run the threaded insert backwards onto the bolt, then use a socket wrench to force the insert into the hole. Handy tricks like this make your low-tech telescope a pleasure to use.*

er so you can easily bolt the handles into place.

T-nuts are simple but hardly elegant. Drill holes in the appropriate places, insert the T-nuts from the reverse side, and hammer them home solidly. They'll work well but as the wood around the holes gets dinged up, you'll wish you had taken a bit more time and used self-tapping threaded inserts. These handy devices are available with a ⅜-inch internal thread in steel, stainless, or brass. Self-tapping threads on the outside vary depending on whether they will be used in plastic, metal, or wood. Check **Appendix E** for the names of suppliers.

Locate the self-tapping inserts or T-nuts on the rocker so that the tire tread is ½ inch from the front of the rocker and the bottom of the wheel 2 inches above the ground when the telescope is resting on the ground. Two inches seems like a lot, but it provides clearance for placement and removal of the handles on uneven ground. The wheels should be on the inside of the handles to keep the assembly narrow and make turning easier.

Self-tapping threaded inserts work great, but it takes some skill to install them. Here are a few of the secrets of threaded inserts.

At the desired location, drill a hole of the size recommended by the manufacturer. Next, relieve the surface plies around the hole by beveling it slightly with a very large drill tip. (To do this, use a ¾-inch drill and turn it slowly by hand. Do not apply power—the drill will remove too much wood.) A three-sided hand awl or a rat-tail file also work. Bevel the top two plies, but no more. Beveling will prevent the outer face veneer from puckering when the insert is driven into the hole.

Install the insert using the "bolt-and-nut" method. (*Do not* try to tap the insert into the hole with a screwdriver in the slot as the manufacturer suggests. The screwdriver blade can slip out of the slot and gouge the wood. The insert invariably goes in crooked, you strip the hole, then the slot, and before you know it you're drilling new holes to start over.) As shown in **Figure 10.15** an ordinary nut on a hex-head bolt will hold the insert from turning. Run the insert backwards onto the bolt up to the nut. Yes, backwards. Self-tapping inserts self-tap better when the slotted top end of the insert goes into the wood first. Use a socket wrench to turn the bolt, and run the insert into the wood until the holding nut is flush with the side of the rocker. Loosen the nut and back out the bolt. You'll find this technique eliminates the screwdriver gouges. Put some epoxy on them before installation and they will never come loose.

To bolt the handles to the rocker, you *could* use fancy bolts with plastic handles, but there's a better way: eyebolts. Two ordinary ⅜-16 by 5-inch long fully-threaded eyebolts connect the handles to the fasteners in the rocker sides. The eyebolts are retained in the handles with ⅜-16 all-metal lock nuts. Countersink the lock nuts into a ⅞-inch diameter by ⅜-inch deep hole in the handle so that the handle can be drawn up tight to the rocker side without interference from the lock nuts.

Eyebolts are cheap and durable, and they won't break off when you crunch a door jamb or smash them against the sides of your vehicle when you're loading your telescope. Don't be cool and use expensive store-bought plastic knobs—

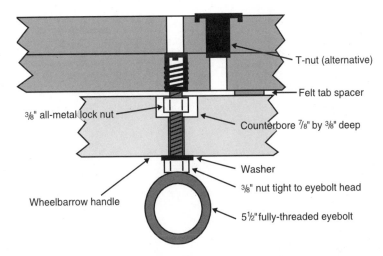

T-nut (alternative)

Felt tab spacer

³/₈" all-metal lock nut

Counterbore ⁷/₈" by ³/₈" deep

Washer

³/₈" nut tight to eyebolt head

Wheelbarrow handle

5½" fully-threaded eyebolt

Fig. 10.17 *Here is how the captive eyebolt works: the nut threaded against the head allows you to turn the eyebolt tight; the all-metal lock nut serves to capture the eyebolt. The counterbore in the handle means the lock nut lies entirely inside the handle. The fit of the eyebolt in the hole is sloppy enough to permit easy alignment with the threaded inserts.*

Fig. 10.18 *No need to wonder where that @#$%&! handle bolt went because it's captured in the handle. An inset all-metal lock nut keeps the eyebolt securely in the handle so you never have to search for it. Little niceties like this means a lot when it's 4 in the morning and time to pack up and head for home.*

Fig. 10.19 It only takes a few moments to attach the wheelbarrow handles and you are ready to transport the telescope. For safety's sake, never move a fully assembled telescope more than a few feet. Once the telescope is disassembled, you can roll the ground board, rocker box, mirror box, and mirror quickly into a vehicle or storage area.

you'll smash the heads off them the first summer. You'll also find that the bolts sometimes bind, and that makes removing the wheelbarrow handles difficult. With plastic knobs, your fingers will suffer. With eyebolts, you put a stick through the eye and you have enough leverage.

The all-metal lock nuts keep the eyebolts captive in the handle. They can't disappear in the weeds when the handles are removed for observing. Don't drive the nuts up tight to the handle. Leave it sloppy so the eyebolt is easy to turn. Also drill the hole for the bolt oversize so you can align the bolts with the inserts on the rocker. Low-tech solutions like these work great.

10.6 Handling "The Big Ones"

Lifting the lower unit of a large-aperture Dobsonian is an easy task with wheelbarrow handles. But "large" is a relative term. "The Big Ones" (30 inches and larger) weigh too much for wheelbarrow handles. Even if you could lift the weight, the rocker box is so wide that your arms aren't long enough to reach both handles. The obvious low-tech solution is to not lift at all. Instead, look for equipment that was designed for some other purpose, like moving a motor boat. For moving heavy telescopes, marine trailer jacks fill the bill perfectly.

To move a 36-inch Dobsonian weighing 500 pounds into and out of a small trailer or a full-size pickup truck requires nothing more than a pair of ramps, a

Fig. 10.20 *You would need Paul Bunyan (or Babe) to move one of "The Big Ones" with wheelbarrow handles. Instead, marine trailer jacks serve as four fold-down wheels for this 500-pound 36-inch telescope. Using a small boat winch, you can ease the rocker up a ramp and into a van, trailer, or pickup truck.*

hand winch, and four marine trailer jacks. It's all off-the-shelf stuff, readily available and affordable. Trailer jacks are made to attach to the tongue of a boat trailer so the owner can raise it to the level of the hitch on his car without breaking his back. There are numerous types on the market. Look for jacks with the following features:

- crank mounted on the side,
- at least 10 inches of lift, and
- pneumatic castor on the foot.

Don't worry about the capacity of the jacks because they were built to handle loads much heavier than a telescope. The side cranks are important because a top crank won't clear the mirror box.

Mount a pair of marine trailer jacks on each side of the rocker. With a jack at each corner of the rocker, you could raise and lower the biggest home-built telescope in the world. Crank the jacks and the telescope is on wheels. When you mount a hand winch (or an electric winch if you *really* want to impress folks) and a pulley to the frame of your trailer or truck, you can off-load the telescope alone.

Attach a pulley to the front end of the trailer or truck bed, and place the winch beside the door of the trailer or the tailgate of the truck. With an "S" hook, connect a stout rope to an eyebolt built into the rocker. Run the rope through the pulley and back to the winch. Set up ramps made of stout 2 by 10-inch timber from the back of the vehicle to the ground. This arrangement lets you stand at the back end of the vehicle where you can keep an eye on the telescope as you crank it on or off.

We recommend at least 10 inches of lift so the ground board won't bottom out at the junction of the ramp and the trailer. If the ground is soft or graveled, it helps to place boards on the ground so the wheels will roll more easily.

Roll the telescope down the ramps by releasing the rope slowly with the winch. Once the telescope is on the ground, lower the jacks and rotate them 90 degrees so that the rocker is free to turn. Set up the truss tube, get out the ladder, attach secondary cage, and you're ready to observe.

Chapter 11
Assembly and Troubleshooting

It's been a long road, but you have made it. At last you are ready to assemble your big Dobsonian for the first time. You are excited and want to set it up as soon as possible so you can have first light. Be prudent—make sure that everything is ready. You don't want to get half-way through assembly only to discover that some big something still must be done. Use the following checklist to make sure that all of the components you will need are ready to assemble.

Optics: The primary and secondary mirrors are on hand. Although you temporarily installed the mirrors to determine the length of the truss poles, for the most part, they are still waiting in safe storage.

Mirror Cell: The tailgate is bolted into the mirror box. The flotation cell and sling are installed. The fan is bolted to the back of the mirror cell and the wiring, switch, and batteries to operate the fan have been installed.

Secondary Cage: The secondary cage is complete. The focuser is in place, the spider and secondary holders are installed, and you have bolted the pole seats to the lower ring. You have a Telrad or finder telescope on hand and ready to attach.

Mirror Box: The pole blocks have been aligned, shimmed, and screwed into place, and the mirror cell is already installed. Except for a few touch-up spots, the wood is finished with the coating of your choice.

Dust Cover: The dust cover is equipped with clips and fits over the opening of the mirror box. This is an important component: its job is protect the primary mirror once it is installed in the mirror box.

Truss Poles: You determined the length of the truss poles and cut them exactly two inches too long. They are all the same length. You have verified that they seat in the pole blocks attached to the mirror box, and you have aligned the pole blocks.

Side Bearings: You have completed the side bearings and bolted them to the mirror box. You did a great job and they are precisely concentric.

Rocker: The pivot bolt joins the rocker and ground board. The side bearing pads and the bottom bearing Teflon pads are in place. You have applied the finish of your choice to the rocker and ground board and—heck, even you have to admit it—they look pretty darn good.

Shroud: You have at least thought about making the shroud, but (if you are male) in typical male chauvinist pig fashion, you are still hoping some woman is going to volunteer to sew it for you. This is stupid. Making the telescope shroud is a macho business called "fabric engineering." Don't put if off.

When the pile of telescope parts in your workshop matches the description above, then you are ready to assemble. If the everything is not ready, then finish the work that needs to be done before you attempt to assemble the telescope. Even the painting. You don't want to drip black paint on the primary, do you?

Read through the instructions carefully. There are several things you'll want to do in advance—center-dotting the primary, installing the secondary in the secondary holder, and so forth. If you get them out of the way a few days ahead, it's a big help.

One final note before you assemble: prepare an area for storing the components afterward. Until now, if the components got dusty, you could hose them down. Once the primary mirror is installed in the mirror box and the secondary mirror is installed in the secondary cage, it's best to store the telescope in a clean, dry, dust-free location.

Take a Saturday or Sunday when you have time to do the whole job in daylight. Review the following instructions carefully. Assemble the telescope slowly and carefully the first time, and then run through it again in the nighttime. After a couple of trials you'll find you can do it in a few minutes. Don't rush it! You will only damage your telescope, hurt yourself, or suffer poor images due to lack of collimation. A few extra minutes invested in making sure that everything is really ready just might save you a lot of grief later on.

11.1 Last-Minute Preparations

If you plan to transport your telescope in a van or trailer, make a ramp to roll the rocker and mirror box in and out of your vehicle. A length of 1x12-inch clear pine works very well. Clear means it has no knots; ask for it by name. Ramps made of clear 1x12-inch pine are strong, wide enough that you won't roll the lower unit off the side, and the boards are flat so they store easily. You'll pay more for clear pine but now's not the time to be frugal! Ordinary 2x10s also work well.

Before first assembly, mark one end of the poles as the split-block end. A convenient way to do this is to wrap a ring of tape around the pole. The reason is wedges that hold the poles in the pole seats will eventually deform the poles slightly. Mark the ends that go into the split blocks so you never have to wonder why the poles are getting so balky.

Get a good quality observing ladder. A six-foot ladder works well with a 20-inch scope, and an eight-footer works well for a 25-inch scope. We prefer wood and fiberglass ladders to aluminum. On a cold night, aluminum is *awfully* cold on the hands.

11.2 Installing the Optics

In the truss tube Dobsonian, you install the optics once and after that they remain in the telescope. While it is certainly possible to remove the optics, every time you lift the primary mirror from its cell, you run a finite risk or dropping or chipping it. However, there should be very few occasions when you will ever do this. Heck, you can even wash the primary in its cell. When you install the mirror, think of the installation as permanent. In the Kriege-style Dobsonian, there is no better place for the mirror to be than secure in its mirror cell.

11.2.1 Center-Dot the Primary Mirror

Before you install the primary mirror, "center-dot" it. This is very important because without a center dot, precise collimation is impossible. All reflecting telescopes can benefit from center-dotting, with the fastest mirrors (those with the lowest f/ratio) benefiting the most.

To locate the precise center of the primary mirror, you will need to make a circular template the size of your mirror from heavy paper. Gift-wrap paper works well. Cardboard is even better. You can use this same template to protect the surface of the mirror when you are adjusting the fasteners and safety clips on the mirror cell. A cardboard template offers better protection in the inevitable "The-wrench-fell-out-of-my-pocket-and-onto-the-mirror" scenario. Also, when you draw the circle, cardboard holds the compass point better than paper.

With the stick compass, draw a circle the diameter of your mirror on the cardboard. Don't simply trace the mirror with a pencil—you need to do it with a compass so you know the exact location of the center. Cut the template with scissors. Enlarge the center hole where the compass point penetrated to accept the tip of a small felt-tip pen.

Gently lay the template on the mirror and carefully align it with the edges. The faster (lower f/ratio) your mirror the more critical it is to be as accurate as possible. With the pen, dot the mirror surface through the small center hole in the template. One small dot is enough: all it does is mark the center of the mirror. Remove the template and save it for later.

In an ideal world, this tiny center dot would be adequate. In the real world, a tiny dot is too small to see when you are looking for it in the dark through a Cheshire collimation eyepiece. It's tricky to eyeball that little dot so far away.

So, you're going to add a clever visual aid to overcome this problem: a hole reinforcement. You know, those little paper circles you lick and put over the hole in the page in a loose-leaf folder to prevent the paper from tearing. They make great center dots.

Go out a buy a package of these little paper rings from your local dime store.

Fig. 11.1 *To locate the exact center of the primary mirror, use a stick compass to draw a circle the size of your mirror on a sheet of giftwrap paper. The pivot of the compass will make a small hole in the paper. Place the paper on the mirror and center it, then using a felt-tip pen, dot the center of the primary through the compass hole.*

You'll only need one for your telescope. Think of the hundred or so that remain in the package as spares—or as a lifetime supply. Color one black with a permanent-ink marking pen. Lick the blackened collimation ring and place it so that the center dot you made on the mirror is in the exact center of the black ring.

Yes, you'll probably smear some saliva on the mirror when you do this. But not to worry, because the central three or four inches of the primary are in the shadow of the secondary mirror anyway. It just looks lousy in the daylight when your friends admire your work. If you fuss around and try to blot the mirror clean, you'll make an even bigger mess. Let it be.

Later, when you collimate the optics, you'll see the dark spot of the Cheshire and the dark ring on the primary. It's quite easy to see the ring as you move it toward the spot because the ring is so large. When the ring completely surrounds the spot of the Cheshire, the collimation is right on. Because the spot will just fit into the ring, if you're off a tiny bit it should be readily apparent. Collimating the primary this way is fast and accurate. The little hole reinforcement sticks to the mirror tenaciously even when the mirror is washed.

11.2.2 Center-Dot the Secondary Mirror

To insure precise collimation, center-dot the secondary mirror as well. A small dot of black paint on the surface of the mirror is all that's needed. Its purpose is to aid you in centering the secondary mirror under the focuser. Using a cross hair collimation tool, you adjust the secondary mirror until the dot on the secondary lies under the cross hair.

Locating the center of an ellipse is quite a trick. To help you, we have included a set of templates (**Figure 11.2**) for the common-size secondary mirrors with their centers marked. Photocopy **Figure 11.2** at 100% and cut out the template you need. Make a pinprick at the center and mark the secondary mirror with a felt-tip pen just as you center-dotted the primary. Over the mark from the pen, place a dot of black paint. (Do not use a hole reinforcement.) Put the secondary mirror in a safe place until the paint is dry.

If you decided to offset the secondary mirror, you will need to place the center dot at the offset center of the diagonal mirror instead of its geometric center. The offset center is 1.414 times the tabulated offset distance, so an offset of 0.167 inches becomes a center offset distance of 0.24 inches. On your template, measure out the center offset distance along the major axis and place the pin-prick there. Orient the template on the ellipse so that the offset center is closer to the eyepiece end of the secondary mirror, that is, the end of the secondary is cut off at an obtuse angle. Even on the biggest telescopes, the center offset is only a fraction of an inch.

If you built your telescope carefully and squared the focuser precisely, the dot helps you set the longitudinal position of the secondary, that is, its positon toward or away from the primary mirror. To do this, you adjust the nuts on either side of the central hub on the spider. If the focuser or the spider is not correctly

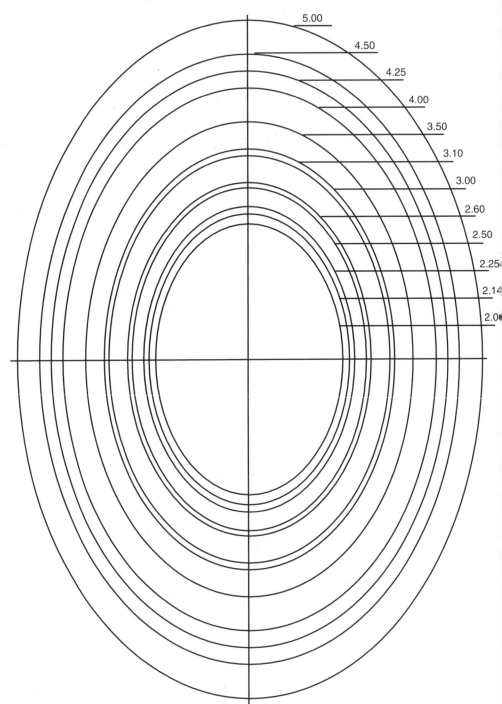

5.00
4.50
4.25
4.00
3.50
3.10
3.00
2.60
2.50
2.25
2.14
2.0

Fig. 11.2 *Photocopy this ellipse template to locate the center of the secondary mirror. To find the offset center, measure the surface offset on the long axis of the mirror and prick a small hole through the photocopy.*

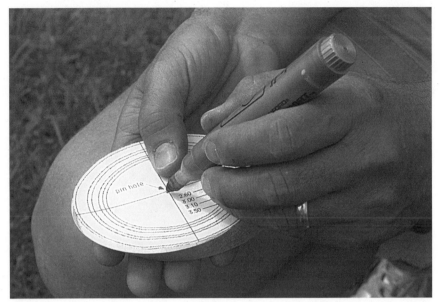

Fig. 11.3 *Center-dot the secondary mirror using the ellipse template you have copied and cut out from Fig. 11.2. Place the template gently on the mirror and align the edges accurately. With a felt-tip pen, dot the mirror through the tiny hole you made with a pen. Permanently mark the center with a small dab of black paint over the ink dot.*

adjusted, then the cross hair collimation tool and center dot will show you the problem and you can correct it.

11.2.3 Install the Primary Mirror

Loosen the nuts on the two bolts that hold the sling. Lay out the sling so it loops loosely around the two lower side pins, where it's out of the way. Make sure that you still have a couple turns of sling on the bolts, though, because you will tighten the sling later.

Loosen the locknuts on top of the mirror clips and turn them aside. They must not block the mirror. Loosen the nuts on top of the two lower off-center pins and rotate them out to allow as much room as possible to install the mirror.

Pat down your pockets and remove anything that could fall out of them and onto the mirror. It is amazing how pens, pencils, screwdrivers, key chains, and all manner of weird stuff will choose this particular moment to exit your pockets and leap toward the mirror. Take a deep breath. Pick up the primary mirror, carry it to the mirror box, and lower the primary into its cell in the box. Check that the mirror is centered and wiggle it a bit to insure that the cell parts are moving freely and have adjusted to float the mirror.

Place the center-dotting template on the mirror. This protects the mirror from falling pens, glasses, wrenches, and drops of perspiration. You can relax now, the

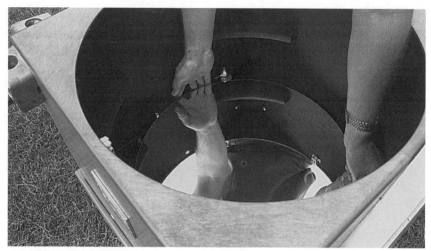

Fig. 11.4 *Before you pick up the primary mirror, pat down your pockets to make sure nothing will fall out. As you lift the mirror, keep your back straight so you don't pop a disk and drop it. Gently lower the mirror into the mirror box and set it in the waiting mirror cell. Shown here is a 20-inch mirror.*

hard part is over!

Next, slip the sling up and over the two lower off-center pins and place it in between the side of the mirror and the pins. Leave it slack. Rotate the two lower side pins inward so there is between ⅛ inch and ¼ inch between the sling around the mirror and the side pins.

Leaving this space is important. The function of the side pins is to keep the mirror captive in the mirror cell when your car is bouncing down the road and when you roll the lower unit around with the wheelbarrow handles. The side pins should not contact the mirror when you are observing. If the primary touches the pins, adjust the sling and or reposition the side pin that contacts it. (We'll say more on this subject later, when you adjust the sling.) Tighten the nuts that hold the side pins.

The mirror clips are also non-contacting. Their job is to catch the mirror if the mirror box is accidentally tipped forward. Set them ⅛-inch to ¼-inch above the mirror surface. Position the mirror clips over the mirror and tighten the lock-nuts.

At this point, the sling should be wrapped loosely around the mirror. On the part of the mirror which will rest in the bottom of the sling, place a short piece of double-sided tape. Press the sling against the tape so there is an equal amount of glass above and below it. The function of the tape is to prevent the sling from slipping off the mirror when the telescope is pointed straight up. Leave the sling loose. You will adjust it after the telescope is fully assembled.

Remove the template from the mirror and place the dust cover on the mirror box. The primary mirror is now installed in the mirror box.

Fig. 11.5 *If your mirror is too heavy to lift safely, don't take chances. Here Larry Wadle centers a 25-inch primary mirror before lowering it into his telescope. The hand-operated winch makes lifting and lowering the mirror both safe and controllable.*

11.2.4 Install the Secondary Mirror

You do not need to have the scope setup to carry out the following installation. All you need is the secondary cage. Remove the secondary holder from the spider.

Begin by attaching a safety cord to the secondary mirror. Obtain a 24-inch length of good quality nylon cord and tie a couple of knots in one end. Place the cord in the center backside of the secondary and cover the knots with a 1½-inch long blob of silicone adhesive. Allow 24 hours for the adhesive to cure.

This safety cord should prevent the secondary from bouncing out of the cell and falling onto the primary, an obvious catastrophe. In addition, by attaching the cord to the spider vane it can catch the entire mirror-holder assembly should you forget to secure the outer mounting nut properly.

We are not aware that *any* secondary mirror has ever slipped out of its cell. However, given the severe bouncing that your Dobsonian telescope may receive on the way to some remote dark site, the possibility is real. Be smart and attach a safety cord.

Once the adhesive has cured, place the mirror in the split sleeve that holds the mirror. Stuff the white polyester batting over the back of the mirror into the

wrap cord around spider vane and tie

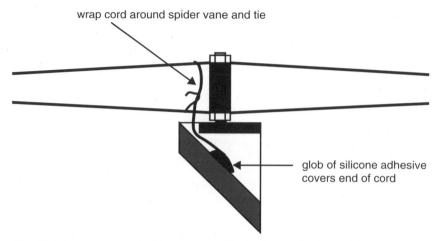

glob of silicone adhesive
covers end of cord

Fig. 11.6 *Protect your optical investment with a safety cord. Glue a 24-inch length of strong nylon cord to the back of your secondary mirror and tie the free end to one of the vanes of the spider. Even if the diagonal mirror bounces out of its holder, it cannot fall on the primary mirror.*

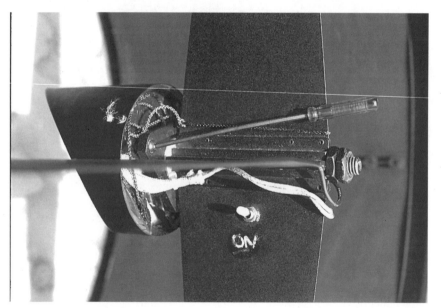

Fig. 11.7 *Look closely at the important details shown here: the nylon safety cord, the optional electric heat rope and its switch, and the Bondhus balldriver for aligning the secondary.*

sleeve. This batting keeps the mirror gently seated in the holder. Run the cord out the back of the holder, and bolt the sleeve to the mounting plate. Install the mirror-holder assembly into the spider hub. Tie the loose end of the cord around one of the spider vanes.

If you live in a wet climate, read ahead in **Section 11.3.2.3**. It's easier to install a heat rope now than it will be later.

Place a collimation sight tube in the focuser and adjust the secondary holder longitudinally so that the center dot is under the cross hairs of the sight tube in the focuser. Tighten the nuts on each side of the spider hub firmly. Check your work to confirm that the secondary mirror is secure in the holder and that the holder is secure in the spider. You don't want anything falling off some night even with a safety cord! It's prudent to inspect the secondary mirror assembly after any trip on a bumpy road.

11.2.5 Adjust the Sling

Your telescope must be fully assembled before you can adjust the sling. Return to this section after you have assembled it. (We'll remind you.)

For proper performance, it is imperative that your mirror be suspended in the sling. Before you initially adjust it, the mirror rests on the two lower side pins. When the mirror rests on these two pins, friction between the side of the mirror and the pins means that you cannot collimate the primary properly, and in addition the mirror deforms slightly around the pins distorting the image.

To check or adjust the sling, point the fully assembled telescope about 15° above the horizon, that is, not horizontal, but low enough that you can comfortably reach the back of the tailgate. Check that the sling is around the mirror properly. In the unlikely event that the sling has fallen off or become twisted, you will need to point the telescope straight up, remove the dust cover, and reach into the mirror box to correct it.

Always adjust the sling from behind the tailgate. Never reach in from the top of the mirror box to do this because it is both risky and unnecessary. You will need two $\%_{16}$-inch wrenches to adjust the sling. To suspend the mirror from the sling, wrap two turns of sling material onto the left sling bolt and then lock it. To do this, rotate the head of the bolt with one wrench and with the other, reach in to the opposing nut on the inside of the tailgate. Lock it tight.

Now go to the sling bolt on the right side. Hold the bolt head with one wrench and loosen the opposing nut. Turn the bolt to draw up the slack in the sling until the mirror lifts off the two lower side pins. Raise the mirror until it almost touches the upper side pin.

Lock the bolt in this position by tightening the opposing nut. Check to determine that the mirror is truly suspended by pushing on the back of the glass with your fingertips. With the scope in a nearly horizontal position, you should be able to push the mirror off the flotation cell triangles quite easily.

If the mirror binds against one of the side pins, loosen the retaining nuts and

Fig. 11.8 *Always adjust the sling from behind the telescope. Using two wrenches, loosen the nut that secures one of the split bolts. Turn the split bolt to lift the mirror off the side pins, and then tighten the opposing nut. Before an observing session, check to make sure the sling is still properly adjusted.*

rotate the offending pin slightly away from the mirror. The mirror must rest in the sling and not on the side pins.

Before each observing session, check that the primary is supported properly in the sling. It takes just a moment and insures that the mirror is going to deliver its best. You may need to adjust the sling once a year to allow for normal stretching. If you live in an area where there is a large change in air temperature from season to season, the sling will shrink or expand slightly, and you may need to make a seasonal adjustment. Checking takes just a few seconds, and adjustment takes just a few minutes. Don't skip checking.

11.2.6 Collimate the Optics

Before you observe, you need to align or "collimate" the optical system. The goal of collimation is to insure that the primary, the secondary, and the eyepiece are correctly lined up to give good images. Of course, the first time you collimate is bound to be the hardest time, but after a few months it will become second nature to you.

You might think that collimation begins with the primary mirror because the primary mirror is the most important part of the optical system, but curiously enough, it does not. Collimation is done backwards, starting at the focuser and working step-by-step back to the primary. The reason is that before you can adjust the

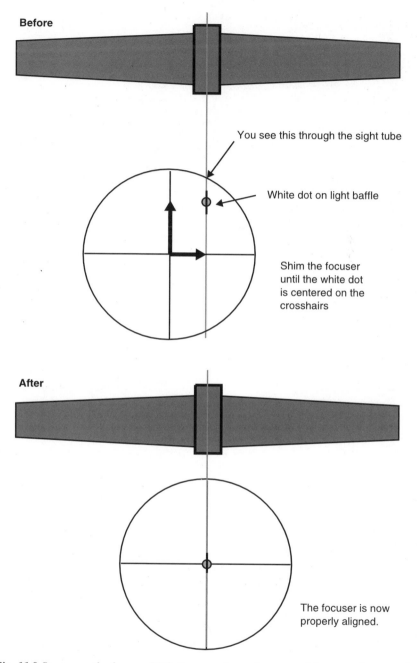

Fig. 11.9 *Square-on the focuser. With the secondary mirror out of the way, place a sight tube in the focuser. Then shim the base of the focuser until the dot opposite the focuser on the Kydex light baffle is centered behind the cross hairs of the sight tube.*

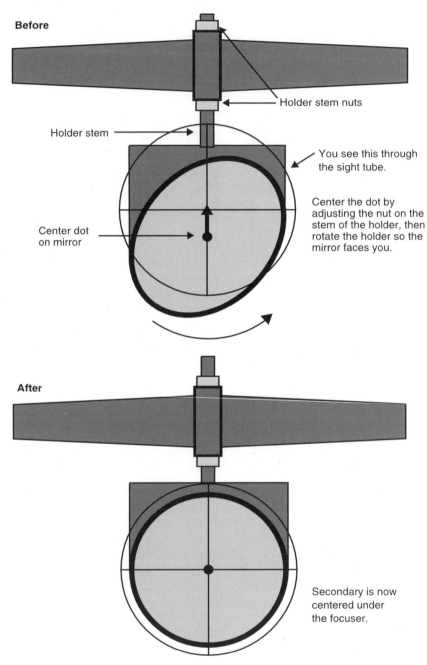

Fig. 11.10 *Then center the secondary mirror. Use the adjustment nuts on the stem of the secondary cell to move the secondary up and down. Rotate the secondary cell on its stem until the outline of the secondary is round. The outline of the secondary mirror should be concentric with the barrel of the sight tube.*

Before

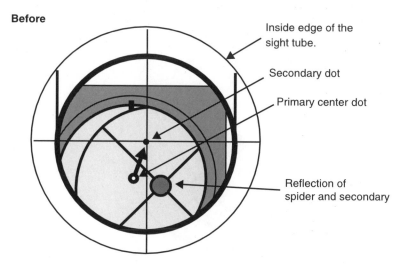

Inside edge of the sight tube.

Secondary dot

Primary center dot

Reflection of spider and secondary

Adjust the secondary tilt and rotation until the primary center dot lies behind the secondary center dot.

After

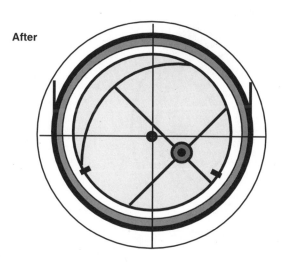

The secondary mirror is now properly aligned.

Fig. 11.11 *Aim the secondary at the primary. Assemble your telescope for this step. Use the adjusting screws on the back of the secondary holder to change the tilt of the mirror. The job is done when the reflection of the center dot on the primary is centered under the black dot on the center of the secondary mirror. Also notice the concentric circles. The sight tube barrel, secondary holder and primary mirror outlines are all concentric.*

Before

Center dot on primary

Black dot reflection
in Cheshire

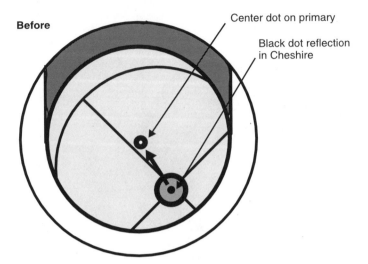

Adjust the collimation
knobs behind the primary
mirror to center the black
dot in the Cheshire
behind the center dot on
the primary.

After

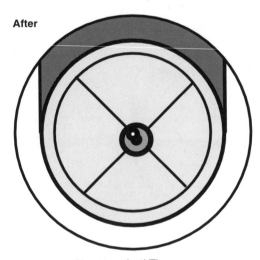

Almost perfect! The
primary needs just a bit
more tweaking to achieve
perfect alignment.

Fig. 11.12 Collimate the primary mirror. Use a Cheshire eyepiece (alignment tool). Turn the alignment knobs on the back of the primary to center the reflection of the Cheshire eyepiece in the primary mirror behind the center dot on the surface of the primary. Verify that the telescope is collimated properly by using the star test.

Fig. 11.13 *Rolling the telescope from storage is a lot easier than lugging it! Vans, hatch-backs, station wagons, pickup trucks and trailers make transporting your large-aperture Dobsonian to great dark-sky observing sites entirely feasible.*

most important optic, all of the less important optical parts must be doing their jobs properly. Think of it that way and it makes perfect sense.

For fast and efficient collimation, you need a set of collimation tools. You can make these yourself or buy them ready made. Those sold by Tectron Instruments are excellent, and they cost no more than a medium-quality eyepiece. A set of collimation tools consists of two devices: a sight tube and a "Cheshire eyepiece."

The sight tube is a long tube with a set of cross hairs mounted at one end and a small peephole at the other. The cross hairs and the peephole are located precisely in the center of the sight tube. The Cheshire eyepiece is a short tube with a peephole in it and a small, flat, perforated mirror tilted at a 45° angle located just inside the peephole.

Square the focuser left and right. Do this indoors on the kitchen table. With a tape measure, check to see if the spider hub is precisely in the center of the cage. If it is not, then center the spider by adjusting the mounting screws on the ends of the vanes. Remove the secondary holder and temporarily install a long bolt or length of threaded rod into the spider hub. Pick a bolt that is the same type as the holder stem, usually it's ⅜ inch in diameter. Insert a cross hair sight tube into the focuser. Take a look. The cross hair should sight to the center of the bolt. Don't worry about up and down, only left and right for now. Loosen the focuser mounting screws and shim it until the cross hair sights are in center of the bolt. Remove the bolt. Draw a short

Fig. 11.14 After rolling the bottom end of the telescope to the location where you want to observe, insert the truss-tube poles. Note that the dust cover remains in place on the mirror box to protect the primary mirror from falling objects until you are ready to observe.

Fig. 11.15 *With small- and medium-sized Dobsonians, you can simply place the secondary cage atop the truss poles and clamp it in place. The setup time for a 25-inch Dobsonian need not be more than 10 or 15 minutes.*

vertical pencil line on the inside of the Kydex directly opposite the focuser while sighting through the cross hair. This is the left and right alignment of the focuser.

Square the focuser up and down. Insert your cross hair sight tube into the focuser from *inside* the cage. In other words, the peep hole is inside the cage and flush with the Kydex baffle. Measure from the top of the upper wood ring of the cage down to the center of the peep hole. Measure the same distance down opposite the focuser and place a small white dot of paint along the pencil line you made earlier. Reinstall the sight tube into the focuser from outside the cage like normal and sight on the white dot. If the cross hairs are not on the dot you will need to shim the focuser accordingly. Use thin strips of wood or plastic until the focuser dot is aimed at the white dot. Congratulations. The focuser is now square to the optical axis of the cage both left and right and up and down. **Figure 11.9** shows what you should expect to see.

Once this adjustment has been made, it's a good idea to check it once a year as part of you telescope's annual maintenance.

Position the secondary mirror. You can do this step indoors, too. First center-dot the secondary as described in **Section 11.2.2.** Install the secondary mirror in its holder and the holder in the spider. Place the sight tube in the focuser and look into it. You will see the secondary holder in some odd position. Turn the holder so you see the face of the secondary mirror. Use the adjustment nuts on the stem of the holder to move it up and down until the center dot on the surface of the secondary mirror lines up precisely with the cross hairs in the sight tube.

Fig 11.16 *Assembling "The Big Ones" is much the same as assembling other Dobsonians—but it simply would not be smart to carry the secondary cage up a ladder and try to rest it on those poles. Instead, there's an easy way. You top the poles with a lightweight connecting ring and attach the secondary cage with the telescope horizontal.*

Rotate the holder so that mirror is exactly face-on. **Figure 11.10** shows you what to look for.

When the secondary mirror is centered under the focuser, the outer lip of the secondary holder should appear to be concentric with the inside edge of the sight tube. However, you cannot trust the eye to judge whether the circles are truly concentric. Instead, rely on setting the center dot under the cross hairs. Check that the secondary is centered each time you collimate the optics.

Align the secondary mirror. In this step, you aim the secondary mirror at the primary mirror, so the telescope must be set up. However, it need not be dark; you can do the alignment in daylight.

Insert the sight tube. Reflected in the secondary mirror you will see the lower end of the telescope, probably off-center in one direction or another, as shown in **Figure 11.11**.

Using the adjustment bolts on the top side of the secondary holder, tilt the secondary mirror until the center dot on the primary mirror is centered behind the center dot on the secondary mirror. This means that the primary center dot also lines up with the cross hairs in the sight tube.

Depending on how the secondary holder is constructed, changing the tilt of

*Fig. 11.17 Dave Kriege stands ready to carry the connecting ring up a ladder and rest it on top of the poles of a 36-inch telescope. The ring is light and easy to handle even at the top of a very tall ladder (see also **Figure 6.21**).*

the secondary mirror may slightly change the centering of the spot on the secondary mirror. When you have gotten the tilt close, recenter the secondary and then readjust its tilt. Make sure that the secondary is both centered and aligned before going on.

The alignment of the secondary mirror can change if you drive over miles of rough road or if you accidentally bump or drop the secondary cage. Recheck its alignment every time you set up.

Align the primary mirror. When everything else is aligned correctly, you are ready to collimate the primary mirror. Insert the Cheshire eyepiece in the focuser. Since it need not be dark to align the primary mirror, it's a good idea to check the optical collimation as soon as you have set up the telescope. It is considerably easier to collimate in the daytime than at night.

When you look through the Cheshire you will see the center dot on the primary at the center of the view. Reflected in the primary mirror you will see the spider and the secondary holder. Within the image of the secondary mirror you will see the reflection of the alignment tool you are using. At the very center you will see the peephole you are looking through as a black dot. See **Figure 11.12**.

Turn the collimation bolts behind the primary mirror to place the reflection of the black dot behind the center dot on the primary mirror. Because the center dot is a hole reinforcement with a hole in its own center, you should be able to put the black dot in the dead-center of the primary. That's it.

Once you have aligned your telescope, the primary usually stays pretty close

Fig. 11.18 *After balancing the connecting ring on the poles, fasten the first pair of poles. Rotate the telescope and fasten the second pair, and so on until the four sets of poles are tied solidly together at the top.*

to proper position from one observing session to the next. Travel over rough roads, the bumps and joggles of setup, and the inquisitive fingers of other telescope builders at star parties can change the alignment of the primary. Since it takes only a few minutes to collimate, check the collimation each time you observe. If it stays dead on, that's great, but if it's gone off, it will take just a few minutes to recollimate.

Star check your collimation. As darkness falls, locate a moderately bright star fairly high in the sky, insert a medium to short focus eyepiece, and check that the star images are round. Roll through focus and verify that the diffraction rings in the out-of-focus star image are uniform and concentric. In the early evening, seeing may distort star images into shapeless blobs, so do the best you can. If the star images are markedly asymmetrical, check that the mirror is hanging properly in the sling, that the side pins are not touching, and that the mirror is not pinched in the support system. If those are okay, recheck collimation.

Throughout the night, check the quality of the star images from time to time. There's no need to get fancy about checking; after you finish looking at a faint galaxy, jog the telescope to a nearby star and check that the image is still round. If the seeing is exceptional and you feel like doing it, star test the optics as described in **Section 4.7**. With a good set of optics, it feels great to see a clean Airy disk and identical star images inside and outside focus. This is how your telescope tells you, "I am ready for any observing challenge you care to try."

Fig. 11.19 With the connecting ring in place, a concrete block holds the 36-inch telescope horizontal. Although a single pole would not be strong enough to hold the out-of-balance telescope down, the eight poles together form a strong truss.

11.3 Setup, Use, and Takedown

Over time you will become familiar with every peculiarity of your telescope. You'll add your own gadgets and gizmos that need special care, and the whole process of assembly will become second nature. The first few times you assemble your telescope, however, should be in the daytime, when you can clearly see what's going on.

Here are a few ground rules to keep in mind the first time and every time you assemble your telescope.

- •*Protect the primary.* Be sure the mirror clips are in place to prevent the primary from falling forward. While you set it up, leave the dust cover on the mirror box.

- •*Protect the secondary.* Be careful how you pick up and set down the secondary cage. Never place the cage where it can slip and fall. A soft, well-laundered old sock stretched over the secondary and its holder keeps it clean.

- •*Protect the truss rods.* The truss rods are precision parts of your telescope. Never use them as pry bars or pile them on the ground where people could step on them and fall.

Fig. 11.20 *It's a cinch to hang the secondary cage on the connecting ring when the telescope is horizontal. Once the secondary cage is clamped firmly in place, the telescope is balanced. You can then easily point it to any celestial object.*

- • *Respect your telescope's weight.* The mirror box and rocker are heavy. Be careful when you handle them. Never put your fingers between moving telescope parts.
- • *Take pleasure in your telescope.* Each time you set up, indulge yourself with a few moments to remind yourself that you have crafted a fine and functional instrument. Enjoy the fit of the components and the smooth routines that you have developed for assembling them. Take pride in an instrument that can take you to the edges of the universe.

11.3.1 Setup

Depending on where and when you use it, the telescope may be stored in a garage or porch, or in the back of a van or trailer. The mirror box should be resting in the rocker/ground-board unit, with the dust cover on it. The secondary cage, truss poles, ramps, wheelbarrow handles, observing ladder, and other bulky items should be stored securely beside the mirror-box, rocker and ground board assembly. Delicate parts like a Telrad, finder, or digital setting circles and eyepieces should be stored in a foam-lined case or box built to hold them.

 1. *Set up the ramp.* If you have been transporting the telescope in a vehicle, set up the ramp. The ramp consists of two lengths of 2x10 pine boards.

Fig. 11.21 *An adjustable tailgate chain lets you fine-tune the balance of your telescope when you change eyepieces. Steel bars bolted to the tailgate frame balance out new finders and other "must have" accessories on the secondary cage.*

2. *Attach the wheelbarrow handles.* Bolt on the handles so that when you lift them, the side of the mirror box that is on top when the telescope is horizontal faces you. If you put them on the other way, the mirror box might swing over the wheels when you lift, and if that happens, the mirror box could slip out of the rocker and crash to the ground. If you lift facing the top side of the mirror box, the taller of the end boards on the rocker will prevent the mirror box from swinging. When you're transporting the telescope, you may elect to leave the handles on while you drive. If this is the case, the secondary cage nestles nicely between the wheelbarrow handles.

3. *Lift and roll.* It's so easy—raise the handles a few inches and roll the mirror box, rocker, and attached ground board out from storage to a flat, level surface. Observe all posted speed limits.

4. *Remove the handles.* Carry the handles back to the storage place or vehicle. If you leave them lying around, you or someone else could trip on them in the dark.

5. *Red Flag Warning:* If you have not installed the primary mirror, do not assemble the telescope. Without the mirror in place, the tube is grossly out of balance. One little push and the whole telescope might come crashing down. And of course make sure the dust cover is on.

6. *Insert the eight truss poles.* These go into the maple split blocks on the mirror

Fig. 11.22 The parts of your truss-tube Dobsonian are laid out ready to load into your car or van. It's hard to wait for a clear night so you can drive to your favorite deep-sky observing site for a night of observing!

box. Gently tighten the knobs. The poles are interchangeable with this exception: two of the poles have shorter lengths of black foam. Place these adjacent to the side bearings. You should have marked one end of each pole as the split-block end; make sure you insert the split-block end of the poles into the split blocks.

7. *Place the ladder next to the scope.* The ladder should be just far enough away so that you can turn the rocker base without having it bump into the ladder.

8. *Prepare the secondary cage.* Check to make sure the locking devices on the secondary cage are ready. If you have cam levers, for example, make sure the cylindrical nuts are even with the ends of the flat-ended bolts. Check that nothing is loose or about to fall off the secondary cage.

9. *Red Flag Warning:* Check that the dust cover is on the mirror box. If there are people around watching you set up, they might have asked for permission to gaze lovingly at your mirror. Don't forget to put the cover back on.

10. *Set the cage on the poles.* With the four wedge-clamp assemblies dangling by their cords, climb the ladder holding the secondary cage and set it on top of the truss poles. As you do this, be very careful! Don't bump the secondary on the pole tips. At night, it helps to hold a small flashlight in your mouth so you can use both hands yet still see what you

Fig. 11.23 *Up the ramp and into the car! Jerry Conroy's 20-inch telescope rides in the cargo space of his compact automobile. With a ramp of 2 x 8 lumber there's no straining and sweating at the end of a great night of observing—you just roll in the telescope and go.*

are doing.

11. *Check the secondary cage.* The wedge-clamp assemblies should be dangling freely from the secondary cage. Make sure the secondary cage is sitting evenly on top of all eight poles. Make sure the focuser is positioned on the side you want it. Untangle, move, or rotate the secondary cage to correct any condition that is not right.

12. *Insert the clamp wedges.* Each wedge goes between two of the truss poles. Push the bolt with the cam lever attached through the hole in the lower secondary cage ring, insert the hitch pin through the hole in the bolt, and then throw the cam lever. You may have to adjust the position of the cam lever on the bolt to achieve a snug fit with the wedge.

13. *Pull on the light shroud.* If you have a light shroud made of Ripstop nylon, you just pull it on like a big sock. Make sure the focuser is racked in completely, and tease the end with the drawstring over the focuser. Align the shroud as you slip it on so the upper drawstring slider faces the ground. Loop the two continuous bungee cords around the top of the two wooden blocks on the mirror box. Thread the two cords with the hooks through the side bearings and then stretch them around the blocks.

14. *Attach the finder or Telrad.* Make sure your finder is firmly attached or that your Telrad is secure and all four clamps tight before you remove the

Fig. 11.24 *After the entire 20-inch telescope has been loaded, there is still room left over for an observing ladder and a case with charts and eyepieces. Even then there is plenty more stuff-it-in space for Oreo cookies and warm clothing.*

dust cover.

15. *Move the telescope around.* As you do this, watch that nothing is slipping, sliding, groaning, moaning, or about to fall off. Check that each pole is seated at both ends. Checking takes moments.

16. *Check the mirror sling.* Point the telescope about 15° above the horizon. Press gently on the middle of the back of the mirror. If the mirror is hanging in the sling, it should swing slightly forward, and settle back when you release the pressure. See **Section 11.2.5** for the details.

17. *Remove the dust cover.* At last, your precious mirror is exposed to starlight. In the daytime, make sure you have pointed the telescope away from the sun. Never ever observe the sun with a big Dobsonian.

18. *You are ready to observe.* Point the telescope, move the ladder, climb the ladder, insert an eyepiece, and focus. First light! Savor the moment.

11.3.2 Using the Telescope

Here are some tips on using your telescope. These are things you need to consider often; they are as much a part of using your telescope as setting it up and taking it down. After a few sessions, you will almost certainly be ready to add your own suggestions to this list.

11.3.2.1 Balance

If your telescope drifts up when you change eyepieces, add a little weight to the secondary cage. There are many ways you can do this. If the balance is really close, put an extra set of batteries inside the Telrad. Whatever you do, make sure you secure all weights firmly to the telescope so they won't come loose and clobber the primary.

If you have added a substantial amount of weight to the secondary cage—a finder telescope or a heavy eyepiece-and-Barlow combination—add weight behind the mirror. Remember that for every extra pound you add to the secondary cage, you need to add about 5 pounds to the tailgate.

For a permanent "fix," obtain strips of steel from your local welding shop. Ask for 1½-inch wide by ¼-inch thick flat stock. If you have a 20-inch scope, ask them to cut you four pieces 20 inches long, and for a 25-inch scope, ask for pieces 25 inches long. Drill two ⅜-inch holes through them to match the location of the two carriage bolts that hold the tailgate to the mirror box via the top rung.

Replace the upper carriage bolts on the tailgate with longer ones to hold as many strips of steel as you need. Don't worry, the tailgate won't fall off when you remove the upper bolts. Placing weights at this location is efficient and unobtrusive, but it's still extra weight you have to move around.

Cloth bags filled with lead shot make excellent counterweights, and you can easily make bags that weigh just what you want. Sew Velcro loop strips to the bags and attach Velcro hook strips to the back of the telescope. Those dandy little one-pound and two-pound weights that runners use in training are another great counterweight, and they have the added benefit of being there when you feel like midnight jogging.

For the ultimate fix, add a length of heavy chain to the tailgate. At the hardware store, purchase a 5 or 6 foot length of very heavy chain, a couple of S-hooks and a small eyebolt that will accept the S hook. Start with a chain that weighs about 10 pounds. Drill a small hole through the top rung of the tailgate and put the eyebolt through it. Check that the eyebolt clears the rocker when the scope is pointed straight up. **Figure 11.21** shows how it works.

When you go observing, attach one end of the chain to the eyebolt with the S hook. Leave the rest free to pile up on the ground as the scope is pointed upwards. By hooking the chain in different positions, you can balance your scope for different situations.

11.3.2.2 Bearing Surfaces

Two or three times a year, apply ordinary silicone car wax (such as Turtle Wax) to the laminates on the bottom of the rocker base and on the side bearings. The wax decreases friction and makes the motion of the telescope smoother.

Clean the laminate surfaces with a solvent, apply the car wax, let it dry to a haze and then buff. After the car wax, spray a little ArmorАll on to the bearing surfaces. ArmorAll is available at most hardware/discount stores in the automotive department. The result is the distinctive buttery feel of a good Dobsonian.

If you observe at dusty sites, the Teflon pads become contaminated with grit. It gets embedded in the pads, and makes the telescope less responsive. To restore the smooth handling of the scope, wipe the dust off the pads and restore the laminate with a damp rag and solvent. After cleaning, apply a fresh coat of silicone car wax and ArmorAll.

If the telescope is stiff in azimuth, add more shims under the central pivot pad on the ground board. Get a friend to help you lift the mirror box out of the rocker and set it on the ground. You can leave the mirror in the mirror box and you can set the box on the collimation knobs on the ground without harm.

When you move the mirror box do not bump the exposed portion of the primary mirror. Remove the azimuth bolt with a socket wrench and slide the rocker aside. Put a couple pieces of cardboard under the center pad. Unfortunately, this is a trial-and-error process. Shim it enough to allow the scope to turn easily, but not so much that it becomes wobbly. Experiment until you achieve the desired friction.

11.3.2.3 Dew

In damp climates, dew may form on the secondary mirror. You can dry the secondary by blowing warm air on it with an electric hair drier or 12 volt heat-gun, but on a humid night the dew will be back in five minutes.

A slick way to prevent dew is to place a length of "heat rope"—nichrome wire wrapped in a woven fiberglass cloth tube—against the back of the secondary between the glass and the polyester batting. For power, run a wire up one of the truss poles to the secondary cage. Important: Use a 12-volt power source. Heat ropes also work great on Telrads and finderscopes. The wire can prevent dewing, but once a surface gets wet, heat ropes usually cannot remove dew.

As soon as you install the heat rope, you'll hear people saying "doesn't the heat from the rope warp the secondary mirror?" Well, of course it does—you're heating the secondary mirror so it cannot be in thermal equilibrium. Tough nuts. At least you're observing through your telescope, while those who are thermal purists are waiting in line to look through your telescope because theirs have dewed up!

11.3.2.4 Ventilation

Without a fan on the tailgate, it can take several hours for your mirror to cool to air temperature. Find a 3-inch 12-volt muffin fan at an electronics surplus place for $10, or new at an industrial electronics store for about $20. To power the fan, a car battery or Porta-Pak works well. Rechargeable gel cell batteries are another option, and you can attach them to the tailgate, thereby eliminating a power cord and the potential hazard of tripping.

Turn on the fan as soon as you set up. While it runs, collimate the primary, have a cup of coffee, and review your observing agenda for the coming night. Turn off the fan. With a fan running for 30 minutes, a 20-inch mirror will have cooled enough to give good images.

If you have electric power at your dark site, point an ordinary house fan at the tailgate and blow air on the back of the mirror. The fan is no cure-all, it simply speeds the natural cooling process. As the night goes on, the mirror will cool further and the images will get better. You probably won't want to observe with the fan on because the turbulent, moving air distorts the images.

If you decide to stop observing and take a nap, turn on the fan again. If the night is getting cooler, the fan will help keep the glass the same temperature as the air. Turn it off when you begin to observe again.

At sites where the humidity is high, point the telescope to the west and leave the fan running when you turn in for the night. The fan and the heat of the rising sun on the back of the primary will both help prevent dew from forming on the mirror.

11.3.2.5 Safety Precautions

While your telescope is set up, it is vulnerable. You can protect it and the people who come to look at it by taking a few simple precautions.

1. *Beware of direct sunlight.* For the sake of safety, make sure the dust cover is on at all times during the day. Large-aperture telescopes can start a fire in seconds if direct sunlight strikes the primary. Unwary observers or even someone just walking by could suffer permanent injury if a focused solar image happened to catch them. Never project the image of the sun onto a piece of paper; the light cone from the primary is so hot it will melt or set the secondary cage on fire. Please be careful.

2. *Beware of wind and rain.* Seemingly mild breezes can push your telescope around like a giant weather vane. If the weather looks threatening, it is often wisest to take down the scope. Put the secondary cage in your vehicle and leave the lower unit in the field covered with a blue plastic tarp. The truss poles are weatherproof so you can leave them out. After the rain stops, wipe moisture and debris off the poles so it doesn't fall onto the primary.

3. *Avoid moving a fully assembled scope with the wheelbarrow handles.* If you lift the handles high, a fully assembled telescope can slip off the rocker and crash to the ground. It's okay to move the fully assembled scope a few feet to a flatter observing spot *as long as you keep the handles low*. If you can, have a friend keep a hand on the scope to steady it as you lift and roll it.

4. *Never wrap your telescope in clear plastic.* In some parts of the country, temperatures under a clear plastic tarp or in a closed vehicle can ex-

ceed 200°F. Clear plastic lets light in, but doesn't let heat out. If you wrap your scope in clear plastic and the sun shines on it, the resulting greenhouse effect can melt the Kydex light baffle in the secondary cage. Even worse, the glue in the plywood may fail, causing delamination.

5. *Let air circulate.* During the daytime, wrap your telescope loosely so that air is free to circulate under it. Blue plastic tarps are great for this and available at most hardware stores. Silver tarps cost more, but they reflect the heat. If you keep your telescope cool, you can start observing sooner after dark.

6. *Before moving your telescope, make sure everything is secure.* Be sure that nothing has come undone or might fall onto the mirror. It is not uncommon for casual bystanders, children, and even skilled observers (who should know better) to "play" with your scope when you're not around. A screw might be loosened and not retightened. It's natural. You probably own the biggest telescope they have ever seen and they want to see how it works. A few moments of inspection can prevent significant damage.

11.3.3 Takedown

Begin takedown by clearing the area where you plan to put the components. If you plan to store the parts in a van or trailer, remind yourself of your strategy for packing them. Set aside the area where you will pile stuff temporarily before loading it for transport. If there are people in the area, ask them to step back so you can move freely. The last thing you need is to be halfway down the ladder carrying the secondary cage when some curious bystander plants himself at the base of the ladder and starts asking you questions.

1. *Put on the dust cover.* You don't want to drop something accidentally onto the primary. Keep the cover on whenever you're not observing.

2. *Remove anything that may fall off.* This includes the eyepiece, the Telrad, the light shroud, and the bag of chips you clipped to the secondary cage. Run the focuser to its lowest position.

3. *Pull off the light shroud.* Loosen the drawstring and release the four lower bungee cords. Work the shroud up the telescope as if you were taking off a tight sock. Be careful not to snag the focuser when you get near the top.

4. *Point the telescope straight up.* Don't even *think* about removing the secondary cage unless the telescope is pointing straight up. If you remove the secondary cage with the telescope at an angle, the lower unit and truss poles might swing rapidly upright, possibly injuring you and damaging the telescope.

5. *Place the ladder next to the scope.* Leave enough room that so the rocker

just clears your ladder when you rotate it in azimuth.

6. *Release the cam levers.* Begin by releasing the cam lever nearest you; also pull out the hitch pin. Rotate the scope in azimuth 90 degrees and release the next cam lever and pin. Do this until all four cam levers and hitch pins are released.

7. *Lift off the secondary cage.* Grasp the lower wood ring of the secondary cage with one hand on each side and shake firmly with a twisting motion. All the clamp assemblies will drop out and dangle by their cords. Confirm that all four clamps are fully released and dangling. If not shake again.

8. *Carry the secondary cage down the ladder.* Take care not to bump the secondary.

9. *Spin on the cam levers.* Run them all the way up on the flat-ended bolts so you won't lose a lever in the field.

10. *Remove the eight truss poles.* If the poles stick a little, release them with a gentle twist.

11. *Attach the wheelbarrow handles.* Be sure to orient the handles so the top of the mirror box will face you when you lift and roll away.

12. *Dry off moisture.* If the components are wet, towel them dry. If the shroud is wet, make a note to put it in a clothes drier at low heat as soon as you get home. Never store any part of the telescope wet for more than a few hours.

13. *Pack the components securely.* In a vehicle, the mirror and rocker/groundboard generally go in first, followed by the rest of the components. After a little practice, disassembly should take you three or four minutes.

A convenient way to store truss poles is in a length of 8-inch diameter tube. Galvanized stove pipe is found at most hardware stores. A lighter alternative is concrete form tube, or Sonotube, the same stuff used for the tubes of small-aperture Dobsonians. In larger cities, you can usually find concrete form tube at building supply centers.

11.4 Cleaning the Optics

Once a year or so, but only if they have become truly soiled, you should wash the optics. There is no need to even take the mirror out of the mirror box—just wash and rinse it in the cell. If dust settles on the mirrors, *do nothing about it.* Dust has no significant effect on the light grasp or resolution of the mirror, and washing the coatings too often will wear them out faster than letting a little dust sit on them. Over the course of a year, however, acidic contaminants and harmful pollutants accumulate on the coatings, and you need to wash them away.

People make a big fuss about washing mirrors, but there is really nothing to it. The secret is that the primary mirror should be washed in the telescope. Taking the mirror out does nothing but create a risk of dropping it. Why take chances? To wash your primary in the telescope, here is what you need:

1. distilled water (3 gallons)

2. hand-pumped spray bottle

3. sterile, lint-free cotton wadding

4. Dawn or Ivory dish detergent

5. roll of paper towels

6. two bath towels.

Proceed as follows:

Step 1: On a level surface, assemble the telescope without its light shroud. Line the bottom of the rocker base with a couple of heavy bath towels to catch the water. Lay them flat so that the scope can be easily rotated without the mirror or the lower collimation knob catching on the towels. (If you have installed an azimuth shaft encoder in the rocker unit for digital setting circles, you should remove it to prevent water damage.)

Step 2: Tilt the scope to about 20° above the horizon. Crawl inside the truss assembly and kneel before the primary. Rinse off the mirror thoroughly using the spray bottle filled with distilled water. The water will run down the face of the mirror floating off the heavy particles of grit and dirt, and drip onto the towels below. Relax—water won't harm the wood inside the mirror box. It will dry in a couple hours.

Step 3: Dissolve *one drop* of Dawn or Ivory dish detergent in one quart of distilled water. If you use more than a drop or two you'll be left with a scummy soap film on your mirror. Rotate the scope straight up so the mirror is as level as possible. Pour the soap solution on the face of the mirror until about two-thirds of the surface is submerged. Because the mirror is a shallow bowl, the soap solution will not run off.

Step 4: Using the sterile cotton wads, swish the soap solution around and around in a circular motion starting from the center of the mirror working outward. *Apply no pressure.* The weight of the wet cotton is all that's necessary. Change the cotton wads frequently. Repeat this process two or three times until the entire surface of the mirror has been swabbed. Soapy water will run down the sides of the mirror, over the sling, and onto the mirror cell, but it won't hurt anything.

Step 5: After the mirror is clean, rotate the scope down to about 20 degrees above the horizon and let the soapy water run off onto the towels. Crawl inside the truss assembly again. Starting at the top and working down, rinse off the soap solution completely by spraying distilled water from the spray bottle on the face of the mirror. Rinse the mirror at least three times.

Step 6: After the rinse water drains off, rotate the scope to vertical again and gently blot off the water droplets with a few paper towels. The key word here is "blot." If you rub, you'll scratch the coating because paper towels are abrasive. Don't try to blot off all the droplets, but get most of them.

Step 7: With a wad of dry sterile cotton, gently wipe off the few remaining water droplets. Start at the top and work down. Change the cotton wad when it becomes too wet.

That's it. You are finished and have a very clean optic. Remove the towels and leave the scope set up until the rocker and mirror cell are completely dry.

The secondary is even easier to clean. Wash it in its holder using the same techniques you used on the primary. Let the water drain out the bottom of the holder. The polyester batting behind the glass will dry out in a few hours. When you wash the mirrors in the telescope, you don't have to remount and collimate everything.

11.5 Troubleshooting

During the initial setup phase of your telescope's life, and at times during the years you use it, you will need to take care of problems that may arise. Not to worry: a few minutes and the problem is fixed.

11.5.1 Too Little Focus In-Travel

The first time you set up your telescope, the truss poles should be 2 inches too long. As a result, it may be difficult or perhaps impossible to run the focuser in far enough to focus with some of your eyepieces. Later, as fashions in eyepieces change, you may acquire wonderful new eyepieces that need to focus even further inward.

This is not cause for concern. You simply need to customize your telescope to the eyepieces you want to use. With 2-inch eyepieces, you must do this by trimming the ends of the truss poles. The idea is to trim the poles until the eyepiece that requires the most inward travel just forms a focused image when the focuser is racked in completely. Trim them with a tube cutter available at most hardware stores for $20. Be sure to shorten all eight poles equally. Remove only ¼ inch at a time until the eyepiece in question reaches focus.

If you have a set of 1¼-inch eyepieces that won't reach focus, you can save yourself a lot of trouble by ordering a low-profile 2-inch-to-1¼-inch adapter. It will probably give you enough inward travel, and you won't have to shorten the poles.

11.5.2 Sticking Poles and Blocks

Wood changes dimension with changes in humidity and temperature. As the seasons change, the truss pole tips may not line up into the pole seats on the secondary cage, making it difficult to install the secondary cage. You can easily correct this by shimming the split blocks until the poles meet as they should.

Place the poles in their split blocks and tighten the black knobs. Examine the poles. If a particular pole does not line up with its opposite member, loosen the two screws on that block and slide a paper or thin cardboard shim under it. Put the shim by the top screw to move the pole tip out and by the bottom screw to move the pole tip in.

Sometimes a split block won't close enough to clamp truss poles securely. Again this is a result of changing humidity and temperature. You have two options: paint the hole smaller or tape the pole bigger.

To reduce the size of the hole, paint the inside of the hole with a few layers of polyurethane. Use a small brush and try to spread it evenly. Avoid fat drips and runs.

To enlarge the pole, wrap a layer of tape around the pole end. Mark the pole and remember which block it mates to. This problem is minor and unavoidable because it is impossible to fabricate blocks that match the change in climate conditions in some parts of the country.

For sticky blocks, coat ends of the poles with silicone spray. When you spray the ends of the poles, do it downwind and well away from the scope. You do *not* want silicone spray to contaminate the optics.

If the foam tubes on the poles slip up and down on the poles, remove them. Apply contact cement to the middle third of each pole and then slip the foam tubes back on. If that sounds too messy, slip a few inches of double-sided tape between the foam and the pole at each end. The fix takes only minutes.

11.5.3 Telescopes and Moisture

Rain and dew (especially the "acid dew" prevalent in urban and industrial areas) are the enemies of your telescope. If the optics dew up or get wet, do not touch the dew. Let it dry naturally or use a hair dryer or portable heat gun to remove it, but never, ever wipe optical surfaces to dry them. You will scratch the coatings.

However, all the other surfaces of the telescope can and should be wiped dry with a towel after a heavy dew. Wiping wood and metal surfaces prolongs the finish and prevents electrolytic action and corrosion.

Furniture polish can be used to protect plywood surfaces and make them look good, but never spray polish on a telescope unless you have first removed the optics. Aerosol mist of any sort contaminates optical coatings. To apply polish, walk away from the scope, spray the polish on a rag, and then polish the wood with the rag.

Large-aperture primary mirrors rarely dew up in use. They dew up in storage. Dewing is a chronic problem in the more humid areas of the country. This occurs when the telescope is kept in a garage or shed, and the mirror becomes cold during the night. The sun rises in the morning and warms the building. The warm, moist air then condenses on the cold mirror. If dewing occurs daily, even low concentrations of pollutants in the dew can rapidly degrade the coating.

Dewing also occurs in trailers and closed vehicles. It helps to leave the doors or windows open so the sun can't heat the air as much, and be sure to park in the shade.

In a garage or shed that has electricity, install a fixture for a 15-watt appliance bulb under the mirror in the rocker unit. Air must be free to circulate around the bulb, and it should not contact any part of the telescope. Don't use more than 15 watts. When the telescope is in storage, leave the bulb on constantly. It warms the primary slightly and prevents dewing.

Chapter 12
Using Big Dobsonians

It's time to change gears. For months you've spent your evenings in the basement or garage working on your telescope. Sawdust, paints, varnishes, and adhesives have matted your hair, and the glare of fluorescent lights has muted your dark adaptation. The telescope is ready for the stars, and it's time for you to become an observer once again, a student of the universe.

12.1 Eyepieces

Assuming you have purchased good quality mirrors and routinely collimate them, it is a mistake to use anything less than the best eyepieces available. Good eyepieces are not cheap. A set of three or four top-shelf wide-field eyepieces cost as much as round-trip air fare to Hawaii. But they are worth it. They will last longer than you will and they have a lot better resale value.

You don't have to buy all your eyepieces at once. You could buy just one a year and in three or four years, you would have all you'll ever need. We don't recommend brand names or specific types, and you'll be best off if you don't put much stock in other people's opinions, either. Hold some star parties with other observers. Compare the different types and brands of eyepieces. This way you can assess which eyepieces are best for you and your scope, and get in some good observing at the same time.

12.2 Filters

Few products perform as well as you would like them to, but two light pollution filters sold by Lumicon are an exception. Their Ultra High Contrast (UHC) and the Oxygen-III (OIII) filters have had an extraordinary impact on visual astronomy. With the ever-growing problem of light pollution from towns and cities of any size, extended objects like nebulae and planetaries are fast becoming invisible.

These two filters wipe out light pollution like nothing else. They transmit light that nebulae give off and block wavelengths that street lights emit. They work well even under dark sky conditions by noticeably improving overall con-

Fig. 12.1 *Above all else, share the enjoyment of your telescope. Take it to beautiful places, invite your family, friends, and colleagues to scan the heavens with you. This picture was taken in the French Alps, by Gilles Meuriot who owns this telescope. He writes, "...everybody was pleased to reach farther into the cosmos when the night came."*

trast. With an OIII filter you can see breathtaking views of the Veil, the Swan, and the Dumbbell nebulae from moderately light polluted skies! So much more is opened up to visual astronomers.

Sadly, nebula filters can do very little for galaxies. The light from galaxies is spread across the entire visible spectrum, so blocking street light wavelengths blocks their light as well. However, at least half the spectacular objects in the sky are nebulae, and these filters do wonders for them.

12.3 Finders

You need a finder, and probably two. One should be a sighting device to help to point the telescope at things you can see. The other should be a telescope to help you pick out the "faint fuzzies."

The field of view with a 20-inch *f*/5 telescope and a 35 mm wide-field eyepiece is just under 1 degree. It is fairly easy to point a big Dobsonian at a 1 degree patch of sky. With nothing more than a reflex sighting device like a Telrad you can unerringly aim a large-aperture scope at anything you can see by eye and it will appear in the eyepiece. Herein lies a decision: Do you want to aim your telescope using only naked eye objects or do you want some magnification to help you find thousands of "faint fuzzies" beyond the naked eye limit?

We think you need both.

There are numerous types of finders on the market. They range widely in field of view, magnification, image orientation, quality, and price. If you don't know it already you will learn that finder telescopes are very personal. A unit that works well for one observer may be despised by another. Like eyepieces, the best way to determine which finder is right for you is to go to star parties. Try out lots of different finder telescopes, talk to the owners for their comments, and get a feel for what's available. Then you can make an informed decision.

12.3.1 The Telrad Finder

The Telrad is a reflex sighting device that can be mounted on any telescope. It has enjoyed nearly universal acceptance. For over 10 years, an amateur astronomer named Steve Kufeld has been making them commercially. Robert E. Cox published an article describing the principle behind these marvelous devices during the 1960s in *Sky & Telescope,* but it took Kufeld to make a commercially viable product.

The Telrad (and the growing number of similar finders) does not magnify the image you see through it; it simply projects three concentric circles in its window. The circles are deep red, and you can adjust their brightness. Put the red circles on the desired spot in the sky and—Bingo!—your monster scope is pointed at the same target. As easy as this is, this is *not* the main reason for its mass appeal. The real reason is that what you see through the window of the Telrad (the ordinary night sky) looks the same as the star chart in your hand. Comparing the bright stars

Fig. 12.2 *The Telrad finder is simple, intuitive, and easy to use. When you look through the angled window you see a red bull's-eye target projected against the stars. Finding is a matter of moving the telescope until the bull's-eye points to the celestial object.*

in the sky with the bright stars on the chart is mentally very simple.

The image is not inverted and/or reversed like it is in some finder telescopes, and you aren't overwhelmed by the "too many stars in the field" syndrome, so you can get yourself oriented. The last thing you need while you're casually star-hopping is an upside down and flipped image. You don't always feel like playing mind games with your finder. Once you learn how to use a Telrad you won't leave home without it.

12.3.2 Finder Telescopes

Observers are remarkably opinionated about finder telescopes. This may be an instance where you love whatever you grew up with, no matter how terrible. In any case, the purpose of a finder telescope is to give you enough light gathering power and magnification that you can see many of the objects that you are looking for, and to help you locate those too faint to see by star-hopping to the exact location. You can then spot the elusive object in the big telescope.

There are lots of finders on the market. We recommend that you avoid the standard straight-through, 8-power units that often come as stock equipment on commercially-made scopes. The image is upside down and very dim. You deserve something better.

The next tier consists of 8x50 finders, 11x80 finders, spotting scopes, and small refractors. These are the heart and core, the classics, what everyone means when they say "traditional finder telescope." The big advantage they offer is that under a dark sky they show you all the Messier objects and many other deep sky objects. If you can aim your scope with a Telrad to the approximate spot in the sky where the desired fuzzy is located, you can probably see it in the finder and center precisely on it. This saves you a lot of time you might otherwise spend searching for but not finding faint objects.

The best advice is to get a unit that has the same scale and shows the same limiting magnitude as your star atlas. For most of us deep-sky freaks this means *Uranometria 2000.0*. This widely acclaimed atlas shows stars down to about 9.5 magnitude. It is very readable for such a detailed atlas. A finder with a three to five degree field of view and a limiting magnitude around nine works well with this atlas.

We recommend the following features in a finder telescope:

An Amici prism. This type of prism bends the light up 90 degrees so you can position your eye behind the eyepiece comfortably, but more importantly, it provides an erect and non-reversed image. That is, the view you see at the eyepiece has the same orientation as the sky and star chart. No mental gymnastics are needed.

A cross hair eyepiece. You only think you can judge the center of the field of view. You need a nice cross hair reticle so you can center it precisely. Illuminated cross hairs are a bonus.

A large dew cap. Or make a large cap for the finder if it doesn't come with one. Once the finder fogs up, you're out of business. Wrap a heat rope around the objective end of the finder to keep it warm and dry.

A set screw for the eyepiece. A small but important feature. No one wants the eyepiece to fall out and land in the grass or on the primary mirror.

Matching mounting rings. Mounting rings are easy to make, but why bother? If the scope comes with mounting rings, that's a plus. Ideally, you would like mounting rings on a quick release base for hassle-free set-up. Where do you get this kind of base? Good question. Several have been advertised in the astronomy magazines over the past few years, but you may have to make your own.

If you are building a telescope that is 20 inches or larger, by all means opt for a finder with lots of light gathering power. This means a bigger objective and more money, but the extra power and image brightness are worth it. If you mount a "jumbo finder" (such as a 4-inch $f/5$ refractor) on the mirror box, rebalancing the telescope won't be a problem, but you'll have to come down the ladder for each faint object you want to find.

12.3.3 Get Two Finders

As general rule, you cannot install any finder except a Telrad or a very light finder

Fig. 12.3 *A good observing ladder is an indispensable accessory for your telescope. Added wheels make it easy to move; rungs added between the regular rungs mean you can climb to exactly the right height.*

Fig. 12.4 *Up the ladder, down the ladder, each time to check some detail on the star map. Bob Ross demonstrates that binoculars prove useful for more than star gazing: they can be useful for star-map gazing too! Photo by Lord Ross.*

telescope on the secondary cage. Doing otherwise makes the telescope top-heavy. However, you can attach almost any size or type of finder to the mirror box. What to do? Simple: Get two finders. Or three. Install as many finders as you need to observe comfortably.

This is no joke. Seasoned observers use both a Telrad (for bright objects) and an optical finder (for locating faint fuzzies). Sometimes they add a jumbo finder to make searching for very faint fuzzies easy. Each finder does its job well and the different types of finders complement each other. You can stand with your feet on the ground and sight through a Telrad even though it's four or five feet above you. You position the ladder, use the most powerful finder telescope to nail whatever you're after, and then climb to the eyepiece.

If you like to star-hop from the top of the ladder, mount a lightweight finder with an Amici prism on the secondary cage. Sure, you'll need to add weights to the back of the mirror box, but you can zip from one object to the next with ease. With "The Big Ones," you need to come down once in a while to reposition the ladder, so mount another finder on the mirror box. That way you can point the telescope without climbing the ladder to your top-mounted finder scope.

12.3.4 Jumbo Finders

The "jumbo finder" comes into its own with Dobsonians in the 30-inch and over range. With your feet on the ground, you spin the scope around and point it approximately where the object should be. You look into the jumbo finder at 25x

Fig. 12.5 *Observer Ron Hill stands besides a beautifully crafted observing desk that can be knocked down for easy transport. Low-level lighting, a chart holder, and a handy book-shelf for atlases and reference books enhance observing pleasure. For dewy nights, the fold-down cover keeps charts and eyepieces dry.*

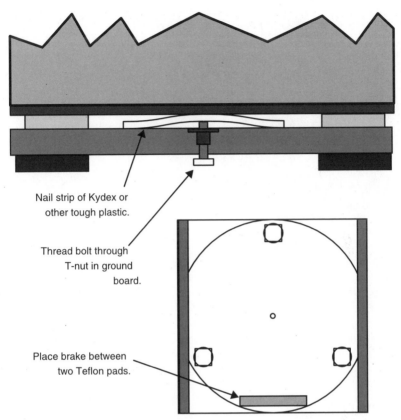

Nail strip of Kydex or
other tough plastic.

Thread bolt through
T-nut in ground
board.

Place brake between
two Teflon pads.

Fig. 12.6 If you observe where the nights are windy, a "breeze brake" is what you need. Midway between two of the Teflon pads on the ground board, add a bolt that presses a length of tough plastic against the laminate on the ground board. The additional friction keeps the telescope from swinging in the wind.

magnification and carefully zero in on a field defined by eleventh and twelfth magnitude stars. When you climb the ladder, your quarry is there. You don't need a jumbo finder for the bright Messier objects, but it sure is handy if you're trying to log some gravitationally-lensed quasars.

12.3.5 Digital Setting Circles

If you don't like to spend time "finding" objects, then install digital setting circles (see **Appendix C**). These consist of two shaft encoders (at $50 each) and a tiny computer costing several hundred dollars. The shaft encoders tell the computer where the telescope is pointing and the computer converts this to right ascension and declination. Digital circles are accurate and easy to use. The newer models have the entire NGC and IC catalogs of deep-sky objects stored in permanent memory! Better units have a "go to" feature that actually tells you which way to

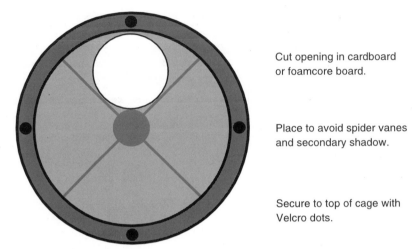

Cut opening in cardboard
or foamcore board.

Place to avoid spider vanes
and secondary shadow.

Secure to top of cage with
Velcro dots.

Fig. 12.7 *On nights when the seeing is poor, an aperture mask may improve lunar and planetary images. Make the mask from black cardboard or foamcore board and secure it on the top of the cage with self-sticking Velcro pads. The mask for a 20-inch telescope will have an opening about 8 inches in diameter; for a 25-inch, about 10 inches diameter.*

push your telescope to reach the desired object. They speed up finding objects, so you can see more in each observing session. Even fanatical star-hoppers enjoy using digital setting circles once in awhile.

12.4 Electrical Accessories

We live in a world full of electrical and electronic devices. Even though your big Dobsonian doesn't *need* any of these gadgets, they sure are handy. After all, how are you going to observe when the secondary dews over? Maybe you really need those electrical goodies after all!

12.4.1 Dew Zappers

In open-tube telescopes, the secondary mirror sometimes dews up as air temperature drops. If dew is an occasional problem for you, blow it away with warm air from an electric hair drier or 12-volt "dew zapper" heat gun available from a variety of suppliers. The 120-volt variety is great when you have electric power at the site, but the 12-volt kind is a necessity at remote sites. You can run them from a 12-volt battery or from a Porta-pak.

12.4.2 Heat Ropes

Use a heat rope to eliminate chronic dewing problems. Heat ropes are surplus heating elements from automatic coffee makers; they cost about $1 each from the American Science Center, a surplus outfit located in Milwaukee and Chicago (see

Appendix E). They are about 2 feet long and consist of a nichrome heating wire in a fiberglass sheath.

Stick a heat rope on the back of your secondary mirror, connect a 12-volt power source like a Porta-pak, and the gentle heat from the heat rope will prevent dew from forming. Heat ropes also work great on the exposed glass parts of Telrad finders. Attach the heat rope with a few dabs of silicone adhesive. Once dew has formed, a heat rope usually cannot remove it; use your heat rope as a preventive measure.

12.4.3 Portable 12-Volt Power

Portable telescopes need portable power supplies. This means batteries. For modest power needs, rechargeable gel cell batteries work best. They have no caustic chemicals to spill, they hold a charge well, and they can be recharged many times. They are used in emergency lighting and children's electric toy cars.

The Porta-pak gel cell has become popular with amateur astronomers. It can supply more than enough current to power a cooling fan, a couple of heat ropes, and digital setting circles for a couple of nights of observing. The Porta-pak and similar gel cells are compact. Mount one on the bottom of the mirror box where it's out of the way, and it will help you balance a top-heavy telescope.

12.4.4 Marine Batteries

If you do not have 120-volt house current at your dark site and you use a lot of electric gadgets, a deep-cycle 12-volt marine battery is the answer. Sport fishermen use these batteries to run their electric trolling motors. They are sold in sporting goods stores and boat shops.

They require an automotive battery charger and some attention and maintenance between observing sessions. In fact, they thrive on tender loving care from someone who likes to charge and discharge them regularly between observing sessions. But they repay the extra time and effort; one big marine battery will run your digital setting circles, your heat ropes, and your cooling fan for months, and your portable computer and your favorite sky-finding software for weeks and weeks. If you need portable power, resign yourself to lug that lead.

12.4.5 Power on the Secondary Cage

Heat ropes, reading lights, and digital setting circles are installed at the secondary cage, where you do not want a heavy battery. If you run a cord from a gel cell on the ground to the secondary cage, you've got an accident waiting to happen.

The solution is to mount the battery inside the mirror box and run the power cord to the secondary cage beside a truss pole. In fact, why not eliminate the cord and use the aluminum truss poles as the conductors?

Run the positive lead from the battery to one of the split blocks. Use an insulated, braided wire for this. Drill a small hole through the block and mirror box into the pole bore. Press the braided end of the wire into the hole so it will fold down alongside the bore when the pole is inserted. Do the same with another block for the negative side. Make sure these two truss poles make no electrical connection.

On the secondary cage, wire the pole seats and snake the wiring out to the heat rope on the secondary mirror. Only the positive insulated wire needs to run the length of the spider vane—the negative wire is attached directly to the end of the same vane. Connect one end of the heat rope to the positive wire and the other to the spider, and the circuit is complete. When you set up your telescope, power is connected to the secondary cage.

Be sure to install fuses between the gel cell and the telescope. Measure the current with everything running and select a fuse with a slightly higher value. If something shorts out, the fuse will blow. Better to blow a fuse than to blow up a gel cell all over the inside your mirror box.

12.5 Keeping Warm and Dry

You've built your dream scope. The following gold nuggets have been gathered from experienced observers around the world to help make observing with your telescope a dream and not a nightmare.

12.5.1 Warm Clothing

Nothing kills an otherwise great night of observing faster than cold feet. In the larger scheme of things, keeping your body warm is more important than optical cleanliness or collimation. Who cares if the telescope isn't collimated if your toes are frozen and your brain feels like mush? All you care about is getting warm!

Forget appearances. No one is going to see you in the dark. Go for comfort and warmth. Lots of layers of loose clothing work best. Polypropylene, wool, neoprene, and down are the materials of choice. High-tech synthetics like Gortex are great for the ski hill where lightweight warmth and style are desirable. For astronomy, don't bother. Buy warm clothes like you expect to gain 50 pounds. Select a coat with spare pockets for eyepieces, heat packs, and filters.

Insulated boots. Buy the best you can afford. When you try them on, make sure you're wearing the same socks you normally wear while observing. If in doubt, opt for a size too large and fill the extra space with big socks. For severe cold, cut polypropylene insoles from a camper's sleeping pad and insulate your feet from the cold ground.

Polypropylene glove liners. The hand that guides the scope is the one that gets cold first. Big mittens keep your hands warm, all right, but then your hands are clumsy. An observing buddy of ours tried polypropylene glove liners. They are thin and flexible—you can turn pages in a book—yet they hold in heat and provide an insulating layer between your skin and freezing metal. They cost a couple of dollars a pair and are sold in stores that sell work clothing. In deep cold, put a heat pack on the back of each hand inside the glove liners.

Neoprene head band. Neoprene is a super insulator. Place a head band made from neoprene over your ears and you'll feel delicious. If you ears are really cold, stick a heat pack under the head band. Within minutes you'll feel toasty.

Neck warmer. Polypropylene neck warmers are sold in ski supply stores. The neck warmer is simply a 6-inch-long cylinder of polypropylene that you pull over your head down to your neck. They seal your neck much better than a bulky, old-fashioned scarf.

Hooded sweat shirt. Buy the biggest hooded sweat shirt you can find. They always shrink in the wash and you'll want to put a couple layers underneath. The hood is the important part. It seals up the back of your neck and head. The bulk of body heat is lost through an exposed head and neck. Keep them covered and you'll stay warm.

If it's really cold, wear a stocking cap under the hood. Wool knit or wool with Thinsulate are the warmest. For an added benefit, make sure the hood is extra big and sloppy. You can pull the hood out and over the focuser and everything gets really dark except the image. No need for a dish towel to block stray light—your hood does it for you.

12.5.2 Keep Warm with Heat Packs

Sheryl Johnson observes with a 20-inch telescope, and on cold nights she feels a little chilly. Anyway, Sheryl hasn't complained about the cold since she started using heat packs. These wonderful little chemical furnaces can make observing infinitely more comfortable.

Heat packs are sold as disposable hand warmers. The active ingredients are iron, water, cellulose, vermiculite, activated carbon and salt. They are odorless, nontoxic, dry and clean. They come in various sizes. For astronomy, the 2x3-inch size is right. You get them in an air-tight plastic wrapper.

To use, remove the heat pack from its plastic bag. Shake it and give it a few minutes to heat up. Place it in a pocket or glove where it will reach a comfortable temperature of 135 to 156°F. Stuff heat packs into your pockets, your gloves, your boots, the back of your neck—wherever you get cold. They really work great. While everyone else is shivering you'll feel guilt and warmth flowing all through your body.

The best place to look for them is at stores that sell to hunters. They use them to keep warm while waiting all day for an animal to show up. David buys a hundred heat packs for 25 cents apiece at the end of every deer hunting season in Wisconsin, when they are on sale.

If you are camping don't throw out half-used heat packs. Toss them into the bottom of your sleeping bag when you turn in for the night. They will keep your feet warm.

Fig. 12.8 Add a handy observing shelf to your ladder and you won't have to make so many trips down for a refresher look at the star chart.

12.6 Eliminating Stray Light

Stray light is the enemy of astronomy. It is insidious, creeping into telescopes and subtly stealing a few tenths of a magnitude even under the darkest skies. In an open tube telescope like the low-profile Dobsonian, stray light reaches the observer's eye via three routes: the middle, the top, and the bottom. The middle is the open tube itself; sky light falls on surfaces that should be dark, and zips into the observer's eye. The top is the open upper end of the tube, where light bounces off the secondary cell, and into the observer's eye. At the bottom, light sneaks around the outer edge of the mirror and into the observer's eye. Luckily, stray light is easy to fix.

12.6.1 The Light Shroud

The light shroud closes the open tube. It is a cloth cylinder of Ripstop nylon fabric. Compared to a solid tube, it weighs almost nothing. Not only does it keep out stray light, but it deflects falling objects that might otherwise land on the primary mirror. Furthermore, it is so thin and breathes so well there are no tube currents, yet it deflects the observer's body heat away from the optical path.

The light shroud is made from "Ripstop" fabric. Available from JoAnn Fabrics, a national chain, Ripstop is a lightweight, black, rip-stopping nylon that is dull on one side and semi-glossy on the other. The dull side faces inward and absorbs light while the glossy side faces out and looks attractive.

Ripstop is nearly impossible to tear. It handles a heavy dew well but it is not

Fig. 12.9 Don't neglect the base of your ladder. Wheels or castors make it easier to move; reinforcing the base makes it more stable. Add extra rungs so you can stand at a convenient height.

waterproof. When you're done observing, just toss it in the clothes dryer on low heat. At about $6.00 a yard (60-inch wide) it's affordable.

Be sure to ask for Ripstop by name. One unfortunate soul bought a generic ripstop nylon to make a shroud, only to discover that the generic material was not colorfast. After a heavy dew, the dye ran out and dripped all over his scope, including his new 20-inch mirror!

The upper end of the shroud is hemmed and contains a drawstring and plastic slider. These nifty little devices can be purchased from Campmor, a camping catalog outfit. They are actually called "Fastex Lock with Wheel" and sell for 25¢ each. You pull the slider tight and the shroud is nicely held on the secondary cage.

At the bottom end of the shroud, by the mirror box, the shroud is held by plastic grommets and bungee cords that wrap around the split blocks. The bungee cords maintain tension on the shroud. Ripstop nylon is strange stuff: it expands when it gets cool or wet, and it contracts when it gets warm. Bungee cord attachments provide the necessary tension for whatever the temperature does.

12.6.1.1 Sewing a Shroud

Sewing an attractive light shroud that fits well can be more difficult than building the telescope. The fabric for the shroud must be draped and pinned over the fully assembled telescope. If you can obtain permission to setup the telescope in the middle of the living room, you'll have plenty of incentive to complete the job quickly. Guaranteed.

Ripstop nylon comes in a 60-inch wide bolt of fabric. Sew together two pieces the length of the truss poles plus an extra foot. Sew the seam on the dull side of the fabric, the side that faces inward. Rotate the scope until it is almost horizontal. Lay the 120-inch-wide piece of Ripstop over the truss assembly with the seam running from the upper pole clamp to the middle of the top of the mirror box, that is, along the top. The other seam will be sewn directly opposite, along the bottom.

With the fabric draped over the truss, gather the two free ends and pin them together underneath. Try to pin a straight seam from the secondary cage to the mirror box. Be patient, it's tough. Remove the pinned tube of cloth from the scope by gently tugging it over the secondary cage. If it won't slide, grasp the lower end and try peeling it off the truss like a banana skin. Sew the pinned seam. Put the shroud back over the scope and, if it's puckered, remove and redo the seam. Keep at it until the shroud fits snugly over the truss tube.

You will notice that pulling the shroud over the truss farther and farther makes it fit tighter and tighter. That's because the truss assembly is conical. This geometry is advantageous because it means you can always pull the shroud on until it's tight.

Once you've got the bottom seam finished and the shroud fits relatively well, the rest is easy. Mark the top and bottom with chalk to locate the hems. The upper end of the shroud needs a hollow hem to retain an ⅛-inch braided nylon cord to be used as a drawstring. Make this hem about ½ inch wide for the cord.

Fig. 12.10 *Black Ripstop nylon shrouds keep stray light out of your eyes and improve the view even at the best dark-sky sites in Texas. Under city and suburban skies, the shroud is an absolute necessity.*

On the bottom, make the hem about 1½ inches wide. Sew a 1½-inch wide basting around the bottom hem to reinforce it for the grommets. Adjacent to the side bearings, split the shroud so it won't interfere with the bearings. Sew a ¼-inch wide hem in the splits to prevent fraying. Install eight plastic snap-on grommets in pairs, one pair for each corner of the mirror box. Do not use metal grommets: they have sharp edges and will cut through the fabric like a sharp knife. Locate each grommet six inches from the corner.

When you first install the shroud, determine the right amount of tension and tie the cord and bungees. To prevent the cord form unraveling, fuse the ends with a cigarette lighter.

The best way to get the shroud done right is to not do it at all. Every town has a someone who specializes in making complex structures (e.g., wedding dresses) from cloth. These talented souls can crank out a beautiful light shroud in short order and they are worth every cent they charge. Ask around and you'll find one. Bring your telescope over and discuss the pictures in this book. Point and say, "This is what I want." Walk away. Pick up the telescope and shroud in a couple of weeks and you'll be ready to observe. It'll be the best money you ever spent.

12.6.1.2 Installing the Shroud

Install the shroud on your telescope before you attach the Telrad or any eyepieces. Lower the telescope to a comfortable height and slip the shroud over it like a sock. Tease the end with the drawstring past the focuser and align the shroud so the drawstring slider faces the ground

Loop the bungee cords at the bottom of the shroud around the wooden blocks on the mirror box. (If you made solid wood side bearings, drill a hole so you can pass the cord through.) After the bungee cords are secure, adjust the tension at the top end with the drawstring and slider.

As the air cools and dew forms, it may be necessary to tighten the drawstring. This is because nylon fabrics expand as the temperature falls. Conversely, it's a good idea to loosen the drawstring and release the four lower bungee cords if you leave the scope assembled in the daylight. In the warm sun, nylon fabric contracts and it will pull on the cords and seams.

If the shroud gets wet from dew, dry it in a home clothes dryer on air fluff or very low heat. Do not use the regular setting or the fabric may melt. Before you slip off the shroud, rack in the focuser all the way, take out the eyepiece, and remove the Telrad.

12.6.2 Focuser Baffling

Even with the shroud in place, short focusers may allow sky light to sneak over the top edge of the secondary cage to the eyepiece. To block this light, you can either baffle the focuser (described in this section), attach an external light baffle (described in **Section 12.6.3**), or both.

A focuser baffle is simply a small disk of dark material such as Kydex attached to the inside of the focuser board. A hole in the disk allows light from the primary mirror to enter the focuser and reach the focus, but it blocks light off the optical axis. For a 2-inch focuser, carefully cut an opening 1¾ inches diameter, center it precisely on the axis of the focuser tube, and secure it with three small screws. When you look through the focuser, you should see that the disk blocks stray light without blocking light from the mirror.

While you're at it, inspect the inside of your focuser's tube. A few years ago, many focusers had tubes that were chrome-plated on the inside! You still see these glittering wonders for sale at amateur swap meets. If the inside of the focuser tube is not already blackened, cover it with a matte black spray paint.

12.6.3 The External Light Baffle

After the open truss, the next worst entry point for stray light is through the top of the secondary cage. Light entering from this source illuminates the field lens of the eyepiece. The result is a very annoying ghost image in the field of view.

Street lights are the worst offenders. It never fails that the object you are trying to view is always placed in the sky such that a street light shines down the tube at just the right angle. The way to beat this is to fabricate an external light baffle.

A baffle is simply a sheet of plastic or cardboard that extends beyond the end of your telescope. It not need go all the way around the opening of the secondary cage because stray light coming from the eyepiece side of the telescope cannot sneak into your view. As a result, an external baffle can extend a foot from the end of the tube and go halfway around the secondary cage.

Leftover Kydex plastic, the same material recommended for the inside of the secondary cage, makes great external light baffles. Cut a piece half the circumference of the secondary cage. Do a good job: taper the upper and lower ends down towards the secondary cage. Apply a strip of Velcro to the shiny side of the Kydex along the bottom edge. Run a mating strip of Velcro along the top inside edge of the secondary cage opposite the focuser. Tip the baffle so that it tilts out and slightly away from the path of the incoming light. If necessary, add a little weight to the bottom of the mirror box to balance the baffle.

12.6.4 The Tailgate Cover

The only significant drawback of an open-frame tailgate is that light can enter around the mirror. If you use the scope on snow or on light-colored concrete, one solution is to put a square of black carpet on the ground. Alternatively, drape black cloth over the back of the mirror box to block stray light. Leave it loose and open around the sides so the air flows freely.

12.7 Telescope Covers

Dust, dirt, and grit are the enemies of telescopes. Dust on the optics clouds your view of the universe, dirt on the truss tubes makes them stick, and grit in the bearings impairs the smooth motion you crave. To a large degree, the best way to clean your telescope is not to let it get dirty in the first place, and that's why you want to make telescope covers.

12.7.1 Mirror Box Dust Cover

Dust is ubiquitous out of doors. Ten minutes after you install the primary mirror, it will already look dusty. The dust cover cannot keep out fine dust, just big chunks. Chunks like falling bolts, Telrads, eyepieces—that sort of thing. Yes, dust cover is a euphemism for "falling bodies" cover.

Think of the dust cover the same way you think about seat belts—they can't protect you if you don't wear them. Always put the dust cover in place when you set up or take down your telescope. Always store and transport the telescope with the dust cover on.

12.7.2 Hat Box for the Secondary Cage

Have you ever gotten down on your hands and knees and looked underneath the furniture? If you did, was there dust on the bottom of the furniture? No. Well, the

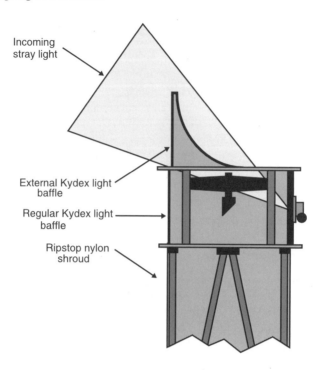

Incoming
stray light

External Kydex light
baffle

Regular Kydex light
baffle

Ripstop nylon
shroud

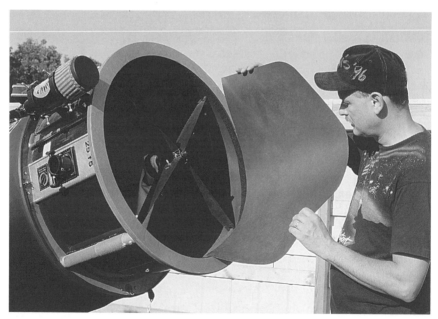

Fig. 12.11 *Make the external light baffle from Kydex left over from the secondary cage, and attach it with strips of Velcro.*

Fig. 12.12 Treat yourself to regular getaways to dark sky sites. As you scan the Milky Way past uncountable numbers of clusters and nebulae, the cares of the world fall away. The time and trouble and expense of building a large telescope will seem trivial compared to the joy of observing on a fine night. Photo by Tim Kimbler.

same logic applies to the secondary mirror in your telescope. Not much can accumulate on a flat surface that faces down at a 45° angle.

However, the secondary cage is another matter. To keep it clean, store it inside a custom made "hat box." The hat box provides excellent protection for both the secondary cage its mirror.

You can make a hat box from a short length of spiral-wound cardboard tubing, better known as Sonotube. It's the same stuff some telescope makers use to make the whole telescope. Get it in a diameter large enough to hold the secondary cage comfortably.

Remove the waxy layer around the bottom edge and glue a disc of ordinary ⅜-inch plywood inside the tube. On the top, make a removable lid. Cut a disc of plywood slightly larger than the outside diameter of the cardboard tube. Glue a one-inch high ring of cardboard around the outside edge of this disc. *Presto*—you've got a dustproof cover that slips easily over the container.

Soak the inside and outside of the plywood and the cardboard with Thompson's Water Seal. It's easy to apply and invisible, and it penetrates the cardboard. The hat box can take lots of abuse, so don't worry about gouges and scrapes. After a couple years of getting banged around, it will still look just as terrible as the day you finished it. Let the hat box dry in the sun for a few days before you store the secondary cage in it.

Fig. 12.13 If you store the secondary cage in a "hat box" made with concrete form tube, it will stay clean, dry and protected.

If you want more storage, make several hat boxes. You can keep eyepieces, charts, books and food in them at star parties. They make great seats for when you're tired. At night they serve as tables to set equipment and charts on. Or a soap box if you want to address the crowd.

12.7.3 Secondary Mirror Cover

Okay. You weren't convinced—you *still* want a cover for your secondary mirror. Well, maybe you have a point. Especially for star parties, because when you leave your scope up for days, some curious Bozo always leaves fingerprints on the secondary.

Here are a few suggestions for secondary mirror covers: a heavy cotton sock, a foam drink cup, a Ziploc freezer bag, a Chivas Regal bag (if it's good enough for a premium scotch, it's good enough for a secondary mirror. They have a cute little drawstring, too).

Have you noticed a common characteristic of all the suggestions above? All are SOFT covers. If they fall off the secondary and land on the primary, there will be no damage.

12.7.4 The Telescope Cover

If you attend a couple of star parties every year, you'll want to leave your telescope setup in the daytime. To protect it from dust, rain, hail, and whatever else, a form-fitting cover is great.

Plastic bags keep the rain and dust off, but when the sun comes out, they make excellent greenhouses. Even if your scope doesn't actually melt in the sunlight, the moisture from the previous night is trapped inside and the heat and humidity might parboil your whole telescope. Making a form-fitting cover is not difficult, and it's certainly much easier than making the light shroud.

Buy some waterproof Ripstop nylon. The type used to make rain-flies for camping tents is best. Select the lightest color you can get, preferably white. Aluminized Mylar plastic also works well. You want something that reflects the heat of the sun so your scope will stay cool inside.

Set up your telescope and attach all the accessories you normally observe with. Nylon generally comes in 60-inch wide bolts, so you'll have to sew at least two pieces together. Cut the pieces a couple of feet longer than the length of the telescope. On Dobs over 25 inches aperture, you may need to sew three or more widths together. Drape enough fabric to go all the way around the tube and pin the pieces together underneath the tube assembly. Remove the fabric and sew it together along the pin lines. You can be somewhat careless since a loose fit works well.

Cut a circle of fabric the diameter of secondary cage and sew it to the top end. Slip this long, skinny sack over the telescope. Gather the fabric around the mirror box and rocker and pin it, just enough to be tidy. You must be able to remove the cover without difficulty. Turn the sack inside out and sew it on the pinned seams, and then cut away the excess fabric. Sew a 1-inch wide hem around the bottom.

Make yourself a simple drawstring to close the cover snugly around the bottom edge of the rocker base. Install plastic grommets through the hem every eight inches around the bottom. Insert a ⅛-inch cord through the grommets and secure with one of those plastic fasteners found on sleeping bag stuff sacks and on backpacks.

Paint a big comet on the side of it, or the pet name of your scope in extra big letters. You can keep your telescope up for weeks at time through all kinds of weather with a custom made cover.

12.8 Dark Observing Sites

When was the last time your eyes were truly dark adapted? Not just "a little dark adapted when you walked the dog last Tuesday evening," or "good enough to see the Lagoon nebula from your suburban back yard," but truly profoundly dark adapted. Is it so long that you have forgotten how exquisitely, wonderfully sensitive human vision is?

Fig. 12.14 *Dark skies are where people are not. This DMSP satellite image of the United States shows that you can find reasonably dark skies within 100 miles no matter where you live. For New York it's the Pine Barrens of New Jersey; for Chicago it's the farmlands of northern Indiana, and for Los Angeles it's the mountains east of the city.*

12.8.1 Why You Need a Dark Site

Most of us live in perpetual artificial daylight. Sadly, the trend towards more and more night lighting seems all but irreversible. Billboards, security lighting, and sports lighting annually waste billions of dollars of electricity to reap very dubious benefits. Much as we may want to, we probably will not reverse this trend in our lifetimes. Instead, we must travel farther and farther away from the cities to observe the deep sky. This is why you are reading our book: so you can build a large-aperture telescope that is *portable*, so you can escape the lights.

Large-aperture telescopes are extremely sensitive to this artificial light, as well as starlight. Deep sky objects appear brighter and more exciting in a dark sky because image contrast is a function of how dark the sky is. Urban light pollution greatly reduces the usefulness of the telescope. It's worth it to spend the extra time to travel to a truly dark site. You really do not know the potential of your scope until you do. Its like driving a Ferrari in the city. You can't let it run flat out until you're out in the country. Words cannot describe the feeling you'll get when galaxies, planetary nebulae and globulars come screaming out of the eyepiece.

Just as large-aperture telescopes are very sensitive to light they are equally affected by the seeing conditions. A large scope cuts through a broader column of air than a small scope, hence you are looking through a larger number of the dis-

Fig. 12.15 *A cover made of waterproof nylon protects your telescope from weather, dust and curious onlookers. Here you see it halfway on.*

Fig. 12.16 *It looks pretty mysterious, but you know that your big Dobsonian is safe inside its cover. By the way, do what we say not what we do: make the cover from light-colored material to reflect the sun's heat.*

torting air cells in the atmosphere. There is no cure for this; if the seeing is poor just grin and bear it and hope for a better night next time.

12.8.2 Site Selection

Get out some topographic maps and look for areas that are remote from cities yet accessible by road. Be realistic. Can you honestly tell yourself you can drive five hours from Denver to 13,000 foot Cottonwood Pass in the Rocky Mountains and still be fresh for observing? It's so dark that Venus casts a shadow. But can you drive back without falling asleep at the wheel? And what if it's cloudy when you get there? Elevation is always nice to get above the ground haze. Get up as high as you can—but remember that high elevations are colder.

Focus your efforts on finding a site on private land. While public lands offer many great observing sites, public land means public access. Private land generally means you'll be left alone. Talk to the owners. Explain what you're doing and ask permission.

Prospect sites by driving to them in the daytime. Bring a compass so that you find sites free of horizon obstructions toward the east, west, and south. In the selected area, look for a big, flat spot. A typical 20-inch $f/5$ telescope has an eight-foot turning radius, so you'll need a level circle of hard ground at least 16 feet in diameter.

As you narrow your selection, assess prevailing winds and look for locations where trees block the wind. Underfoot, note how smooth the ground is, and look for barriers like wire fencing that are invisible at night. Close your eyes and walk around the area. If you can do this without stubbing your toes or tripping on stuff, you have a good spot.

Mark the exact place where you will set up the scope at night. In the dark it is sometimes impossible to locate a spot you found with no trouble during the day. You can stick little flags in the ground, or spray-paint it, or both.

The first couple of nights, go to the area without a telescope. Lie down and do some naked eye astronomy. How bright is the Milky Way? What is the limiting magnitude? How bad are the mosquitoes? Is foliage a problem? Do the bushes have thorns? Does the grass soak your shoes if there is a heavy dew? Are radio towers visible? Are you under the stacking pattern for a major regional airport?

If you can select places like these and have clear skies when you go there, then you have been blessed. Enjoy it, but remember, there is no perfect dark site.

12.8.3 Human Factors

Observing alone at a remote dark site can be scary. The slightest noises and shadows can play upon your imagination. Raccoons and skunks will come visiting. But most of us can deal with that. It's the crazies, the kids on alcohol or drugs, the couples looking for a place to park and the local police who really disrupt observing.

Whenever you observe on public land you are at the mercy of all these folks.

And they *will* come. They will come with their headlights on. They will come expecting the place to be deserted. They will be just as afraid of you as you are of them. The reason these types show up at your dark site is to avoid the authorities. The reason the authorities show up at your dark site is to catch these types. You look suspicious to both: you are caught in the middle.

The lesson is this: if you must observe on public property, take along a friend. Leave a note telling a loved one where you went.

Memorize a couple of the standard answers we use with people that show up uninvited. For example:

Kids drinking beer: "Have you kids seen the sheriff? He was supposed to meet me here half an hour ago. He wanted check up on me and have a look through my telescope." Clears them out every time.

Lovebirds: "It's a telescope. Would you like to have a look?" Some of them will want to and you'll all have a good time. Those who refuse and stay want to get physical with each other no matter who is around. So turn your radio up loud to a news channel. Turn on your car headlights. They will leave.

Drinkers: Their night vision is shot and they are obnoxious. They can't see anything in the eyepiece because of the alcohol they have consumed. Try this line: "Well, I'm doing research for the University on quasars. You'll have to leave because I've got to get back to work. Keep your headlights off when you go because they ruin my sensors." They'll leave if you get rude enough.

The police: Law enforcement people we've run into at night have been very understanding. They may approach with a hand on their gun until they see that you are harmless. Everything's fine once they turn off the headlights. Give them a brief tour of the universe through your scope and they'll get back to their work. They are looking for the people described above. Some day perhaps you'll have a chance to say, "They went that-a-way."

No matter where you observe, make sure your vehicle has a strong battery. Knocking on a farmer's door at 3AM is no fun for you *or* the farmer. Picture yourself standing on the doorstep trying to explain that you are an astronomer and your car back in the woods needs a jump start. It's a tough sales job.

Chapter 13
Epilogue: Making a Small Telescope

An epilogue is an addition that rounds out and completes a story, and that's what we do here. For well over 300 pages, we have talked to you about building large telescopes. To round out this book—to complete the story—we show you how to build a small telescope.

This is necessary because now may not be the best time for you to build a large-aperture telescope. After all, you have to be more than a little obsessed to expend the kind of time and effort that a large telescope requires. The time, money, and skills simply aren't always available. Despite the obstacles, however, your desire to build a telescope with your own hands and use it to explore the heavens remains as strong as it ever was.

In these pages, we show how you can build an 8-inch *f*/6 Dobsonian that's a great performer. It is smaller, lighter, and less expensive than a big telescope, and to be quite frank, these features make it a lot more practical for many people. But should you ever decide to build a huge Dobsonian, you can apply the lessons you learn from building this 8-inch to the construction of your dream telescope.

Please understand that we don't want to talk you out of building a big telescope if that's what you want to do. Both of us use our own big telescopes on a regular basis, and we certainly understand and appreciate why you want a big one. For light grasp and resolution, you can't beat aperture. However, we would not be honest if we failed to point out that the deepest observing satisfaction often comes from being out under the stars with a modest telescope or pair of binoculars.

Small Dobsonians are perfect for nights when you have only a few minutes to quench your celestial thirst. If things get busy at work and you don't have the energy to observe, you'll feel a lot less guilty if your telescope is a small one than you will if it is a big one. Better yet, you can carry your 8-inch outside for ten minutes of before-bedtime skygazing, which is simply not possible with a big telescope. It's easy to use. It's easy to store. It has a wide field of view. No ladders. It's perfect for comets, Milky Way cruising, open clusters, but still big enough for celestial events like occultations, grazes and satellite transits.

Besides, even after you have built your dream telescope, you're going to

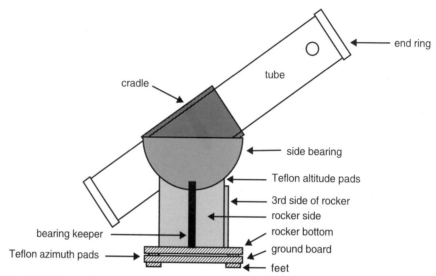

Fig. 13.1 *Our small Dobsonian resembles the classic instruments of the 1980s, with a solid tube and a relatively tall rocker, but the big side bearing and low-friction laminates we recommend add up to exceptionally smooth performance.*

need a small, portable instrument for the nights you don't feel like setting up a big one—so why not build the small one now? Well, enough preamble—let's talk telescopes!

13.1 Materials You Need

Collect the following parts and materials to build an 8-inch *f*/6 Dobsonian telescope:

> One 8-inch *f*/6 aluminized and overcoated primary mirror. (We might add here that if the telescope-making bug has bitten you badly, you might want to consider grinding, polishing, and figuring the primary mirror. Although making a big mirror is not a task for a novice, tens of thousands of people have made an excellent 8-inch mirror on the first try.)

> One 1½-inch minor-axis secondary mirror (sometimes just called a "diagonal"), aluminized and overcoated.

> One commercially-made spider. The length of the spider's legs should be appropriate for the tube you are using, that is, they should be designed for a 10-inch inside tube diameter.

> One commercially-made secondary-mirror holder. Be sure you order one with the same diameter as the minor axis of the secondary mirror.

> One commercially-made low-profile 1¼-inch focuser. Fully racked in, it should have a maximum height of 1¾ inches. Stay away from focusers for 2-inch eyepieces; they are tempting, but eyepieces made for

these are too heavy for small telescopes.

One 5-foot length of 10-inch diameter cardboard concrete form tube. This material is sold at home-improvement centers, builders supply houses, and places that sell ready-mix concrete under the trade names Sonotube, E-Z-Form, and a variety of others. It usually has a plastic or wax coating that you will have to remove.

One-half sheet of ½-inch high-quality HVHC plywood to make the mirror cell, cradle and rocker.

One half-sheet of ordinary ¾-inch plywood for the side bearings and ground board; SVSC is adequate.

One sheet of Kydex in the color of your choice. In this design, Kydex forms the outer covering of the telescope's tube, so your choice of color matters.

Ebony Star counter-top laminate. You will need enough to cut one circle 16-inches diameter and two strips 30 inches long by 2 inches wide.

Teflon plastic for the bearings. You will need three pieces 1½ x 1½ inches and four ¾ x 1 inches by at least ⅛-inch thick. This is not much Teflon, so call your local plastics supplier (in the Yellow Pages under "Plastics") and ask if you can obtain scraps or cutoffs.

In addition to these components, you will need paint, a small 12-volt fan, contact adhesive, silicone adhesive, wood glue, finishing nails and assorted small bits of hardware. Read through this entire section before you do any work so you understand the whole process in advance.

13.2   Overview of Construction

The construction sequence for a small telescope is similar to building the big ones; the overall plan is to build the telescope around the optical system. Here is the construction sequence that we recommend.

Step 1. Order the mirrors, focuser and spider. Purchase the length of concrete form tube.

Step 2. Construct the primary mirror cell and install the primary mirror in it.

Step 3. Temporarily install the spider, secondary mirror, and focuser in the tube.

Step 4. Position the primary mirror cell and verify the spacing by focusing on a distant object outside.

Step 5. Trim off the excess tube behind the mirror cell.

Step 6. Remove the spider, secondary mirror, focuser and primary mirror from the tube.

Step 7. Paint the inside of the tube black and wrap Kydex around the outside of the tube.

Step 8. Construct the cradle and side bearings.

Step 9. Locate the balance point of the tube assembly. Use this value to calculate the depth of rocker.

Step 10. Construct the rocker and ground board.

Step 11. Put on the finishing touches and savor "first light" with a look at your favorite celestial objects.

13.3 Begin with the Tube

Before you order the primary and secondary mirrors, read **Chapter 5** for advice. It is almost always possible to order small mirrors "off the shelf," so you will probably receive yours in a few weeks, at most. For a primary mirror of 12 inches aperture and under, we suggest that you order a full-thickness one. Although you could argue that a thin mirror might cool faster, in the smaller sizes even full-thickness mirrors cool rapidly, and the extra weight helps to balance the tube.

Purchase a length of concrete form tube roughly 2 inches larger in diameter to keep any tube currents away from the light path. The tube should be roughly 12 inches longer than the focal length of your primary mirror. For an 8-inch *f*/6 mirror, the tube would be 10 inches diameter and 5 feet or even 6 feet long. You won't need so much, but form tube is inexpensive and the extra length means that you can select the better end and discard the extra. Concrete form tube is sold in 2-inch increments based on the inside diameter (which would be the outside diameter of the concrete casting), so 10-inch tube is available.

When you buy it, concrete form tube often looks terrible. The surface doesn't matter since you will cover it, but make sure you get a piece that hasn't been squashed out of round. If your tube has a heavy wax coating on the inside, paint may not adhere. The remedy is simple. Carefully peel away one layer of paper from the inside of the tube. The resulting surface is rough and makes an excellent light absorber when it is painted black.

With the tube in hand, order a secondary mirror, holder, and spider. We recommend a standard straight-vane spider; they are tough and don't get out of adjustment. There is nothing wrong with two-vane, three-vane, or curved-leg designs, but the four-vane types are best for portable instruments.

While you are waiting for the optics and other components to arrive, paint the inside of the tube flat black. Ordinary flat-black primer paints work the best. Attach a small paint brush to a thin wood stick at a 90° angle. Start in the middle and work out to the ends. The stick should be at least half the length of the tube.

When the components arrive, decide which is the better end of the tube. Drill holes for the focuser 8 inches from the end you have selected. There is no need to be concerned about vignetting the incoming starlight because the tube is already oversize to handle potential tube currents. Drill holes for the bolts that hold the legs of the spider; place them so that the secondary holder lies directly opposite the center of the focuser.

Refer to the manufacturer's directions to install the secondary mirror in the secondary holder, and to place the secondary holder in the spider. Next, install the focuser and spider in the tube. This installation is temporary, so don't fuss trying to make everything perfect: the purpose is for you to check the location of the optics. Double check that the secondary holder is located directly opposite the draw tube of the focuser.

13.3.1 The Primary Mirror Cell

For mirrors up to 10 inches aperture, there is little danger that the mirror will flex under its own weight. You can simply fasten the mirror to a plywood disk with three dabs of silicone adhesive. The disk that carries the mirror is called the mirror disk; it is the same diameter as the mirror.

To align the mirror, three collimation bolts opposed by stiff springs allow you to tip and tilt the mirror disk relative to a second plywood disk, the tube disk, that fits inside of the telescope tube. The tube disk will be screwed securely inside the bottom end of the telescope tube.

The collimation bolts are ¼-20 by 3-inch long carriage bolts; the springs that oppose them are stout 1 inch long by ½ inch diameter compression springs that are slipped over each bolt. Each bolt is held captive by a wing nut. By turning the wing nuts, you can tilt and tip the mirror disk to send the reflected cone of light up to the secondary mirror.

Cut the two disks out of ¾-inch plywood. The mirror disk should be the same diameter as the mirror; you can trace the outside of the mirror with a pencil. The tube disk is drawn with a compass to match the inside diameter of the tube. Cut out the disks with a saber saw or band saw, then do the final trim to round with a router or power sander. Check the fit of the tube disk. It should just slip into the tube.

Set the mirror disk on top of the tube disk and center it. Drill three ¼-inch holes 120° apart through both disks for the collimation bolts. The holes should be drilled about 1 inch in from the edge of the mirror disk. Drill the holes through both disks at the same time so they line up precisely. After you drill the holes, mark them so you know which corresponds to which. To permit the bolts to pass freely through the tube disk, redrill its three holes to ⁵⁄₁₆ inches.

Finally, to provide air circulation, drill a 2-inch diameter hole through the center of both disks. These holes can be bored with a spade bit or a small hole saw. After all the holes are drilled, seal both wood disks with gloss polyurethane to insure a good bonding surface for the silicone adhesive you will use shortly. Paint or varnish the outer side of the tube disk to match the finish you plan to apply to the rest of the telescope.

To assemble the mirror cell, insert the carriage bolts through the mirror disk and seat them completely by tapping their heads with a hammer. Slip a metal washer over each bolt, then the compression spring, and then another washer on top of the spring. Slip the tube disk over the bolts, place a washer on each one, then thread the wing nuts onto each bolt. Tighten the wing nuts until the springs are par-

Fig. 13.2 *The mirror cell is made from two plywood circles and a handful of stuff from the hardware store. The miniature fan guarantees that the tube will cool even though the bottom end is closed.*

Fig. 13.3 *To install the primary mirror, assemble the mirror cell and then squeeze three blobs of silicone adhesive over the heads of the collimation bolts. Center the mirror on the disc.* **Do not press it all the way down.** *The silicone between the mirror and bolt heads has to stay about ⅛-inch thick or the mirror is distorted when the plywood changes dimensions. When the blob is thick it stays resilient. When the adhesive cures, these pads hold your mirror securely.*

tically compressed.

To aid air circulation, mount a miniature 12-volt muffin fan, the type used to cool computer chips, in the central hole of the tube disk. Orient the fan so that it will pull air from behind the mirror and blow it out the bottom end of the tube. You can leave the fan running while observing if you find that the images are better. We have never had any problem with vibration from these little fans. Finally, screw a 3-inch metal door handle to the back of the tube disk so that you will have a good grip when you install the cell in the tube.

Stop for a moment. Vacuum around your work area and the blow all the accumulated wood chips and grime off the mirror cell. From now on, you want to work clean so the aluminized mirror surface remains free of dust and dirt.

Place the cleaned mirror cell on the work surface with the mirror disk facing up. You may need to place blocks under the cell to hold it level. Squeeze blobs of silicone adhesive about 1 inch diameter and ⅜ inch thick over the head of each collimation bolt. Silicone adhesive sticks tenaciously to clean glass, metal, and polyurethane.

Align the rim of the mirror with the edges of the disk and lower it onto the silicone blob. Wiggle it slightly to get good adhesion, and gently shim the mirror until it is just ⅛ inch short of contacting the heads of the collimation bolts. This leaves about ⅛ inch open under the glass. Let the adhesive cure undisturbed for 24 hours.

Our experiments have shown that three silicone blobs will hold a mirror 8 inches diameter in complete safety for many years. However, if you are nervous, you can wrap a couple turns of duct tape over the edge of the cell. Make sure that the gummy adhesive on the tape stays well away from the aluminized surface. Frankly, we don't recommend taping because the silicone will outlast tape by many years, but if you cannot sleep worrying about your mirror, then it is best to set your mind at rest.

Since the tape obstructs the flow of air around the mirror, cut six slits each at least two inches long between the bottom of the glass and the disk. With the big central hole in the two plywood disks, air can circulate around the mirror.

13.3.2 Locate the Mirror Cell

At this point, the focuser, spider, secondary mirror and secondary holder should be installed in the telescope tube, and the primary mirror should be mounted on the mirror cell. Knock together a simple cradle about 18 inches long with a simple "V" block at each end. This handy device allows you to rest the tube on a surface without worrying that it will roll. Prepare a work surface—a backyard picnic table will serve—in a place outdoors where you can see some "targets" several hundred yards or more distant. Recruit someone to serve as your helper, and you're ready to go.

Place the tube on the cradle and aim the open end toward the distant target. Rack out the drawtube about ¼ inch. With a ruler, measure from the top of the fo-

Fig. 13.4 *The handle attached to the bottom of the mirror cell makes it easier to install the cell in the telescope, and it gives you a secure place to hold the tube. Note the fan mounted under the handle.*

Fig. 13.5 *Kydex plastic makes an excellent finish for the tube: it is both tough and waterproof. Recruit a helper. Coat both surfaces with contact cement and allow the cement to dry, then slowly roll the Kydex onto the tube.*

cuser to the center of the secondary mirror. Subtract this distance from the focal length of the primary mirror. The remainder is the distance from the secondary mirror to the surface of the primary. Measure this distance from the center of the focuser along the side of the tube. Carefully insert the mirror cell with the mirror into the back end of the tube and push it forward until the front surface of the primary is roughly under the mark on the tube. (After doing this, you will appreciate the handle on the back of the cell.)

Look into the focuser and have your helper tip and tilt the cell until you see the open end of the tube reflected in the mirror. Don't worry about collimation; you just want an image. Place a low-power eyepiece in the focuser. Have your helper move the mirror cell back and forth until the image of the distant object comes into focus. Wedge the cell lightly in place with paper shims. Try all of your eyepieces to be sure that the eyepiece with most "in travel" reaches focus when the focuser is racked in. Mark the location of the back of the tube disk on the inside the tube.

If you cannot wait another moment, drill small holes and drive three small wood screws through the tube and into the side of the tube disk to hold the cell in place. Do a rough collimation, and that night prop up your telescope to do some temporary stargazing. Be careful not to let the tube fall—that would ruin your fun!

When you are satisfied that the mirror cell is correctly placed, remove the cell. Draw a line around the tube, and cut the tube to length with a hand saw or saber saw. Here's a simple way to draw the line perpendicular to the axis of the tube: wrap a large sheet of paper around the tube and adjust it so that the edge of the paper meets itself after one full turn around the tube.

13.3.3 Finish the Tube

There are lots of ways to complete the tube. You could leave the outside of the concrete form tube as is: as long as the cardboard stays dry, the images won't suffer. If you live in a humid region where dew would wet the tube, then you could seal the outside of it with varnish or paint. You'll likely have a spiral stripe down the tube, and paint tends to look at bit "hairy" on concrete form tube, but you'll have a fully functional telescope tube for very little money.

For a really good-looking job, you can wrap the tube with a sheet of Kydex plastic. Kydex is an excellent covering for cardboard tube: it is waterproof, it comes in bright colors, and it has a "hair cell" texture that hides fingerprints. If you prefer a glossy look, you can bond the textured side in and leave the smooth, shiny side out. Not only is Kydex inexpensive, it is easy to work with.

Start by removing the focuser, spider, diagonal mirror, and mirror cell. In the process of getting everything right, you may have drilled a few extra holes, but the Kydex will cover them. Sand off any bumps on the tube or protruding layers of paper around the holes.

Some concrete form tubes have a heavy wax coating on the outside. If yours does, remove as much as you can by peeling away the top layer of paper, or wash the tube with a paint solvent. A sure method is to evaporate the wax with a heat gun, or propane torch—but do this outside and take pains not to overheat the wax and start a fire! In the southwestern states, you can put the tube out in the hot summer sun and the wax will vaporize. If you bought a tube that was somewhat too long, experiment on the leftover piece. When you can no longer scrape away any excess wax with your fingernail, enough of the wax is gone and you are ready to begin.

For wrapping the tube, recruit a helper because it involves more than one person can easily do. Warn your helper that you expect to be crabby and snappish, but you really do need their help.

Measure and cut a sheet of the Kydex an inch longer than the tube and an inch wider than the circumference. Make sure its edges are straight and square. On the side of the tube opposite the focuser hole, draw a pencil line down the length of the tube to help you align the edge of the Kydex. Align one edge of the Kydex on this mark and do a trial wrap to make sure that the edges of the Kydex overlap slightly. If the line is not parallel to the length of the tube, you may find you have too much Kydex at one end and not enough at the other when you wrap it around the tube.

To apply contact cement, purchase several inexpensive foam brushes. You will need at least two because once contact adhesive dries, the brush is ruined. Paint a strip of contact adhesive a couple inches wide along the length of the tube beside the pencil line. Paint another strip of adhesive along the mating edge of the Kydex. Be sure you apply the adhesive to the side you want mated to the tube. Let these strips of adhesive dry to touch.

Here is where the helper comes in. Align the edge of the Kydex sheet with the pencil line on the tube. Starting in the middle, press the Kydex to the tube. Work toward both ends. You want to join the edge of the Kydex along the line parallel to the long axis of the tube. With the Kydex firmly attached to the tube this way you can control the final wrapping better.

Place the tube with the Kydex sheet attached at the end on the floor or a large table. Paint the entire outside of the tube and the mating surface of the Kydex with contact adhesive. Let both surfaces dry to touch. Carefully wrap the Kydex sheet around the tube pressing it firmly with your hands.

Go slowly. Roll and wrap only a couple inches around the tube at a time. Force out air bubbles, and make sure you have a good bond over the entire surface.

When the tube is fully wrapped, there should be about an inch of overlap. With a pencil, trace the location of the final seam on the length of the tube. Paint adhesive between the free end of the sheet and the mating surface on the tube, being careful not to go beyond the pencil line. Any adhesive applied beyond the pencil line won't be covered by the Kydex and will look bad. When the adhesive is dry to touch, press it down. Trim away the excess Kydex at the ends of the tube with a sharp knife, and then step back to admire your work. Be sure to give your helper plenty of thanks. Imagine doing the whole thing by yourself!

Fig. 13.6 *The optical tube assembly sits on your workbench balanced on the side bearings. When you complete the rocker and ground board, your telescope will be ready for a night under the stars.*

13.3.4 Add End Rings

The exposed ends of your cardboard tube need reinforcement and protection, so you have an excellent opportunity to try a little creative scrounging. You can make end rings from plywood, metal or plastic. Metal baking dishes, plastic food containers, and a wide variety of other common items like hub caps make excellent end rings. The trick is to keep looking until you find something that fits.

You may get lucky at the junk yard. The chrome-plated end rings used on the telescope in the pictures came from a pair of Ford Motor Company hub caps. They fit perfectly over the tube ends, so the centers were cut out. The front ring can be cemented permanently with silicone adhesive, but you should attach the back ring with small screws so that you can remove it. These hub cap end rings look professional and cost only a dollar apiece.

13.3.5 Assemble the Tube

Assembling the tube takes only a few minutes. Redrill the holes for the focuser and spider through the Kydex, then install the focuser, spider and diagonal at the front of the tube. Vacuum up the scraps from drilling.

To install the mirror cell, remove the bottom end ring. Drill four or five small pilot holes through the tube and into the tube disk. Secure the cell with 1 inch long drywall screws. Be careful when you drill the holes: the primary mirror is in there. Vacuum up the scraps from drilling.

With the mirror cell in place, install the bottom end ring. Use screws. You need to be able to remove this ring so you can slip the tube into the tube cradle.

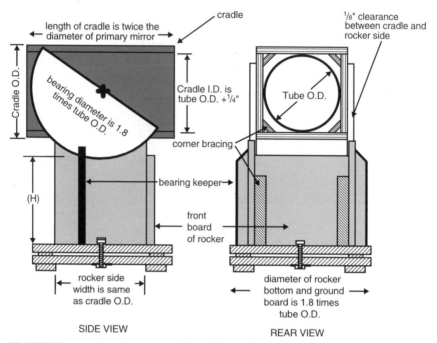

Fig 13.7 *The mounting is constructed from ½-inch hardwood plywood. The tube cradle is sized to fit the optical tube assembly; the large side bearings are attached directly to it. This mounting is quite compact: the rocker stands only about a foot high.*

13.4 The Mounting

Compared to building a large telescope, mounting a small Dobsonian is a cinch. The mounting consists of four easy-to-build plywood components:

    1. tube cradle,
    2. side bearings,
    3. rocker, and
    4. ground board.

The tube cradle is a squarish box that holds the tube; the side bearings are mounted on the sides of the tube cradle. The rocker is a three-sided box that supports the tube in the cradle on Teflon bearings, and the ground board is the flat piece on the ground that everything turns on.

13.4.1 Build the Tube Cradle

Because the optical tube assembly of your telescope has a round cross section, it needs a foundation with flat sides where you can attach side bearings. That's the job of the tube cradle, a four-sided structure that goes around the tube. You will

Fig. 13.8 *Use a carpenter's framing square to check that the rocker box is precisely square before you put it aside for the glue to set. If the rocker is out of square, the tube cradle may bind against the rocker.*

use high-quality ½-inch HVHC plywood for the cradle.

Begin by accurately measuring the outside diameter of the tube. The inside dimension of the tube cradle should be ¼-inch larger than the outside diameter of the finished tube. It can then be inserted easily into the cradle and held firmly with cardboard shims. The length of the cradle is relatively arbitrary. We recommend a 16-inch long cradle for an 8-inch telescope with a cardboard tube.

Since your tube may be a bit larger or smaller, be sure to figure your own dimensions. As an example, however, an 8-inch *f*/6 Dobsonian with a tube 10 inches outside diameter would have a cradle 10¼ by 10¼ inches inside by 16 inches long. Because two of the cradle sides must overlap at the corners, two of the sides must be 11¼ inches wide and the other two must be 10¼ inches wide.

Lay out the four sides on a sheet of ½-inch HVHC plywood and cut them out. Be as accurate as you can when sawing. Each pair of opposing sides should have exactly the same dimensions and should be perfectly square. Bond the corners of the cradle with wood glue and a few small finishing nails. Clamp or set heavy weights on it to improve the strength of the joint. The finishing nails serve to keep the pieces aligned on the slippery glue while it sets; they add little to the strength

Fig. 13.9 *Finish the rocker and ground board with a pleasing wood stain and several coats of polyurethane varnish. This mounting is remarkably compact considering that it carries an 8-inch telescope.*

of the joint. Wipe off any excess glue with a damp rag.

Before the glue sets, double-check the corners with a framing square. If the cradle isn't square, the tube will get squashed out of round, the side bearings will not be aligned so that the scope will wobble in the rocker, and the bottom end of the tube will scrape the inside of the rocker when you try to aim the telescope at the zenith.

After the glue sets, reinforce the corner joints with corner braces that run the length of the rocker and are an inch or two wide. If your cradle is square, the 45° saw cuts on the braces will mate perfectly in each corner. Glue them in by wedging scrap sticks of wood between opposite braces. Be careful that you don't push the cradle out of square.

Install a metal door handle on the top of the cradle to make it easy to carry the cradle-tube assembly out to your back yard. Insert the telescope tube into the cradle. Rotate the tube so the focuser is at a 45° angle from vertical. This is the most comfortable all around viewing position. Press cardboard shims between the tube and the cradle to secure the tube firmly.

Congratulations. The optical tube assembly of your telescope is ready for photons.

13.4.2 Side Bearings

The side bearings are large semicircular disks cut from ¾-inch ordinary SVSC plywood. For a 10-inch diameter tube, we recommend bearings 18 inches in diameter. Note that the side bearings are ¾ inch thick but that the rocker sides are ½ inch thick. The greater thickness of the bearing spans the ⅛-inch clearance between the inside of the rocker and the outside of the cradle.

Using a stick compass, draw an 18-inch circle on the plywood. Mark the center where the nail in the stick compass pricked the wood so you can recover the center later. Cut the disk with a saber saw or band saw, staying as close to the pencil line as possible without crossing inside it. Trim the disk to final dimensions with a router.

Draw a line parallel to the face grain of the plywood across the disk through the center point. The direction is important because the plywood is stiffer along the grain, and the bearings will look more attractive too. Then with a square, draw a short line perpendicular to the diameter at the center point so that you can locate the center of rotation of the bearing after you cut the disk in two.

Cut the disk in half along the pencil line to yield the two side bearings. Apply a ¾-inch strip of Ebony Star laminate to the outside circumference of each side bearing with contact adhesive. File or sand the edges of the laminate smooth.

The two side bearings must lie exactly opposite each other on the tube cradle. On each of the opposite sides of the cradle, draw a pair of pencil lines from opposite corners. These lines cross at the exact center of the cradle. Place the exact center of the bearing at this point and then rotate it until one end is flush with the top surface of the cradle. Recheck that the center of the bearing is still at the center of the cradle, and then attach the bearing to the cradle with a couple of wood screws. Flip the cradle over and attach the other bearing the same way.

Do not glue the bearings to the cradle. If they wobble in the rocker when you assemble the telescope, you can readjust them.

13.4.3 Locate the Tube Balance Point

Balancing the tube is much easier on a small scope than it is on the big one, but to do it right, everything that you plan to attach to the tube assembly should be attached now. The optics must be in place, your heaviest eyepiece, your finder or Telrad, and miscellaneous things like dew heaters must all be installed. Slip the

Fig. 13.10 *The center of the side bearing coincides with the center of the tube cradle, and the ends of the flat side of the side bearing cross from one side of the tube cradle to the other.*

completed tube into the cradle, removing the bottom end ring if necessary, and then reattach the bottom end ring.

Place the assembly on the floor. The telescope will rest on the side bearings. Slide the tube back and forth in the cradle until it sits level on the bearings. The balance point is then precisely at the center of the cradle and tube assembly and at the center of radius of the side bearings. Balancing is easy; the hard part was attaching all the extras.

To secure the tube, slip in cardboard shims. Do not glue the tube into the cradle because you may need to rebalance it in the future.

13.4.4 Construct the Rocker

Before you build the rocker, the optics must be installed in the optical tube assembly, the cradle and side bearings must be finished, and the balance point of the whole system must be determined. The reason is that before you can build the rocker, you need to determine how tall it should be.

With stick or ruler, measure from the center of the side bearing circles to the bottom of the tube. Just to be on the safe side, add two inches to the bottom length. The two inches give you an extra bit of clearance if you need to find a different balance point to accommodate a new 3-pound eyepiece that you couldn't resist. The height of the rocker sides is this height minus the radius of the side bearings.

Fig. 13.11 The best reasons for building a small Dobsonian are its portability and ease of use. Long after you have built the enormous telescope of your dreams, you'll still use "The Little One" often.

For example, if the tube balances 18 inches from the bottom end, add 2 inches clearance for a total of 20 inches. With side bearings 18 inches in diameter, the radius is 9 inches, so 20 minus 9 equals 11 inches. The height of the rocker sides to the bottom of the arc should be 11 inches.

The rocker sides should be the same width as the tube cradle, 11¼ inches. Cut out two pieces of ½-inch HVHC plywood 11¼ inches wide and longer by the radius of your side bearings than height to the bottom of the rocker arc; this would be 11 plus 9 inches, for a total of 20 inches. You need the extra length to locate the point of a compass so that you can draw an arc that matches the side bearings.

Get out your trusty stick compass and set it to the radius of the side bearings (9 inches) plus the thickness of the Teflon pads you are going to use. If the pads are ⅛-inch thick, then set the compass to a radius of 9⅛ inches.

Next, locate the center of the arc at the top of each of the rocker sides. From the center of the bottom edge of the rocker side, measure up 20⅛ inches and mark a point exactly in the center of each rocker side, that is, 5⅝ inches in from both sides. Place the compass at this point and draw an arc across the face of the plywood. Check the line against your side bearings to confirm it falls inside the arc by the thickness of the Teflon pads. Cut out the arcs on each side. The center should be the same you used for your compass point. If you don't have a router, then cut the arc·with a saber saw. After you make both rocker sides, set them on top of each other and check that the length, width, and arcs match perfectly.

Cut the front board of the rocker from ½-inch HVHC plywood. The front board connects the sides that support the side bearings, and substantially reinforces the rocker. To make the rocker as rigid as possible, the front board should reach the top of the rocker arcs; measure this directly off the rocker sides. Its width should equal the distance between the outer surfaces of the side bearings plus a small clearance. Since the cradle is 11¼ inches wide and the side bearings are each ¾ inches thick, the total is 12¾ inches. Allowing another ⅛ inch for clearance between the outside of the cradle and the inside of the rocker sides, the front board should be 13 inches wide by at least 11 inches tall. Cut it out and check that your cuts are accurately square.

Begin assembling the rocker by gluing the rocker sides to the front board. Tap in a few finishing nails to hold the assembly together. While the glue sets, place the assembly on the flat surface of your work bench and check that the bottom edges contact the bench evenly. If they do not, twist the assembled sides until they do, and then recheck the squareness of the three sides with a framing square. When the assembly is both square and rests flat on the bench, clamp the joints until the glue sets.

Reinforce the inside corners of the rocker with braces as you did for the cradle. Be sure these braces do not interfere with complete rotation of the tube assembly. If they extend too high, the cradle may hit them and prevent you from aiming the scope straight up.

The three-sided rocker box sits atop the round rocker bottom, a disk of ½-inch HVHC or ¾-inch SVSC plywood. We recommend that you make the rocker bottom 1.8 times the tube diameter. For a 10-inch tube, this would be 18 inches. Since the ground board should be the same diameter, when you cut and trim the rocker bottom, make two disks. One of them can serve as the ground board. For a good-looking telescope, use the better-looking face of the better-looking disk for the top of the rocker bottom.

Place the three-sided rocker assembly atop the rocker bottom disk so that the sides are equidistant from the center. To help, draw two lines across the disk through the center. After you draw the first line, use a framing square to draw a second line perpendicular to it. These lines are a good visual aid to centering the rocker sides. The three sides should just fit onto the bottom disk.

Mark the location of the rocker box on the bottom disk, then remove the box and drill two pilot holes per side through the disk. Replace the box and run 1½-inch long drywall screws through the bottom of the disk and into the end grain of the sides. If you set the telescope on the box, you should be able to swing it from vertical to horizontal without its binding or striking anything.

When you are confident the fit is good, remove the screws, apply wood glue to the bottom edges of the sides, and reassemble the rocker. Tighten the drywall screws until the heads are flush with the bottom of the disk. If any screw head protrudes from the surface of the underside of the disk you will need to file it flat or remove it after the glue has set.

With contact cement, attach the Ebony Star laminate to the bottom of the rocker assembly. Brush the cement on both surfaces, allow it to dry until it is no longer tacky, and then bring the surfaces together. They will bond instantly and permanently. Trim off any excess with a router and a laminate trimming bit.

Drill a ½-inch hole through the center of the rocker bottom for the pivot bolt.

Cut out four Teflon rectangles for the altitude pads. They should be ½ inch wide and 1 inch long. Attach the pads to the ends of the rocker arcs using tiny finishing nails. Use a nail set to recess the heads of the nails halfway through the Teflon so they will not score the laminate bearing.

Set the tube-cradle assembly on the rocker and check the movement. Note that the side bearings tend to slide sideways off the rocker pads. To prevent this, attach a short wood strip 1 inch wide and ¾ inches thick to each rocker side. These "keepers" should rise 1 inch above the bottom of the rocker arc. Glue a felt covering on their inside surface to prevent scratches on the faces of the side bearings.

13.4.5 Make the Ground Board

The ground board is easy to make since you have already cut out the disk. Make three feet each 2 inches square by ¾ inch thick from a hardwood and glue them to the bottom of the ground board 120° apart. Their outer edges should lie about ¼ inch from the outside edge of the ground board.

Drill a ½-inch hole through the center of the ground board for the pivot bolt. Construct a ½-inch welded nut-and-plate as described in **Section 10.3.2**, and attach the plate to the underside of the ground board.

At this point, all of the construction is done. Stain, varnish, or paint the rocker and ground board top and bottom, inside and outside. Allow the finish materials to dry thoroughly before you continue.

Cut three pieces of Teflon 1½ inches square; attach them to the ground board directly over the feet using tiny finishing nails. Recess the heads so they cannot scratch the bottom bearing. Wax the bearings with car wax.

Connect the ground board to the rocker with a ½-inch hex-head machine bolt. Don't tighten this bolt; leave about 1/16 inch clearance under the washer so that the rocker is free to rotate.

If you have not done so already, slip the tube into the tube cradle. Turn the tube so that the focuser is 45° above horizontal, and then shim the tube securely into the tube cradle with pieces of cardboard.

Finally, set the tube-cradle assembly into the rocker and, come nightfall, enjoy "first light" on the star of your choice.

Appendix A
Wood as a Structural Material[1]

Plywood is a glued panel made up of relatively thin plies, or layers of wood, with the grain in adjacent layers running along and across the sheet in alternation. The outside plies are called face veneers or face and back plies, and the inner plies are called core veneers or centers. The plies with grain running perpendicular to the face and back plies are called crossbands. Plies may vary in number, thickness, species, and grade of wood.

A.1 The Properties of Plywood

Plywood is a generic term. There are a multitude of different plywoods on the market. Before we can discuss using it as a construction material, we need to do some homework. As the name suggests, plywood is made from thin layers or "plies" of wood. Understanding the properties of plywood comes in extremely handy at the lumber yard when you buy materials for your telescope. It's fun to learn about and the knowledge will help you get the "right stuff."

Compared with sawn lumber, the chief advantage of plywood is that its properties are nearly the same along the length and across the width of the panel. In addition, plywood offers greater resistance to splitting, and, of course, plywood comes in versatile large sheets. The type, number and size of the layers, different grain directions, and various filler ply species all affect the mechanical properties of composite plywood panel.

All wood has a tendency to warp when its moisture content changes, as the result of uneven shrinking and swelling along, across, and through the grain of the wood. In plywood, this tendency is largely eliminated by balanced construction. The plies in a panel are arranged in pairs having the same thickness and kind of wood on either side of the core. Panels may be balanced with three, five, seven, or even larger odd numbers of plies. Balanced construction is highly important in panels that must remain flat.

1.This appendix is an adaptation of selected text from *The Wood Handbook*, a publication of the U.S. Department of Agriculture.

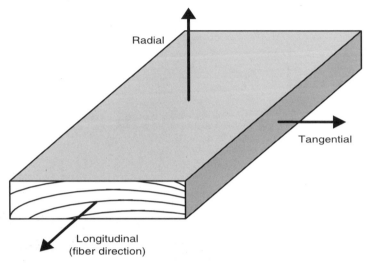

Fig. A.1 *Wood has different mechanical properties in each of its axes. In the longitudinal axis, along the direction of the wood fibers, it is strongest, and in the tangential axis, which runs across the grain, it is weakest. By combining thin plies running in different directions, plywood offers far more uniform mechanical properties than board lumber.*

Plywood panels may be as little as ⅛ inch thick, or as much as 1⅛ inch thick. The core may be thick or thin wood veneers, strips of solid lumber, or even a grossly deficient material such as particleboard. When you select plywood, never judge a panel by its face veneers. Two sheets of "oak" plywood may have radically different mechanical properties because the internal plies may be made from different species, and the number, thickness, and grain direction of the internal plies may be different.

A.1.1 Types of Plywood

Broadly speaking, two classes of plywood are produced: construction and industrial, and hardwood and decorative. Construction and industrial plywood has traditionally been made from softwoods such as Douglas fir, southern pine, white fir, larch, western hemlock and redwood. Most construction and industrial plywood used in the United States is produced domestically. The bulk of construction and industrial plywood is used where strength, stiffness and speed of assembly are more important than appearance.

Hardwood and decorative plywood is made of many different species of hardwood. Well over half of the hardwood and decorative panels used in the United States are imported. It is normally used for furniture and cabinet panels where appearance is more important than strength.

Plywood panels are marked according to type and grade. The two basic types of plywood are interior and exterior; the grade refers to the quality of the surface veneers. The two plywood types differ in their "glueline durability," or the water

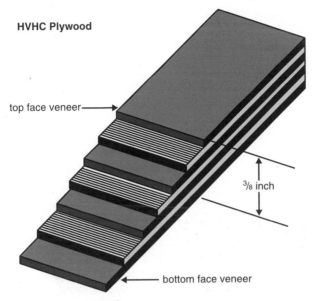

HVHC Plywood

top face veneer

³/₈ inch

bottom face veneer

Fig. A.2 *Hardwood veneer hardwood core (HVHC) plywood is made from many thin layers of hardwood bonded together. The internal laminations are free of defects, and the face veneers often have beautiful grain patterns.*

resistance of the glue that bonds the plies. Unless you intend to leave your telescope permanently outside uncovered and exposed to the elements, interior plywood works well in telescope construction.

The grade of the panel depends upon the quality of the veneers used, particularly of the face and back. A panel labeled A-A grade means the face veneers on both sides are made from the best quality wood with few to no imperfections that might detract from the appearance. The grain is smooth, tight, and full length. Plywood of A-A, A-B or A-C grade finds applications in furniture construction (stereo stands, book cases, kitchen cabinets, etc.) where a pretty face is more important than strength. Where only one side matters, you can save money by purchasing A-C plywood. Use the A side facing out and hide the C side.

A sheet labeled C-D has one side that looks poor and one that looks horrible. C-D plywood is used extensively in building construction for walls, floors, and roof sheathing where strength counts more than appearance.

A.1.2 The Stiffness of Plywood

The stiffness of different plywoods can be compared by knowing the number and thickness of the different plies that make up the panel and knowing the species of each ply.

The following formulae pertain to flexure of plywood that causes curvature of the plane of the plywood sheet. The modulus of elasticity in flexure or bending is equal to the average of the moduli of elasticity parallel to the span of the various plies weighted according to their moment of inertia about the neutral plane. That is, for

$$E_w \text{ or } E_x = \frac{1}{h} \sum_{i=1}^{i=n} E_i h_i$$

and

where E_{fw} is the modulus of elasticity of plywood in bending when the face grain is parallel to the span; E_{fx}, the modulus of elasticity of plywood in bending when the face grain is perpendicular to the span; E_i, the modulus of elasticity of the ith layer in the span direction; I_i, the moment of inertia of the ith layer about the neutral plane of the plywood; and I, the moment of inertia of the total cross section about its centerline.

For the sake of simplicity, assume all the layers are of the same thickness and wood species. Then the formulae reduce to:

$$E_w = \frac{1}{2n} \ [(E_L + E_T)n + (E_L - E_T)]$$

and

$$E_x = \frac{1}{2n} \ [(E_L + E_T)n - (E_L - E_T)]$$

where n is the number of layers (n is always odd so the panel is balanced), E_L is the modulus of elasticity of the veneer parallel to the grain, and E_T is the modulus of elasticity of the veneer in the tangential direction.

What do these equations mean? In simple terms they mean the stiffness of plywood depends on the species, thickness, and face grain direction of the plies. Here are some simple rules that can come in very handy when you're building a telescope and want maximum performance. Focus on understanding these important properties of plywood:

The thicker the panel, the stiffer it is. In other words, if you use more layers of the same thickness, the thickness of the panel increases and so does its stiffness. Under a load, the deflection of a plywood panel depends on the inverse cube of its thickness. This means that a small increase in thickness produces a dramatic increase in stiffness: ½-inch plywood is eight times stiffer than ¼-inch plywood.

This rule can save you money. If you bond two panels of cheap, weak, ¾-inch plywood together, you'll have a panel that is much stiffer than one layer of expensive ¾-inch hardwood plywood with 13 layers, such as Baltic Birch. Doubling up the thickness of the rocker sides and bottom with two layers of ¾-inch plywood makes the rocker unit eight times more rigid than the same unit made from single layers of plywood. The downside is that it weighs twice as much.

The moduli of elasticity of the wood species that make up the different layers determine the stiffness of the panel. Layers made from hardwood with higher moduli of elasticity are stiffer and resist bending better.

For high-stress areas that must be light in weight, like the secondary cage and focuser board, select plywood made from hardwood with larger moduli of elasticity. For the stiffest construction, use solid hardwood plywood like Baltic Birch, where all the layers are made from a species with a high modulus of elasticity. At the very least, pick a plywood for high stress areas that has face veneers made from hardwood.

The number of layers in a panel of a given thickness does not appreciably affect the stiffness of the panel. No matter how many layers there are, the grain of half of the layers run parallel to the span and half perpendicular to it. So a ¾-inch plywood panel with 13 thin layers has roughly the same stiffness as a ¾-inch plywood panel with 7 fat layers providing they both use the same species of wood for all the layers.

This means that using expensive multi-layered hardwood plywood for non-stress components is a waste of money. Use plywood made from cheap softwood for the ground board, rocker bottom, and side bearings. For high-stress areas, remember that it's the species and the total thickness and not the number of layers that makes material like Baltic Birch plywood so great.

Placing the face veneers of the panel parallel to the span provides greater stiffness than aligning the face veneers perpendicular to the span. The face veneers of a plywood sheet always run in the same direction, and since the face veneers are usually the wood with the highest modulus of elasticity, placing them on the faces makes the stiffest panel.

On rectangular components like the mirror box, rocker sides, and focus board, run the grain of the face veneers parallel to the longer span. The resulting part will be much stiffer.

By aligning the face veneers perpendicular to one another, you can make a panel that is equal in stiffness in both directions. When you double-up panels of ¾-inch softwood plywood for a square component like the rocker bottom, bond them so that the face veneers of the top layer are perpendicular to the face veneers of the bottom layer. The result is a rocker bottom that has equal stiffness in either direction.

A.2 Selecting Plywood for Telescopes

It is clear that an understanding of the nature of plywood depends on breaking it down to its smallest structural components and taking the sum of their properties. Now that we are familiar with these characteristics of plywood, we can put our knowledge to practical use. We can select and arrange commercially available plywood panels to the best possible advantage. In general, here's what you'll find:

- Hardwood Veneer Hardwood Core Plywood (HVHC)
- Hardwood Veneer Softwood Core Plywood (HVSC)
- Softwood Veneer Softwood Core Plywood (SVSC)

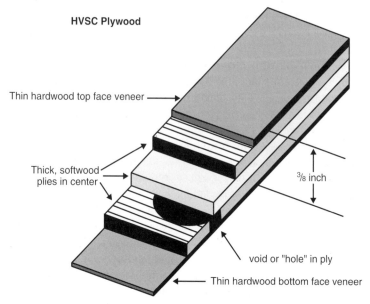

HVSC Plywood

Thin hardwood top face veneer

Thick, softwood plies in center

³/₈ inch

void or "hole" in ply

Thin hardwood bottom face veneer

Fig. A.3 *Hardwood veneer softwood core plywood looks attractive but its beauty is only skin deep. The hardwood veneer is very thin, and the core is filled with inexpensive softwood plies that are thick and may contain splits, cracks, and voids.*

• Particle Board

Each of these has a place in telescope making, though there are large differences in their properties.

A.2.1 Hardwood Veneer Hardwood Core Plywoods (HVHC)

HVHC plywood is the strongest, hardest, heaviest, and most expensive plywood. All the inner cores as well as the face veneers are made from hardwood species. Their high density is provided by solid grade core which in turn provides high strength per unit weight. HVHC is the plywood of choice for stress-bearing areas like the rings of the secondary cage and the focuser board.

HVHC also has very high resistance to splitting when nailed or screwed near the edge of a panel, as you must do along the corners of the mirror box. Compared to plywood with hardwood face veneers and softwood inner cores, the end grain of the inner plies is quite hard and solid, and can be finished as is. Because the inner plies have few patches and internal voids, the end grain can be finished smooth and beautiful.

You will not find HVHC at your local lumberyard or home improvement center. You will have to search. Call the customer service numbers in **Appendix E** for stores that sell it. If you are the type of telescope maker who wants to build a really good-looking first-class telescope, these plywoods are worth the effort. However, if you care only about what you see in the eyepiece and not at all about

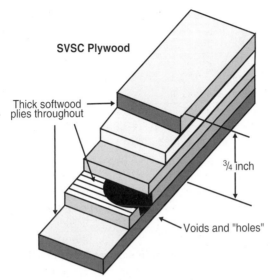

SVSC Plywood

Thick softwood
plies throughout

³/₄ inch

Voids and "holes"

Fig. A.4 *Softwood veneer softwood core plywood is made of thick layers of softwood. In a high-quality telescope, it is suitable only for crude structures such as the ground board and the rocker box. A "home-improvement" panel of 3/4-inch plywood usually contains only five plies.*

the structure that holds the optics together, select one of the less expensive plywoods with hardwood face veneers and softwood inner cores.

Baltic Birch HVHC Introduced to the United States in 1967, Baltic Birch has long been the plywood of choice for cabinet makers and furniture and specialty wood crafters. Because the quality of Baltic Birch is second to none, it is a great choice for plywood telescopes. Baltic Birch is made from many layers of birch hardwood each just a bit under ¹/₁₆ inch thick. The exposed edge of Baltic Birch, with its many fine plies, gives this material high visual appeal.

Baltic Birch is made with one-piece face, back, and inner plies that stand up under great stress and strain. It has tremendous tensile strength and screw-holding power, yet it machines easily and fasteners can be driven close to the edge without danger of splintering. Compared to a comparable panel of ordinary birch veneer plywood made from soft woods with a thin veneer of birch on the surface, Baltic Birch is more than three times as stiff.

Because Baltic Birch plywood has twice as many layers and every layer is birch, you can use thinner Baltic Birch in high stress load-bearing areas than you could with ordinary birch plywood. This keeps your telescope lighter and stronger. Standard birch ply may look pretty, but it isn't any stronger than construction grade.

No matter how you cut it, Baltic Birch always has a void-free edge because there are no joints in the veneers. Even when exposed raw, the edge creates a highly decorative treatment and gives the finished product greater individuality. The

Table A.1
Plywood Summary

Brand or Kind (4x8 Sheet)	Type and Number of Plies ¾"	Cost in Dollars per Square foot**
FinnPly-AA birch (5x5 sheet)	HVHC-13	$3.00
ApplePly-AA (maple face-alder core)	HVHC-13	$2.85
Baltic Birch-AB (5x5 sheet)	HVHC-13	$1.85
Oak veneer-AB	HVSC-7	$1.93
Oak veneer-AB*	HVSC-5	$1.60
Maple veneer-AB*	HVSC-7	$1.60
Birch veneer-AB*	HVSC-7	$1.60
Pine-BC*	SVSC-5	$1.00
Marine-AB	SVSC-13	$2.60
Aspen-BC (fir core)	SVSC-7	$1.00
Douglas Fir-AC	SVSC-7	$1.27
Clear Pine-AI	SVSC-7	$3.55
CDX (²³⁄₃₂" wall sheathing)	SVSC-5	$0.88
#45 particle board	wood particles	$0.56

*Plywood typically found in "home improvement" stores (core unknown).

**Cost in dollars per square foot: Retail prices December 1997. Wood is a commodity. Prices vary tremendously over even short periods of time. When demand is high, prices rise for all brands and types of plywood.

color is uniformly light, attractive, and takes a stain well.

Three grades of face veneer are available. All three grades are the same inside, so it depends on how picky you are and how much finish sanding you want to do, and how much you want to spend. Baltic Birch plywood is usually sold in 60 by 60-inch sheets. Have you ever tried lifting something like this? Unless you're an ape, your arms won't be long enough.

Since the collapse of the Soviet Union, the availability and quality of Baltic Birch has become more variable. It now pays to check each sheet before you buy it. Don't let some grunt running the fork lift select your plywood for you. Remember, you can get whatever you want. Don't settle for whatever you get.

Finn Ply HVHC Finn Ply, a product of Finland, is made from the same trees from the same forest as Baltic Birch from Russia. It is identical to Baltic Birch in every way except its batch quality: every batch of Finn Ply is consistently perfect. Every sheet off every truck load is manufactured to this same high standard. It too comes in 60 by 60-inch sheets. Good material and highly recommended.

ApplePly HVHC ApplePly is a brand name and has absolutely nothing to do with the wood species used in the plies. It is one of the best solid hardwood plywoods on the market for telescope making. As with the other HVHC plywoods, the laminations of Apple Ply are 1⁄16-inch thick hardwood. But instead of birch for the core veneers, Apple Ply uses Western Red Alder cut from the coastal areas of the Pacific Northwest. Because the core veneers are alder, ApplePly weighs less than Baltic Birch and Finn Ply. The result is a solid hardwood plywood with an outstanding strength-to-weight ratio—the perfect choice for a high quality portable telescope.

What's great about ApplePly is that its face veneers are as smooth as a 1⁄20-wave primary mirror. Okay, that's an exaggeration—but this plywood is incredibly smooth and beautiful. No sanding is needed. The veneers are AAA grade. Not only are the face veneers of the highest quality, you can select from seven different species of them. Maple is the most common, but you can special-order oak, birch, pecan, hickory, ash and pine. Best of all, it's sold in 4 by 8-foot sheets.

A.2.2 Hardwood Veneer Softwood Core Plywoods (HVSC)

When someone tells you they built their telescope or stereo cabinet out of birch, oak, or maple plywood, don't believe it. Plywoods sold as such may have surface veneers made from a hardwood species, but the core veneers are a poor grade C softwood. HVSC may look pretty but it doesn't have the structural strength or end-grain strength that the solid HVHC plywoods have. When you cut into HVSC, you'll find numerous voids that always seem to end up in the most unsightly places, like all around the rings in the secondary cage.

In the last few years, the core veneers in HVSC have become more and more lousy. Legislation protecting endangered animal species, huge foreign demand for U.S. wood, and continuous domestic development have reduced the number of trees available for harvest. To stay competitive and save money, the mills use whatever they can get to fill the space between the thin hardwood face veneers. Traditionally, one of the fir species was used for this, but now luan mahogany from the tropics is fast becoming the wood of choice for the inner cores. Luan is a softwood, roughly equivalent in strength to strong cardboard. Luan is the Spam of plywoods.

If you are on a budget, HVSC plywoods perform adequately, but be prepared to fill a lot of edge voids with wood putty. Also, to minimize splitting, predrill all nail and screw holes that are close to the edge of the piece.

A.2.3 Softwood Veneer Softwood Core Plywoods (SVSC)

SVSC is what comes to mind when you think of plywood. Walk into any lumber yard and ask the clerk where the plywood is located and he will direct you to stacks and stacks of this yellowish stuff. All the face veneers and inner core plies are made from the same species of softwood.

The better quality SVSC plywoods are made from various species of fir. Four by eight foot sheets of softwood plywood can be obtained with A-B or A-C face veneers. The inner core veneers are all C or D grade, with the ubiquitous "holes" characteristic of grade C and D plies. Don't sell softwood plywoods short. These plywoods are strong, inexpensive, and can be used to build nearly everything imaginable.

Softwoods like Douglas fir, for example, have a high modulus of elasticity and can provide all the strength or stiffness you need at about two thirds the weight of a typical hardwood plywood. Use SVSC where you aren't concerned about the low surface hardness of the face veneers, the cosmetically poor face grade, and voids and holes in the end grain. SVSC is perfect for the ground board and rocker bottom in a big Dobsonian telescope.

Marine SVSC Plywood This is a hybrid. It is a softwood veneer softwood core plywood designed for the boat industry. It is a solid *softwood* plywood. As with hardwood plywoods, the face veneers and the inner core veneers of marine plywood are all $1/16$-inch thick and come from a softwood species such as Douglas fir, bonded with a water-resistant exterior grade glue. The plies are typically grade B, so there are no voids in the core and you won't find holes when you saw into it.

Marine plywood is tough stuff with a high modulus of elasticity, and it is the least expensive of the multiple veneer core plywoods. It is an excellent choice for someone wanting a high strength-to-weight ratio. Because the face veneers as well as the core veneer are all softwood plywood, marine plywood is a good choice for any component that doesn't get dinged a lot or where appearance is not critical.

A.2.4 Particle Board

Blend sawdust with glue and what do you get? Particle board. Softwood fibers (particles) are bonded under heat and pressure to form a panel. The bonding agent is either a phenolic resin or urea formaldehyde. Pound for pound, phenolic resins provide superior strength properties, allow for a tighter, smoother finished edge, and are much more water resistant than urea formaldehyde resins. Unsealed urea formaldehyde resin particle board swells up and dissolves over time if it is exposed to moisture. If you must use particle board in a telescope project and you observe in a humid environment, then make sure you get phenolic particle board and seal it thoroughly.

Unless you are really strapped for cash, we do not recommend building a telescope from particle board. The savings over plywood is not worth the extra weight of the particle board, and particle board is a real hassle to assemble. Nails and screws pull out easily, and it doesn't hold up to knocks and bumps as well as softwood plywood. In addition, if particle board gets wet, it gets soft and dissolves. Yuk!

Table A.2
Mechanical Properties of Wood

Group	Species	Specific Gravity	Modulus of Elasticity	Surface Hardness
HARDWOOD SPECIES				
Alder	Red	0.41	1.38	590
Ash	Black	0.49	1.60	850
	Blue	0.58	1.40	
	Green	0.56	1.66	1200
	Oregon	0.55	1.36	1160
	White	0.60	1.74	1320
Aspen	Bigtooth	0.39	1.43	
	Quaking	0.38	1.18	350
Basswood	American	0.37	1.46	410
Beech	American	0.64	1.72	1300
Birch	Paper	0.55	1.59	910
	Sweet	0.65	2.17	1470
	Yellow	0.62	2.01	1260
Butternut		0.38	1.18	490
Cherry	Black	0.50	1.49	950
Chestnut	American	0.43	1.23	540
Cottonwood	Balsam Poplar	0.34	1.10	
	Black	0.35	1.27	350
	Eastern	0.40	1.37	430
Elm	American	0.50	1.34	830
	Rock	0.63	1.54	1320
	Slippery	0.53	1.49	860
	Hackberry	0.53	1.19	880
Hickory, pecan	Bitternut	0.66	1.79	
	Nutmeg	0.60	1.70	
	Pecan	0.66	1.78	1820
Hickory, true	Mockernut	0.72	2.22	
	Pignut	0.75	2.26	
	Shagbark	0.72	2.16	
	Shellbark	0.69	1.89	1390
Locust	Black	0.69	2.05	1700
Magnolia	Cucumbertree	0.48	1.82	700
	Southern	0.50	1.40	1020
Maple	Bigleaf	0.48	1.45	850
	Black	0.57	1.62	1180
	Red	0.54	1.64	950
	Silver	0.47	1.14	700
	Sugar	0.63	1.83	1450
Red Oak	Black	0.61	1.64	1210
	Cherrybark	0.68	2.28	1480
	Laurel	0.63	1.69	1210
	Northern Red	0.63	1.82	1290
	Pin	0.63	1.73	1510
	Scarlet	0.67	1.91	1400

Table A.2
Mechanical Properties of Wood

Group	Species	Specific Gravity	Modulus of Elasticity	Surface Hardness
	Southern Red	0.59	1.49	1060
	Water	0.63	2.02	1190
	Willow	0.69	1.90	1460
White Oak	Burr	0.64	1.03	1370
	Chestnut	0.66	1.59	1130
	Live	0.88	1.98	
	Overcup	0.63	1.42	1190
	Post	0.67	1.51	1360
	Swamp Chestnut	0.67	1.77	1240
	Swamp White	0.72	2.05	1620
	White	0.68	1.78	1360
Sassafrass		0.46	1.12	
Sweetgum		0.52	1.64	850
Sycamore		0.49	1.42	770
Tupelo	Black	0.50	1.20	810
	Water	0.50	1.20	880
Walnut	Black	0.55	10.7	1010
Willow	Black	0.39–1.01	8.8	
Yellow Poplar		0.42	1.58	540
SOFTWOOD SPECIES				
Bald Cypress		0.46	1.44	510
Cedar	Alaska	0.44	1.42	580
	Atlantic	0.32	0.93	350
	Eastern Red Cedar	0.47	0.88	900
	Incense	0.37	1.04	470
	Northern White	0.31	0.80	320
	Port-Orford	0.43	1.70	630
	Western Red Cedar	0.32	1.11	350
Douglas Fir	Coast	0.48	1.95	710
	Interior West	0.50	1.83	660
	Interior North	0.48	1.79	600
	Interior South	0.46	1.49	510
Fir	Balsam	0.35	1.45	400
	California Red	0.38	1.50	500
	Grand	0.37	1.57	490
	Noble	0.39	1.72	410
	PacificSilver	0.43	1.76	430
	Subalpine	0.32	1.29	350
	White	0.39	1.50	480
Hemlock	Eastern	0.40	1.20	500
	Mountain	0.45	1.35	680
	Western	0.45	1.33	540
Larch	Western	0.52	1.87	830
Pine	Eastern White	0.35	1.24	380
	Jack	0.43	1.35	570

Table A.2
Mechanical Properties of Wood

Group	Species	Specific Gravity	Modulus of Elasticity	Surface Hardness
	Loblolly	0.51	1.79	690
	Lodgepole	0.41	1.34	480
	Longleaf	0.59	1.98	870
	Pitch	0.52	1.43	
	Pond	0.56	1.75	
	Ponderosa	0.40	1.29	460
	Red	0.46	1.63	560
	Sand	0.48	1.41	
	Shortleaf	0.51	1.75	690
	Slash	0.59	1.98	
	Spruce	0.44	1.23	660
	Sugar	0.36	1.19	380
	Western White	0.38	1.46	420
Redwood	Old-growth	0.40	1.34	480
	Young-growth	0.35	1.10	420
Spruce	Black	0.42	1.61	520
	Englemann	0.35	1.30	390
	Red	0.40	1.61	490
	Sitka	0.40	1.57	510
	White	0.36	1.43	480
Tamarack		0.53	1.64	590
IMPORTED WOOD SPECIES				
Balsa		0.16	0.49	
Bulletwood		0.85	3.45	3190
Luan	Dark Red	0.46	1.77	780
	Light Red	0.34	1.23	460
Mahogany	Honduran	0.45	1.50	800
Rosewood	Brazilian	0.80	1.88	2720
	Indian	0.75	1.78	3170
Teak		0.55	1.55	1000

We have included the relative values for specific gravity, modulus of elasticity (i.e. stiffness), and surface hardness for many types of wood for those who want to try other varieties of hardwood. The actual units of measurement don't matter. The numbers should only be used to compare the relative properties of different woods to one another.

Hardwoods are much harder, stiffer, and denser (higher specific gravity) than softwoods. Now compare different species within a each group. For example, sugar maple is stiffer, harder, and denser than any other maple. The mechanical properties of the maples are considerably better than the aspens which are another hardwood species often used in plywood. A maple board is nearly twice as dense, a lot stiffer, and three times harder. Now you know why bowling alleys prefer maple for the lanes.

It is clear that there is a vast range in mechanical properties between different species, and even between the location of growth. When purchasing lumber and plywood you should ask the dealer what type of wood you are getting and if it's been adequately dried. Know what you are buying. Wood that is dried to 15% moisture content will assemble easily and remain dimensionally stable. A telescope made from "wet" wood will buckle and warp. At the lumber yard you can buy whatever you want. Don't settle for whatever for you get.

* Table from the *Wood Handbook: Wood as an Engineering Material*. Forest Products Laboratory, 1987. U.S. Dept. of Agriculture.

Finally, particle board develops a stress memory over time. Bend a piece of particle board and it stays permanently bent even after the load is removed. Although it acts elastic in the short term, particle board does not recover from long-term deformation.

As you head off for a good lumber yard, you'll find more than thirty different species of hardwood veneer plywoods and nearly a dozen different species of softwood plywoods on the market. The selection is vast. You can build your telescope to whatever stiffness and appearance you desire. Many of these hardwood veneer plywoods will make excellent components. A cherry plywood mirror box and rocker with maple split blocks would look very attractive.

Appendix B
Grinding, Polishing, and Figuring Large, Thin Mirrors

By Robert Kestner

If you have made telescope mirrors before, this appendix should help you apply techniques you already know to the problems particular to making big, thin mirrors. The techniques described work well for mirrors of 16 inches aperture and larger. However, if you have never made a mirror before, don't start with a large, thin mirror. It's a much bigger job than you realize. What you will learn from first making several small mirrors is directly applicable to making a big, thin mirror, and it will serve you well.

There are many different ways of making telescope mirrors. The methods I describe are those I used years ago when I was an amateur astronomer working at home, grinding and polishing entirely by hand. When I offer information on working mirrors by machine, it's for your information, and is not essential to making a successful mirror by hand.

B.1 Obtaining a Mirror Blank

For the purpose of large amateur telescopes, thin primary mirrors are advantageous. They are less expensive than thick ones, and they are lighter (more important than one might think at first), and they attain thermal equilibrium more rapidly.

In the past, amateur astronomers and telescope makers believed that the glass for a telescope primary mirror had to be at least one-sixth as thick as its diameter. Nowadays, however, most large primary mirrors have thickness-to-diameter ratios of 1:12, 1:16, or even 1:20, and they still provide excellent optical performance. Using the techniques set forth here, thin mirrors aren't much more difficult to make than full-thickness ones.

Obtaining glass is often a problem. The classic Dobsonian solution was a plate-glass porthole. Today's choice is a low-expansion glass like Pyrex. However, only a few companies want to supply thin Pyrex blanks to amateurs. You may find it necessary to buy a rough-cut blank from a glass company that is not used to selling one-of quantities over the counter, although this has been changing for the better. Exercise tact and understanding in dealing with these places. Outline what you do need as well as what you don't need from them. Remember to tell them what it's for. People love telescopes; promise them a look through it, and they will bend over backward to help you.

B.1.1 Porthole Glass: the Classic Choice

The mirrors in John Dobson's classical sidewalk telescopes were made with plate-glass portholes from large ships. There are still a few places where you can obtain Navy-surplus plate glass portholes. Start with surplus stores in seaport cities. Portholes of 16-inch and 18-inch diameter made of 1-inch thick glass are still out there for prices ranging from $10 to $300. A reasonable price is whatever you're willing to pay.

Plate glass has a coefficient of thermal expansion three times greater than does Pyrex. It should be worse for telescope mirrors than Pyrex. Yet when a plate glass mirror is used in a solid tube with a closed back, as is the case in the classic Dobsonian, after initial equilibrium in the evening, temperature change in the mirror box is quite slow. The telescopes that I've observed with most have plate glass mirrors mounted in closed-back tubes, and I can testify that the figure change after the mirror reaches equilibration is quite small. As a result, the performance of a plate-glass mirror can be quite good. Perhaps more important, however, is that portholes come round and 99% of them have surfaces flat enough to start grinding the curve right off, so making the mirror is far easier. Without porthole glass, the original sidewalk telescopes would have been far more difficult to construct. That's the good news.

Now the bad news: Plate glass is more difficult to figure than Pyrex, and the strain in portholes is sometimes a problem. Figuring plate glass is more time-consuming and more tedious than figuring Pyrex. After you attack a plate glass mirror with a warm pitch lap, it's several hours before you can tell anything about the progress you've made figuring. That means when you're doing the final figuring, you'll wait three hours before you can test the mirror to see what needs to be done next. Considering the time it takes to get going again, it can take all day to work the mirror twice. If you plan on making a porthole mirror, don't let this stop you: just recognize it is plate glass, and allow for it.

In today's Kriege-style Dobsonians, where the back of the mirror is directly exposed to the air, temperature changes in the air affect the mirror rapidly and directly. Under such conditions, I have been annoyed with the changing figure of plate glass mirrors. For modern Dobsonians, Pyrex is a must.

B.1.2 Pyrex Sheet Glass

Pyrex sheet glass is available from companies which do not normally sell to amateurs. Corning makes Pyrex in many different thicknesses; the thickest currently available is 2⅛ inches. As the world market for low-expansion glass grows, man-

Fig. B.1 *Yes, it's big! Dan Bakken is here shown with a 41-inch f/4 mirror that he ground, figured, and polished by hand—with help from friends. The mirror is used in a big truss tube telescope that Dan calls "Hercules."*

ufacturers in Russia, Japan, and Europe may begin to offer low-expansion sheet glass and blanks suitable for mirror making. Therefore, when I refer to Pyrex, please understand that I mean Pyrex or any equivalent low-expansion glass. At present, Corning is still the major supplier in the United States.

The manufacturers sell sheet glass to their distributors in large squares. If you can't find Pyrex sheet glass locally, get quotes from United Lens in Anaheim California or from United Lens in Southbridge, Massachusetts. Also search the current crop of astronomical magazines for amateur astronomers and specialty firms selling sheet glass and mirror blanks. Optical shops that make large mirrors may also be willing to sell you a suitable piece of glass, or even a blank cut to the proper size and shape for the mirror.

Pyrex is available in coarse annealed and fine annealed grades. I recommend fine annealed for telescope mirrors. Coarse-annealed glass might suffice for a telescope mirror, but I've never tried it. It's probably a hit-or-miss thing—most blanks would work but a few would fail because of stresses locked in the glass.

The asking price for blanks cut from sheet glass varies quite a bit. For a piece 16 inches in diameter by 1⅝ inches thick expect to pay at least $225, and for a piece 24 inches in diameter by 1⅝ inches thick about $500, but this could change.

When you order your glass, specify the rough-round diameter and the thickness. The company will cut a square piece big enough to yield the diameter you want. They cut off the corners several times until the piece is round. What you'll get is a glass blank with 16 to 60 sides with the surfaces only roughly flat—to less than ⅛-inch. Insist they cut lots of sides—you don't want just eight.

Sometimes a rough-round blank turns out to be longer across one diameter than the others by more than ¼ inch. Usually the suppliers scribe a circle on the glass, then cut up to it. If they do a sloppy job, the blank could be out of round. Although I have no solid evidence, I would worry that an out-of-round mirror could be astigmatic. Specify that you want the blank as round as they can reasonably make it.

Once you have such a rough-round blank, you'll need to contrive a way to grind it round and flat. Grinding it round is not very difficult, especially if your piece has enough sides—it's just a matter of spending time to grind off the high spots. I use a hand-size piece of tile and #80 grit. Grinding by hand against the outer edge of the glass, I've made a 16-inch blank with 32 sides presentable in 2 or 3 hours.

Preparing the front and back surfaces is not as easy. You must grind both flat—especially the back. This can take 20 or 30 hours per surface on a 16-inch if you're working by hand, but the grinding time may be much less if the blank comes with smooth flat or nearly flat faces.

If you can, buy the glass from a company willing to diamond-generate the blank. They can Blanchard grind the back flat, edge it round, and generate the radius. For a 16-inch blank, this will cost from $100 to $200. Work of this sort is usually done at your risk; if the glass breaks, you pay anyway. Although complete failure of a disk is rare, do not be surprised if you find a few chips.

B.1.3 Choose the Mirror Diameter

I recommend that you start your big, thin mirror career with a 16-inch. By keeping your ambitions relatively modest, your chances of success are much higher than if you start with a mirror over 20 inches diameter. Even though I have made amateur telescope mirrors twice as big as 16 inches, my observing friends and I agree that we could live the rest of our lives observing with a 16-inch without the slightest regret.

In choosing a blank size, you should think of the outer ¼ inch of the mirror as lost to a turned-down edge. You won't be alone in this—most big mirrors have a clear aperture at least ½ inch smaller than the blank diameter. If you find a 16-inch porthole, think of it as a finished 15.5-inch mirror. If you're ordering Pyrex, add ½ inch to the aperture of the mirror you want.

B.1.4 Choose the Mirror Thickness

The thickness you choose, if you have a choice, is determined by the diameter of the mirror and how you plan to mount it in the telescope. For mirrors under 19 inches, use glass at least 1 inch thick. If you're buying Pyrex, use the thickest sheet Pyrex that you can get, that is, 1⅝ inches thick. For mirrors in the 20-inch to 25-inch range, I strongly recommend 2⅛-inch thick Pyrex.

Some years ago, a friend and I made a 25½-inch mirror on a 1⅜-inch thick Pyrex blank. The mirror had a relatively shallow *f/*6 curve. We paid strict attention to preventing flexure every step of the way, especially during testing when the mirror hung in a vertical sling. I remember wishing that we had the extra ¼ inch thickness of a 1⅝-blank. When the mirror turned out well, it seemed good to have an extra-thin 25½-inch mirror that could be lifted with little trouble.

Especially for a 16-inch or 18-inch mirror, 1⅝-inch thick Pyrex allows you a lot more latitude for error. For mirrors around 30-inches in diameter, working by hand, use 2⅛-inch thick Pyrex. If you plan on working your mirror with a machine, I strongly recommend that you search for a thicker blank. In machine-grinding, you will run into problems that require a stiffer blank than does hand-grinding.

B.1.5 Choose the Focal Ratio

After the aperture, choose the focal ratio. From the standpoint of an optician, the most important factor is that a long-focus mirror is easier to make than a short-focus one because it departs less from a sphere. From the observer's standpoint, long focal lengths make big telescopes too big! Unless you have an army of people to help you set up, do not let the focal length get out of hand. Don't minimize the problems of simply using a telescope longer than ten or twelve feet—observers spend a lot of time clambering up and down the ladder.

However short-focus mirrors—or at least good-quality mirrors with short focal lengths—are very difficult to figure. As a practical consideration, the deep curve of the short-focus mirror removes valuable thickness from your glass.

My recommendation is to play fairly conservative, and choose a focal ratio between *f/*5 and *f/*6. For most observers, *f/*5 is probably the best choice.

B.2 Grinding Tools

In addition to finding a blank for the mirror, you will need several more disks as grinding tools. They may be solid glass tool or plaster tool covered with hard ceramic tiles. Your taste, energy, and what you can get your hands on usually dictate the type of tool you use.

B.2.1 Solid and Built-up Tools

You can cut a solid glass tool from sheet glass, or build it up by laminating plate glass disks from ¼ to ½ inch thick. Buy plate glass circles from a retail window glass company. For the face of your tool you need a piece thick enough not to be ground through during rough grinding. If you grind through the top layer, the gap between the layers may hold stray grains of abrasive and increase the chances of getting scratches during fine grinding.

The tool need not be as large as the mirror. A tool that is ¾ of the mirror's diameter—a subdiameter tool—works well and you will be able to use it without much difficulty. Do not use a tool smaller than ¾ diameter because these have less tendency to produce a spherical curve during fine grinding. Although a 16-inch tool works well with a 16-inch mirror, for mirrors larger than this, I recommend a subdiameter tool simply because full-size tools are heavy and awkward.

Make the tool thick enough that it does not bend while you are grinding. A tool that is too thin presents two problems. First, if any astigmatism occurs in the mirror, it will not completely grind out because the tool will flex to conform to the astigmatic contour of the mirror. Second, the weight of your hands on the back of the tool will force the center of the tool to grind harder on the center of the mirror, generating a curve that is deeper than spherical. This is seldom a severe problem, but it means you will spend extra time polishing out the center.

How thin is "too thin?" With a 16-inch solid glass tool, troubles start with an edge thickness of about ½ inch. A friend of mine used a 16½-inch glass tool only slightly thicker than this and made a good mirror. However, I used a 20-inch glass tool that was ⅝ inch thick at the edge, and I got a bad case of astigmatism. I ended up pitch-blocking a ¾-inch-thick, 16-inch diameter porthole to the back of the tool to stiffen it, and then it worked well. The glass for a 20-inch tool should be at least ¾ inches thick, and larger solid glass tools should be proportionately thicker.

To make a tool for grinding the back of the mirror blank flat, glue several layers of ¼-inch plate together. If you grind through the first layer, just glue another layer to the tool. For gluing the layers together, aquarium cement is readily available and it sticks well to glass.

B.2.2 Segmented Tools

For mirrors with the curve already generated, a ceramic tile grinder is the answer. Tile tools are made by blocking ceramic tiles onto a solid support. The support can be aluminum, glass, or plaster. Segmented tools can be made flat or curved to fit a generated blank. Whatever type of support you use, it is important that the curve of the tool match that of the mirror. In this respect, plaster is most practical because the cast plaster surface matches the curve of the mirror.

Almost any unglazed ceramic tile will do for the grinding surface. They should be around 1 to 2 inches square or round, and of uniform thickness. Attach them with epoxy. It helps if you can buy an epoxy softening agent to make it less brittle. Epoxy or hard pitch is used to glue the tiles in place. Epoxy is the better choice; the problem with pitch is that the tiles may fall off in rough grinding.

To make a plaster support, wrap a metal or cardboard dam a couple of inches high around the mirror. Smear soap over the surface of the glass to keep the plaster from sticking, then pour on the plaster. If the mirror has a curve, the plaster will match the curve. The plaster should be at least twice as thick as the equivalent solid glass tool.

I highly recommend Kerr Dental Plaster, Vel-Mix Stone Pink, for making plaster tools. This stuff hardens like stone! See **Appendix E**. Add water to the dry plaster and stir the mixture until it is slightly thicker than cream. Allow the plaster to dry thoroughly before proceeding. Before applying tiles, give the disk a thin coat of epoxy to protect it from penetration by water.

To epoxy the tiles to the support, start by smearing a release agent on the mirror so that excess epoxy will not stick. (You should buy the epoxy manufacturer's release agent—but I suspect grease or soap would work just as well.) Place the tiles in the desired pattern on the mirror. Space them at least ¼ inch apart. Pick up the tiles one by one, spread a thick coat of epoxy on them, and set them back on the mirror surface epoxy side up. It is a good idea to get a friend to help you, because you must coat the last tiles with epoxy before the first ones set. After all the tiles have been coated, carefully lower the support onto the tiles and let the epoxy cure.

This is a very general discussion of making tools. Experiment with the materials and use your smarts to keep you out of trouble.

B.3 Prepare a Suitable Work Area

Ideally you should have two work areas—one for grinding and another for fine grinding and polishing. Because they require cleanliness and temperature uniformity, fine grinding and polishing are best done indoors. Working with abrasives larger than 30 microns is messy so, if the climate and the time of year permit, rough grinding can be done outdoors.

My favorite way to work mirrors by hand is with the mirror face up on top of a 55-gallon drum weighted with 300 pounds of sandbags. You can buy drums at salvage yards, and sandbags can be bought at nurseries. Make a plywood top for the drum and add cleats to hold the mirror in place. A reasonable substitute for a barrel is a sturdy counter—remember, you don't have to walk around the barrel if you rotate the glass regularly.

Another substitute for a barrel is so obvious it's usually overlooked: the floor. I have ground and polished mirrors on my knees with the mirror on a sheet of plastic on a thick carpet on the floor. The grinding goes well and the only trouble is figuring out how to walk again after 30 minutes of polishing on your knees. Tool companies sell strap-on rubber pads that protect your knees from bruising when you kneel—these are really helpful.

B.4 Preparing for Grinding

Before you begin rough grinding, the mirror blank must be faced on both front and back surfaces. The idea is to work both surfaces into nearly flat figures of revolution. To prevent chipping, bevel the edge of the glass on both the front and back faces before you begin. During grinding, you must rotate the blank and the tool frequently to prevent astigmatism.

B.4.1 Facing the Blank

To face the blank, place the mirror face up on a deep, soft support such as an old piece of shag carpet. The carpet will get wet and be ruined. Wet the mirror face, add a few teaspoons of #60 Carborundum, place a weight on the back of the tool if you wish, and grind like mad. When the grit stops making lots of noise, splash it off, rotate the mirror ⅓ of a turn, and start again. After three or four wets, wash it off and look at the surface of the mirror. The high spots will be ground and the low spots will be untouched. This will give you some idea of what you're up against. Look at the tool to see how it started.

At this point, you should decide which face of the blank is to be the front and which the back. The front surface is usually the one that will grind out to the very edge with the least amount of work.

On the front side, grind with #60 Carborundum until the ground area reaches the edge of the blank. If any low areas remain at the edge, they could be left unground when you generate the curve. The work of grinding the curve takes place in the middle of the mirror, so if the edges are not in contact at the start of grinding, they will not be in contact at its end. Grossly high and low areas will cost you an enormous amount of work because you must grind the entire surface down to meet them.

At this stage, however, the back of the blank is the more important side. The reason you are grinding is to ensure that the back is completely free of astigmatism. If the back is not flat, the blank will flex preferentially along one axis and it will be difficult or impossible to avoid astigmatism on the mirror surface. You can tolerate a few isolated low spots on the back if they are small, but you absolutely cannot tolerate a blank with a cylindrical back.

The key to avoiding astigmatism is rotating the blank on its support frequently. If you rotate the blank with every wet and the tool with every stroke, the two surfaces can do nothing but become long-radius spheres or flats. As you continue grinding, keep track of the convexity and concavity of both surfaces with a straightedge. If the mirror starts going convex, concentrate grinding on the center with short strokes. If the convexity persists, grind with the mirror on top for a while. You don't want the back to become convex, but it's okay if it goes a few thousandths on an inch concave. When you actually come to do it, you'll find it's not a difficult thing to manage.

When you're satisfied with the flatness of the back, fine grind it through #120 and #220 grits. When you start grinding with the fine abrasives, low areas will have coarser pits on them and you'll have no trouble recognizing them. Even if you had the back Blanchard ground, grind it flat with #220. Blanchard grinding may give a perfectly plane surface—but you're better off safe than sorry.

B.4.2 How to Prevent Astigmatism

To prevent astigmatism, you must support the mirror properly. The best way for an amateur working at home is to support the mirror on a piece of deep shag carpet and rotate it frequently. The carpet supports the mirror and allows you to rotate it with no trouble. When the carpet loses its resilience, replace it with a new piece. The carpet goes between the mirror and the work surface which must also be flat. Plywood or particle board is flat enough, but watch for warping when the wood becomes wet.

Rotate the mirror on the carpet frequently while working. This prevents non-uniformities in the support from warping the mirror and showing up as astigmatism. Every time you complete one turn around the barrel, rotate the mirror about a third of a turn in the opposite direction. Randomly vary the amount of rotation you give it. If you don't have a barrel, then rotate the mirror almost constantly in the direction opposite the direction you are rotating the tool. As simple as it sounds, this method is quite effective.

With a machine, the mirror usually cannot be rotated on its support frequently enough. This is why you want mirrors worked on machines to be thicker than hand-worked mirrors. An 18-inch disk of 1⅝-inch thick glass is the thinnest I would recommend for machine grinding.

There are several ways to support the mirror on a grinding and polishing machine turntable, but first make sure the turntable itself is rigid. If your machine does not have one, make a rigid turntable by casting a 5-inch thick disk of Kerr dental plaster.

You can block the mirror directly to the turntable with pitch. This is somewhat risky because the pitch can deform the mirror. A more manageable way is to pour pitch on the base ⅜ inch thick. Groove it like a pitch lap with spaces about ½ inch wide. Cover it with a single piece of paper, and set the mirror on the paper and tape it down. Be careful not to tape it too tight. Let the mirror sit on this lap for 24 hours. The back of the mirror should press the pitch to its exact shape. Each time you take the mirror off for testing, put it back in the same orientation and let it sit again for a few hours. Do not let the grooves in the lap under the mirror press together. When they start closing in, regroove them. Although this method is not without its troubles, it is widely used in the optical industry.

Another method is an 18-point flotation system. I've never used this method for support while working a mirror, but there are those who swear by it. Preventing the mirror from rocking under the lateral forces of grinding and polishing presents a tough mechanical problem.

B.4.3 Bevel the Edge

In all optical work, it is important to maintain a bevel on the edge of the mirror at all times. If the edge becomes sharp, a slight tap can knock off a large conchoidal chip. A smooth 45° bevel about ⅛-inch wide prevents this.

An easy technique for beveling the edge is to stroke a double-sided coarse/fine Carborundum stone, the sort used for sharpening wood chisels, against it. Dip the stone in a bucket of water, hold the coarse side at a 45° angle to the face, and stroke gently down and and away from the edge. Continue until the bevel is about

⅛ inch wide. Always work wet. As the bevel widens, you can use more pressure and a longer, firmer stroke, but during the initial stages of forming the bevel, keep the pressure light to prevent chips. When the bevel is fully formed, smooth it using the fine side of the stone.

Another technique is to grind with a loose slurry of #80 Carborundum using a hand-held piece of sheet iron 3 inches wide and 6 inches long. After the bevel has been shaped with #80, smooth it with #120.

During rough grinding, check the bevel frequently, and if it becomes narrower than ³⁄₃₂ inch, restore it to its original width.

Later, during fine grinding, continue to monitor and keep the edge properly beveled. Because it can now serve as a hiding place for coarse grit and glass chips, after I finish #220, I fine grind the bevel through 12-micron grit. This prevents scratches during the later stages of fine grinding and polishing.

B.4.4 Grinding Strokes

During grinding, you will normally place the mirror on the bottom and move the tool back and forth across it. In the normal, or center-over-center, stroke, the tool extends about ¼ of its diameter over the edge of the mirror at the end of each stroke.

When I refer to the W stroke, I mean any stroke that goes forward and backward while progressing left to right or right to left on the mirror. Take four or five back and forth strokes while the center of the tool moves from one side of the mirror to the other. In a long W stroke, the center of the tool nearly reaches the edge of the mirror, and in a short W stroke, the back and forth motion is just 1 or 2 inches. By "short stroke," I mean a stroke with the tool on top that hangs over front and back, left and right only 1 or 2 inches.

Although it does not matter during rough grinding, in fine grinding it is a good idea to vary the stroke length randomly by a small amount to avoid digging a zone or groove where the strokes start and stop.

B.5 Rough Grinding

Once you have ground the back of the mirror flat through #220 abrasive and the face is ground to the edge, you're ready to start roughing the curve into the mirror. I recommend #60 Carborundum for rough grinding.

If the mirror is light enough to grind on top, grind much the same way you would a smaller mirror, with the mirror on top. Your goal is to grind the curve from the middle out, timing it so that you reach the desired focal length at about the time the curve reaches the edge of the disk. This method removes the least amount of glass.

Start grinding by concentrating the center of the mirror on the edge of the tool. Use a long W stroke, occasionally stroking the center of the mirror out to 2 to 3 inches from the edge of the tool, progressing around the circumference of the tool rotating the mirror.

If the mirror is too heavy to handle on top, grind with a tool that is ¾ of the mirror diameter or smaller. Rough grind the curve by concentrating on the center. John Dobson says, "Rough grinding is a caveman job, so do it like a caveman. Eat well, sleep well, and work like hell." Rough grinding a large mirror by hand is a lot of work.

As grinding progresses, monitor the focal length of your curve. On a sunny day, splash water on the mirror, and then focus an image of the sun on a piece of cardboard and measure the distance to it. Alternatively, you can keep track of the sagitta of the curve. Calculate the sagitta of the mirror for the focal length you want. Find something that can act as a gauge equal to the sagitta of the mirror. The time-honored gauge is a drill bit. Check the progress of rough grinding by slipping the gauge under a straightedge spanning the mirror.

Rough grinding produces a grossly hyperbolic curve. You cannot leave the mirror grossly hyperbolic because it makes fine grinding difficult. Therefore, as the curve nears the desired depth and reaches the edge, take shorter strokes to produce a smoother, more spherical curve. Short strokes also help move the curve toward the edge. Placing the mirror face up and grinding with a moderately long W stroke moves the curve toward the edge, and also tends to produce a more spherical surface. If you have timed it right, a reasonably spherical curve will meet the edge at the same time the mirror reaches the desired focal length.

B.6 Fine Grinding

Everyone has heard stories about fine grinding taking hundreds of hours. Correctly done, however, fine grinding takes about 2 hours of continuous work at each grade, and is, in fact, one of the more manageable jobs in making a large mirror.

Fine grinding rather naturally divides into two stages: up to and including #220, and after #220. Until you have completed #220, concentrate on controlling the radius and getting a smooth curve. After #220, the primary mission is to prevent astigmatism. Switch from your rough grinding area to your polishing area after #220, not before. #220 is still gritty stuff.

The fine grinding compound you start with depends on the condition of the mirror and fit of your tool. If you rough ground the mirror with #60 grit, start fine grinding with #120 grit before going on to #220. If you had the curve generated, you can start with #220 providing the tool makes a reasonable fit to the curve. If you are working with a machine, you can use 30-micron grit.

B.6.1 Fine Abrasives

For abrasives smaller than #120, I prefer aluminum oxide to Carborundum because it has less tendency to cause large pits in the fine-ground surface. There are several sequences of fine abrasives you can take, but as long as you are careful to have ground away the last abrasive pits from the previous grade before going on, you'll be all right.

More than anything else, the abrasive sequence depends on the quality of the abrasive you use. Most abrasives you buy have a few larger abrasive particles that continue to generate new pits even as the old ones are ground away. Although #320 is mostly #320-size grit, for example, there are some grains closer to #220 size grit in it, plus a lot of much smaller stuff. There is nothing wrong with this— but you must compensate for it by taking small steps between abrasive sizes. The standard sequence of #80, #120, #220, #320, #400, #600, and #305 emery goes in rather small steps.

On the other hand, if you use a really high-quality abrasive like Microgrit, you're in luck. 30-micron Microgrit really is 30-micron grit. As a result, you can take much larger steps. For example, with Microgrit, you can follow #220 with 30-micron, 12-micron, and 3-micron or 5-micron grits. The 3-micron Microgrit is considerably finer than the #305 emery.

Depending how you happen to employ the grit, the amount of abrasive you need can be quite small or large. If you clean the glass after each wet and start the new wet by sprinkling new abrasive on the wet glass, you may consume as little as a tablespoon of fine abrasives in fine grinding. If you suspend the abrasive in water and pour it onto the glass, a cup of abrasive may not be enough. For #120 and #220 abrasives, the amount varies mostly with the amount of work you need to do. On the average, I've found 2 to 3 cups of #120 and 1 cup of #220 will do for a 16-inch mirror. If you are purchasing a supply, however, get twice to three times what you estimate you will need. The cost of abrasive is small when you consider the frustration of running out in the middle of fine grinding.

B.6.2 Getting a Sphere

Even the most careful finishing up job with #60 grit leaves the mirror surface somewhat hyperbolic. After #60, your main interest, besides removing the pits, is getting the mirror spherical. At this point, the support of the mirror is not at all critical, but using a piece of carpet under the mirror allows you to rotate it easily.

Start by sprinkling the abrasive on the face of the wet mirror, as you did with smaller mirrors—not too much and not too little. Place the tool down edge first, and start grinding. Walk around the barrel, rotating the tool as you go, and also rotating the mirror the opposite direction every time around.

A solid tool traps abrasive in the center of the mirror and resists grinding the mirror spherical. When you first start a new and smaller abrasive, this shows up as a tendency to form a large bubble in the center, especially in the larger sizes. Remedy this by using short W strokes and stirring the abrasive under the mirror every minute or so until the bubble is gone. To stir up the abrasive, just run the center of the tool clockwise around on the 50% zone of the mirror once or twice while you spin the tool clockwise. You will need to spread the abrasive like this all through fine grinding.

Because they do not trap grit as solid tools do, a segmented tool will generate a spherical curve a good deal more rapidly. With segmented tools, a short W stroke serves well.

Don't let your fingers touch the bevel of the mirror while rotating it during grinding and polishing. Instead, grab it closer to the center of the edge. Touching the bevel heats the glass, causing it to expand. Then you grind or polish the expanded edge off, resulting in a turned edge when the mirror cools.

When the grit ceases to work, stop grinding, separate the mirror and the tool, add more water and grit, and then continue. If you have a segmented tool, you can slop some water and grit on from the side without separating the mirror and tool.

Never drag the tool off on the edge of the mirror—this can depress, or "roll," the edge of the mirror. Instead, pull the tool three-fourths of the way off, then tip the tool slightly and lift it off. With the finer abrasives this can be difficult. A solid tool won't want to separate, and lifting tends to pull the tiles off a tile tool—so take it easy.

As the curves of the mirror and tool approach spherical, you will feel the difference—the motion of the tool over the mirror becomes tighter and more uniform. You're done with #120 when the surface appears evenly ground and no #60 grit pits remain. Don't be fooled by some of the larger pits that #120 can make itself. These will come out in #220.

B.6.3 Going on to #220

As you begin #220, start to pay a little more attention to preventing astigmatism, especially if the tool is thin. Rotate the mirror on a regular basis, and be sure you're grinding it on a soft backing. Keep on mixing the abrasive between the mirror and the tool. In most respects, #220 is the same as #120, except that #220 is your last chance to remove any large pits and the surface *must* be spherical when you finish #220.

To control pits, follow the progress of the ground surface with a loupe magnifier. If you've had your curve generated, inspect the surface carefully for any remaining generator marks. If you do a cursory examination, they may fade into the ground surface only to reappear when you polish.

Remember that #220 can leave some isolated pits a bit larger than the normal #220 pit size. You distinguish these from leftover #120 pits because they will not stay in the same place after a spell of grinding. It is not uncommon for a telescope maker to grind for many extra hours trying to remove these pits when, in fact, they are caused by the abrasive and will be removed by the next grade. However, never let this be an excuse for failing to remove all the pits from the previous abrasive. (If you have any doubts, make a map of the two dozen largest pits and check to see whether they are still present after two further wets.) After a couple of hours of #220, the mirror should be ready for the light test.

This test is a very sensitive gauge of the evenness of the ground surface. Place the mirror face up on a table and arrange a strong light several feet behind it. Stand a couple feet back from the mirror and lower your eye until the angle between the light, the mirror, and your eye is sufficiently large that the surface takes on a reddish shine. The finer-ground the surface, the less extreme the angle.

Study the surface: it should appear evenly lit from edge to edge as you move your head from side to side. If the center of the mirror looks dull or hazy, it has not been fully ground by the latest abrasive. That also suggests that the mirror's curve is not spherical—so continue grinding until it is. This test gives information only on the overall smoothness of the ground surface, and is not a test for isolated pits.

Keeping the edge properly beveled becomes more important as fine grinding progresses. After finishing #220, I like to fine-grind the bevel through 12-micron grit. This prevents scratches in the later stages of fine grinding and polishing. To fine-grind the bevel, obtain a small piece of sheet metal or brass 2 to 3 inches square. Tape the sheet to a block of wood the same size, leaving the metal surface exposed. Start grinding the bevel with a slurry of #220 grit, rounding it off. When the bevel is round with #220, you can continue to fine grind the same way through the finer abrasives, or wait and grind the bevel with smaller abrasives as you use them on the mirror.

When you complete #220, it is a good time to get an accurate reading of the mirror's focal length. The #220 fine-ground surface, when sprayed with water, forms a sharp image of the Sun that can be accurately focused and the focal length measured.

B.6.4 The Fine Side of Fine Grinding

The next abrasive is #320 or 30-micron aluminum oxide. These two grits are about the same size. I prefer to suspend fine abrasives in water, although sprinkling very small amounts of abrasive on, then smearing it around with your wet fingers is also an acceptable method. For suspending abrasives, I use 1 tablespoon of grit to ½ cup of water. This goes a long way when using a solid tool. More of the abrasive mixture may be needed with a segmented tool.

In the finer stages, grind at a slow, steady pace. Be very careful not to let the surfaces get dry. Monitor whether the tool and mirror show a tendency to stick, or if it's hard to push the tool center over center. These symptoms mean that the surfaces are not spherical. Keep the abrasive stirred and use a shorter stroke until you have remedied the situation and reestablished spherical figures.

Conscientiously apply the astigmatism precautions described in **Section B.4.2**. Pay close attention to cleanliness. Wash thoroughly between abrasive grades, and take care that you do not carry grit from one stage of grinding to the next in your clothes.

As the abrasive gets smaller, the danger of getting a scratch from the tool increases. Lower the tool onto the mirror gently and without lateral motion. If the bevel at the edge of the mirror is coarser than the current abrasive, the rough surface can cause scratches and harbor abrasive grains. Fine-grind the bevel as you proceed to finer grits.

Continue keeping track of pits. Also monitor the uniformity of grinding by applying the light test often. It is an effective way to spot slow action at the edge or center of the mirror. In the later stages of fine-grinding, the light test can be done without a special light because room light will be sufficient. With your eye at the correct angle, an evenly illuminated mirror surface means uniform grinding and spherical surfaces. One note of caution: rubbing the mirror with your hand will produce an extra shine at that spot. With fine abrasives, dry the mirror with a clean lint-free towel and do not rub the surface with your hands before performing the light test. Continue checking through the finest abrasive.

If the tool and the mirror suddenly stick together tightly, you must act quickly to separate them. Splash water on the exposed mirror, then place a 2 x 4 on the edge of the tool and hit it hard with another 2 x 4. Be careful the tool does not fly off and land on the floor. Do not use a hammer. Although I have seen hammers used, if you slip and hit the mirror, it will break. If the tool and mirror refuse to separate, soak them in warm water.

Segmented tools seldom stick unless you leave the tool and the mirror together until the water dries off. Never leave the mirror and tool together, and never allow the tool and mirror to run dry during grinding. With segmented tools, a small amount of Ivory dish soap added to the fine abrasive gives a smoother grinding action.

In grinding with abrasives smaller than 9 microns, the edges of the tiles on a segmented tool may seem to dig in and cause scratches. If the problem occurs because the tool warps when it sits unused, then don't give it any time to warp. Go directly from the next-to-last to the last abrasive—just clean up and continue working.

To prevent scratching with the very finest grades, order Microgrit abrasive mixed with talcum. This material is signified with a T after the micron size number, so Microgrit 5T is 5-micron grit with talcum. If you are having scratch problems with 3-micron grit, it is okay to finish the mirror with 5-micron grit. However, don't finish with anything coarser than 6-micron grit or #305 emery.

If the scratching problems persist, try using the abrasive in a thicker slurry, like cream, or add more soap, or both. As a last resort, bevel all the tile edges on your grinder. If nothing works and the scratches are not too numerous, forget them. They will only flaw the mirror cosmetically.

B.7 Preparing to Polish

Polishing is carried out using a tool coated with pitch, a viscous tar-like substance that serves as a carrier for a polishing agent, usually a metal oxide in optical work. During polishing, the pitch deforms into intimate contact with the optical surface to be polished. And because pitch is a viscous liquid, it flows and deforms as you polish, thus allowing you to control the location and speed of the polishing action.

The type, hardness, and behavior of the pitch are important in determining the closeness of the fit between its surface and the mirror, so mirror makers spend a lot of time talking about pitch. A lot has been said about various methods of hardening and softening it, but my own experience is that for a large lap, pitch of almost any reasonable hardness will do the job. I started making mirrors with straight resin pitch tempered with a very small amount of turpentine. These laps were extremely hard. Later I used some very soft pitch laps. They worked well too. Very hard laps can be a nuisance because they tend to act up a lot and begin to cut sporadically. Soft laps cut evenly, but they may lose their shape and need a lot of maintenance. On the whole, you're better off when the pitch is a little on the soft side.

B.7.1 Pitch

The best pitch I know is Swiss black tar pitch, which comes in different hardnesses. I recommend Adolf Miller pitch #73. Oddly enough, this material is less expensive than many of the pitches packaged for sale to amateurs. See **Appendix E**. Making one lap for a 16-inch mirror takes 1½ kilograms of pitch, so for a 16-inch mirror, I would order three kilograms. If you mess up one lap, you'll have extra on hand for another lap or to make smaller laps for figuring.

B.7.2 Polishing Agents

For efficient polishing, choose one of the many cerium-oxide-based polishing agents such as Rhodite 15, which I like for figuring. Don't go nuts trying to find these particular brand names. There are lots of good polishing agents available from many different suppliers sold under many different names. Most of them will be quite suitable to your needs.

The only thing you should avoid are the "super-fast" polishing agents used for metal mirrors. These cut much too fast for a glass mirror. Don't worry about inadvertently purchasing these materials since I do not know of any being sold outside of the optical industry.

B.7.3 Pitch Base

The pitch lap requires a base, that is, something solid to support it on. It can be flat, or it can have the radius of the mirror on it. Traditionally, amateur telescope makers have used the grinding tool as a pitch base. The base may be anything that is thick enough not to flex, including glass, aluminum, or plaster. What you choose is up to you. For polishing 16-inch and 18-inch mirrors, I have found that 1-inch-thick portholes make excellent pitch lap bases. If you try plaster, I recommend Kerr Dental Plaster for its rigidity.

The diameter of the lap, like that of the grinding tool, is to some extent up to you. The closer it is to the diameter of the mirror, the more likely it is to produce a spherical figure. However, it's not always practical to make a full-sized lap. For starters, a full-size base for the lap is not always available. There is also the problem of managing a big lap. Furthermore, the larger the lap, the more work is needed to push it back and forth on the mirror, and the harder it is just to lug around from one place to another.

For mirrors in the 16- to 20-inch range, I recommend a full-size lap unless a suitable pitch base is not available. For mirrors above 20 inches, I recommend using a 75% sub-diameter lap. Laps larger than 20 inches may become too hard to push for one person working by hand, but don't let this stop you if you have a strong build. It is not impossible to push a 24-inch lap by hand. I do not recommend a lap smaller than 75% of the mirror diameter for polishing.

Sub-diameter laps tend to overcorrect the figure of the mirror, that is, they tend to produce a mirror that is spherical in the center and parabolic at the edge. This is a manageable problem—even desirable at times. It may prove to be a new experience to amateur telescope makers who are used to parabolizing from a spherical surface. Some time ago, I polished a 25-inch *f*/6 mirror with an 18.75-inch lap. The figure came out as predicted: the center spherical and the edges parabolic. This was corrected to a full paraboloid quite easily using various sizes of figuring laps.

Figuring laps can be much smaller than 75%, and are covered below.

B.8 Making the Pitch Lap

If you've never done it before, pouring a pitch lap for a big mirror can be quite an adventure. If you are not careful as well as a bit lucky, you may end up doing it all several times before you pour a good lap. Pouring a big lap requires a good sense of timing. Once you have done it, successfully or not, you'll have some feel for the timing, and making the next lap becomes much easier. Do not be overly discouraged if it takes you three tries to make your first successful pitch lap.

Start by heating the pitch on a hot plate. A five-pound metal coffee can is an excellent container to heat the pitch in, and it will hold enough to make a 20-inch lap. Heat pitch slowly. One or two hours is not too long. If you try to heat it in 15, or even 30 minutes, the pitch at the bottom of the can will probably overheat.

While it is heating, stir it now and again. Use this time to clean the pitch base so the pitch will stick firmly to its surface. With masking tape, make a dam at least ½ inch high around the base. Be sure that the tape is strong enough to separate from your base when you strip it. If the tape tears and leaves a residual layer of paper on the lap when you try to strip it off, you'll need to shave the paper away so the edge of the lap is pure pitch.

Prepare the area, too. Place the base on a solid and level surface. Be sure to have a water slurry of polishing agent handy and also a little Ivory dish soap or Windex. A propane torch and matches are useful but optional. It is a good idea to keep a bucket of cool water nearby in case you get hot pitch on your hand. Immerse it in the water and get right back to work.

When the pitch melts, pour a little on something and let it cool to make sure it is as hard or soft as you want it. Before you pour, check that the pitch is hot enough to stick to the base. When it's is the right temperature, it is viscous but will still flow freely.

Smear water, polisher, and two or three drops of soap or a small squirt of Windex on the surface of the mirror. Then light the propane torch and set it safely aside. Start by pouring a layer of pitch on the base to a maximum depth of ¼ inch. This will take about three-fourths of the pitch in the can. Even if the base is convex, the pitch will flow into a more or less uniform layer.

Immediately after the pour, flash the flame of the torch over the surface of the pitch to pop all the bubbles and leave it smooth.

At this point you should have about one-fourth of the pitch still left in the can. Don't set it back on the hot plate: let it cool a little for the next step. Just how cool to let it get is one of the tricky parts of this job. You don't want it so cool that it won't pour from the can, and it can't be too runny or you won't be able to mold it on the lap. When I do the first pour, if the pitch is hotter than I planned, I set the can on the cement floor to cool. If the temperature is right, I set it on a wooden counter so it will hold its heat. As you will learn, the timing of the entire operation depends on the temper and temperature of the pitch.

Monitor the viscosity of the pitch as it cools. Pull the tape off just a little and observe how rapidly the pitch flows. If it does not run at all, you are behind schedule. If the pitch starts running quickly off the base, wait a bit until the pitch becomes more viscous. When the pitch lap has cooled to just the right temperature, pull the tape off the base. The pitch should just run over the edge a small amount.

Immediately after you pull off the tape, slowly pour more of the somewhat-cooled pitch right in the middle of the lap. If everything is at the right temperature, the new pitch will form a convex crown on the lap that will flow toward the edge. If there is time to do so, flash this with the torch to break the bubbles. By the time the crown reaches the edge zones of the lap, it should have cooled enough so it won't stick to the mirror. Pick up the mirror, which should still be wet on the face with polisher and soap, wet its back so you can see through it, and prepare to lower it onto the pitch.

Start lowering with the mirror tilted slightly so that it contacts one edge of the lap first. Then, as you continue lowering, slowly bring it level with the pitch. The purpose of this exercise is to allow the air between the mirror and the surface of the pitch to escape. If you simply plop the mirror flat onto the lap, you will trap a large air bubble in the center. If you end up with a large space in the center, let the pitch cool, then separate the mirror, heat more pitch, and pour another crown. This extra step is often necessary for big laps and fast mirrors.

Move the mirror around on the pitch and keep doing so as the pitch cools. If you stop, it will stick. Keep working until the lap is in contact with the mirror all the way to the edges. If you have a full-size pitch base, some of the pitch will ooze off the base and the mirror may sink into the pitch. Keep moving the mirror around as you push down the excess pitch. Be sure the lap stays in contact with the mirror right to the edge.

When the pitch hardens, you can stop moving the mirror and lift it off. If you lift too soon, the lap will deform. If the lap base is glass, let it cool on its own after you separate it from the mirror. A sudden splash of cold water can break a thick glass disk.

If the mirror you are making is too heavy for you to pick up and turn over, place the pitch lap face down on the mirror instead of pressing the mirror on the lap. Again, bring them together at an angle to avoid trapping an air bubble and deforming the surface of the lap.

If the whole thing fails, you may be able to save it by heating up more pitch and pouring another layer on top of what you have, then pressing the mirror on it. It is a good idea to heat the lap with the torch until it's tacky before pouring on new pitch so the layers will stick together.

B.8.1 Grooving the Lap

To allow the pitch to flow and conform to the mirror surface during polishing, the surface of the lap must be cut into facets. One method for doing this is to cut grooves into the face of the solid lap. For pitch laps under 20 inches, a groove spacing of 1½ inches works very well.

Start by marking the center of the lap with an "x" using a single-sided razor blade. Arrange the squares off-center with respect to the x. With a straightedge, incise parallel lines 1½ inches apart to mark where the grooves will go. Then, using single-sided blades, deepen the cuts into grooves in the pitch. Run cold water on the lap to remove pitch fragments. Make several cuts down one side of a line, then rotate the lap 180° and make more cuts down the other side until the grooves are about ¼ inch wide and deep. Because pitch shatters into tiny fragments, this job is very messy. It is best done outside, in a large sink, or on lots of newspapers. Pitch sticks to everything, and it is very hard to get out of clothing. You have been warned.

B.8.2 Pressing the Lap

Once the lap is grooved, you have to be sure it is "pressed" to conform precisely to the curve of the mirror. If you have a hard lap, warm it slowly in water or in the sun. Wet the mirror with the polishing slurry, then press them together. Put 40 or 50 pounds on top and leave the lap on the mirror for an hour or two. Be sure to keep them wet; if the water dries up, the lap and mirror will stick together. If you have a softer lap, press it without heating it beforehand.

If you plan to stop polishing for a few hours, place a single sheet of waterproof paper on the face of the mirror, and then dry the lap and place it face down on the mirror. Plastic-coated butcher paper works well for this. Be sure the lap is dry or it will wrinkle the paper.

Don't let the lap sit on the mirror for more than a day or two or the squares will start to press out. For periods of pressing between polishing spells in a single day, just remember to keep the mirror wet.

B.9 Polishing

What probably concerns you most is how much time it will take you to polish the mirror. Using the techniques described below, it takes about 10 hours to polish out a 16-inch porthole with a full-size lap, and about 12 hours to polish out a 25-inch Pyrex mirror using an 18¾-inch diameter lap. These times may seem much too short compared to your previous experience or stories that you have heard, but the polishing techniques you will learn below are very efficient and they produce an extremely even cut. Even though the lap friction is high, the strokes are slow. When you are done, the mirror will have an excellent polish with a smooth surface texture.

I do not want to make these polishing procedures seem easy. Polishing requires more physical work than any other stage of mirror making, even more than rough grinding. When I say 10 hours to polish a 16-inch mirror, that's not two afternoons—that's 10 hours actual polishing time. Efficient polishing is so physically demanding that I am only able to polish two or three 15-minute sessions a day.

If you use a machine, you cannot expect such short polishing times. Normal machine polishing techniques are not as efficient as hand polishing. But don't get me wrong about machine polishing: it is the only way to go. I work with polishing machines all day in my profession and precision optical production would be impossible without them. I simply want you to know that it's no disadvantage to make a mirror with your own hands.

To start with, mix your polishing agent with water. The amounts of water and polishing agent can vary quite a bit depending on the brand of polisher you use and how thick you like it. Start with 2 or 3 tablespoons to a cup of water. If you realize it's too thick, you can always add a little water later. After mixing it, strain the mix through cheesecloth to remove lumps.

Use a squirt bottle to apply polisher and water. They come in many shapes and sizes. You can find suitable bottles in sporting goods stores—they are used by distance runners and other athletes—or make your own by punching a small hole in the top of a plastic bottle and shaking it to sprinkle polishing mix on the mirror. A second squirt bottle filled with water is also helpful.

Before starting, "wet press" the pitch lap. Squirt some polishing compound on the mirror so it's nice and wet, then set the lap on the center of the mirror. Load about 40 pounds of weights on the lap and let it press for 15 minutes. Be sure the mirror does not dry. Remove the weights and you're ready to start.

Begin by moving the lap around to stir up the polish on the mirror. Add a little water if the mix feels thick. Move the pitch lap slowly back and forth in your polishing stroke. I recommend the W stroke. Starting with a left overhang, push the lap forward over the front of the glass, then pull it back. Work slowly—each stroke should take one to two seconds. After four or five strokes, you'll be over on the right. Continue the forward and back motion while progressing to the left side. At the same time, walk slowly around the barrel. Rotate the lap every now and then, or continuously, as you prefer.

You can vary the length of the stroke from about 2 inches to 5 or 6 inches, depending on the size of the mirror and lap. With a sub-diameter lap, use strokes of moderate length or you may end up overcorrecting the mirror. With a full-size lap there is little danger of this, and you can use longer strokes.

B.9.1 Judging the "Feel" of the Lap

How well the polishing lap works depends on several things. You can judge its performance by the "feel" of the lap. Most of all, you want the lap to be extremely hard to push. It should move only with a great deal of effort on your part, and then it should move slowly and smoothly across the mirror. If the lap is working right, it should cause blisters on your hands after a while. I recommend wearing leather gloves, or taping the tender areas of your hands, or both. You may have trouble finding gloves that don't slip on the back of the lap. I've found that snug gardening gloves work pretty well even when they're wet.

It is during "smooth-and-tight" polishing that the mirror polishes quickly. When the lap and the mirror are in full contact and you are pushing hard, the action generates heat, and the glass polishes rapidly and evenly. Amateur optical workers often seem to be afraid when the lap begins to function correctly. They can tell that something is happening, so they stop and add more water, or more polisher—anything to make the lap slide easily again. This is the wrong thing to do: when the lap feels smooth and tight, keep working.

Often it just slides around on the mirror with little or no drag at all. This loose and skiddy action is all wrong. Or the lap may slip and stick suddenly again, feeling anything but smooth. These problems often happen when the pitch is too hard, or too cold from your work area, or just cold from the night before. It may also be that the lap is not grooved well enough or has not been pressed into contact. Quite often laps slide around for no discernible reason.

The remedy that works almost 100% of the time is heat. Warm the lap by submerging it in hot water at about 43°C for 5 minutes. Sunshine or a sunlamp also work well. Be careful not to heat the lap too much if you have soft pitch. You may have to heat a lap only once to cure it, and it may work well from that day on, or you may have to heat a lap every day you polish. Learn to love your pitch lap and cater to its idiosyncrasies.

Another remedy that sometimes works is walking around the barrel in the opposite direction. For some reason, pitch laps seem to have a preferred direction (either clockwise or counterclockwise) that they like to go. They also seem prone to changing direction at any time. Experiment with the phenomenon as polishing progresses.

How much pressure? Lots: a slipping lap may result from lack of pressure on your part. Lean into it!

Sometimes the lap works so well that one person can't push it. Usually reversing the direction you are rotating things helps. Walk around the barrel the other way, turn the mirror the other way, turn the lap the other way, etc. Making the polisher slurry either thicker or thinner may work. As a last resort you can lubricate the lap by adding 2 or 3 drops of Ivory dish soap to the water in your squirt bottle. In any case, you want to get the lap working right or the mirror will take many more hours to polish.

B.9.2 Holding the Lap

Once you get things working right, you will have to experiment with different ways of holding the lap to push it back and forth. I always grab the lap over the far edge at 10 o'clock and 2 o'clock and rest my forearms across its back. This allows me to put my weight on the lap and also gives me a smooth transition between forward and backward strokes. Another method is to put one hand on the far side of the lap at 12 o'clock, with that forearm resting on the back of the glass, and the palm of the other hand on the near side of the lap at 6 o'clock. It does not seem to harm anything if you have to hesitate between the forward and backward stroke.

Don't push the lap rapidly back and forth. It should take 1 or 2 seconds to complete one stroke. This may sound fast, but it is actually quite methodical. If it helps, time your strokes to the slow count of "one hippopotamus, two hippopotamus" spoken in a rhythmic way. (Each hippopotamus takes one second to say.)

When the mirror starts running dry, squirt on polishing agent and water. Be sure to rotate the mirror every time you complete one revolution around the mirror. Turn it the opposite direction from the direction you are walking around the barrel. Vary the amount of rotation a little each time, and vary the point at which you stop to rotate the mirror. Keep things random so errors do not accumulate.

Don't try to polish more than 15 minutes at one time or you may end up hurting yourself. I recommend you do about 15 minutes, then rest for 30 minutes, then do another 15 minutes, then quit for the day. After you get used to it, you can determine your own pace.

Don't let all this deter you. Once you master it, you will think nothing of setting up a routine of polishing two or three times a day for 15 minutes. Stick to it and you'll get a 16-inch mirror polished out in 2 weeks. Don't expect to polish out your first big mirror in the minimum amount of time. Go easy, especially before the muscles and tendons of your arms and shoulders toughen up. There is nothing wrong with taking twice as long on your first mirror as you will on your second.

B.9.3 Completing the Polish

Your first goal is to completely polish out the mirror before going on to figure it. If you make the assumption that the mirror will get polished out while figuring, more often than not, you'll end up with a figured mirror that is not fully polished.

When you check the polish, there is one thing you can count on: if the mirror looks at all underpolished, it is probably worse than you think it is. The 25-inch *f*/6 mirror that I mentioned earlier looked very nearly polished after 7 hours of work, but it took another 5 hours to polish out completely.

The first and most widely known method of checking polish is to focus light on the surface. In the initial stages of polishing, use a penlight that focuses a bright beam. Shine it on the cleaned surface of the mirror—right away you'll see the beam there. As polishing progresses, you can monitor which parts of the mirror are polishing quickly and which parts are lagging behind. Keep records, and continue to check the parts that originally lagged behind until they are fully polished.

When the above method shows the whole surface is polished, put black paper under the mirror to give a dark background. Focus an image of the filament of a bright desk lamp on the surface of the mirror with a positive lens such as a magnifying glass. Angle the light at approximately 45° so that the back of the mirror is not illuminated under the area being observed. If there is still haze on the mirror, you will be able to see the image of the filament on the surface.

When the glass is fully polished, no haze is left and it will be virtually impossible to see the surface. You may find it difficult to get the surface clean enough so you are not just looking at dust or oil or leftover polishing agent. If it passes naked-eye inspection, examine the focused beam on the surface with a magnifier such as a low power eyepiece. Experiment with different angles from which you observe the surface. Do not consider polishing completed until it is completed.

B.10 Testing

The secret of making a fine telescope mirror is knowing, on the basis of optical tests, its true figure. When you know the figure on the surface of mirror, you can work toward changing it to the figure you want—a paraboloid—by applying corrective polishing strokes.

If you are unfamiliar with mirror testing, you need to learn about it. Testing is discussed in numerous books, including *Amateur Telescope Making 1* and Texereau's *How to Make a Telescope*. At this stage, you will begin to appreciate how useful previous optical experience with small mirrors is for making your large, thin mirror. In the directions that follow, I have assumed that you have prior experience testing telescope mirrors or have the help of someone who done so before, and that you understand the basics of testing in this manner.

B.10.1 Test Stand and Testing Tunnel

To test the mirror, you need a test stand and testing tunnel. Even if you plan to do the final figuring by testing the mirror in the telescope, you will benefit from learning how to evaluate its shape indoors under controlled conditions. It certainly does not hurt to have several independent methods of testing.

The environment for indoor testing is extremely important. You need a stand to hold the mirror and a tunnel to maintain a uniform air path between the tester and the mirror. If you try to get away with a second-rate test environment, you may end up with a second-rate mirror.

The test stand allows the mirror to hang freely in 'a sling while resting lightly against a piece of shag carpet. The sling can be made of any flexible material that bends to the circumference of the mirror. Nylon or plastic straps are often used; I prefer old automobile seat belts. The strap must not support more than a 180° arc, or it will squeeze the mirror. Conversely, the strap should not support less than 180° or the bottom of the mirror will bear too much of its weight.

The mirror hangs vertically in the sling with its surface about ¼ inch forward of the front edge of the sling. If the edge of the mirror is wider than the strap, center the strap on the mirror's edge. The carpet just hangs behind the mirror against the sturdy frame of the test stand itself. I use the same piece of carpet in my test stand that I polish on. The mirror should tip back ever so slightly and rest lightly against the carpet.

If you plan to do the critical testing "in the lab," take steps to shield the optical path from air currents. Your test area must be twice the focal length of the mirror. In planning it, allow space for the test stand, the tester, and for you. If you are fortunate enough to have a room long enough to accommodate testing, you may be able to close it off by shutting the windows, blocking air vents, and sealing cracks under doors. If the test path traverses more than a single room, you will need to close all the doors and windows in the test area, and block any large airways that lead to other parts of the house.

To isolate the air in the test path, construct a light wooden scaffold between the test stand the tester and enclose the light path with large sheets of plastic. Paint stores carry big plastic drop cloths, and hardware stores sell rolls of plastic 12 feet wide.

B.10.2 Begin Testing Early

When the mirror looks shiny enough to test, test it. This gives you a chance to verify that it's good enough to continue working on. Although 15 minutes of polishing may permit some testing, a polish suitable for evaluating the mirror will probably require an hour.

Setup a Foucault tester with a 100 or 150 lines per inch Ronchi screen. This is an excellent quick check for overall quality and smoothness of figure. Edmund Scientific Company offers excellent Ronchi screens for sale.

When you first look at the mirror with a light source at center of curvature (without the Ronchi screen in place), you may see a dark region on it, often the edge or the center. If you have been proceeding according to this article's instructions, very few of you should have to deal with this. Normally the mirror will look more or less evenly illuminated after an hour of polishing.

In the event that you do see a dark region, however, the low "shine" on that area indicates that it is polishing more slowly than the rest of the surface. It is difficult to describe what degree of darkness is serious at this stage. Examine the region closely to see if it was properly fine ground. If you see more than a few large pits, return to fine grinding and remove them. If the dark region is free of pits, insert the Ronchi screen and see if there is a gross zone within the unpolished area. If there is, you may need to regrind the mirror. Dark spots can be caused by using a stroke that is too long or because the tool flexed during grinding. The shape of the mirror then deviates so greatly from a sphere that the polishing lap cannot (and will not) reach all of it.

If the dark region is not severe, it may polish out. Keep track of that area during the next 4 or 5 hours of polishing to see if the problem remedies itself. If it remains dark and shows little or no progress, you will need to return to fine grinding to correct the problem.

Assuming, however, that the mirror is more or less uniformly illuminated, estimate how much it departs from a sphere. If you have been using a full-sized lap, it is very likely you have a nearly spherical surface. If you have been working with a sub-diameter lap, the mirror will probably have a progressively longer focus toward the edge. If you are working plate glass and you tested immediately after polishing, the still-warm mirror will appear overcorrected. Allow it to cool for 3 hours or so, and it should settle down and look more nearly spherical.

If the figure looks reasonably spherical, you're on Easy Street! Just test the mirror for astigmatism (described below) and then continue polishing. If the curve is an overcorrected sphere, i.e., a parabola or hyperbola, check the uniformity of the curved Ronchi bands. Even though the Ronchi lines are bowed, they should be free of sharp breaks or zigzag areas indicating a gross zone. If you see bad zones on the mirror and several hours of polishing produce no appreciable improvement, you need to rethink your fine grinding techniques and try again from #400 or 12-micron abrasive.

If the mirror shows appreciable overcorrection, you'll want to determine approximately how overcorrected it is. For a tester with a stationary source and moving knife edge, the difference between the focus at the edge and at the center of a parabola tested at the center of curvature is:

$$\Delta R = \frac{r^2}{R}$$

where r is the radius of the mirror (i.e., half of the diameter), and R is the radius of curvature, or twice the focal length. For a 16-inch $f/5$ mirror, the focus difference is:

$$\Delta R = \frac{8^2}{160} = 0.400 \text{ inches.}$$

If the mirror exceeds this by more than 10% or 20%, take some steps to reduce it early in the polishing process. Crude measurement of the correction is a simple matter especially if you are experienced with a Foucault tester: just measure the difference of the focus of the edge zone and of the center. If the light source on your tester moves with the knife edge, divide the focus difference, ΔR, by 2. For a 16-inch $f/5$ tested with source and knife moving together, the focus difference should be 0.200 inches.

You can even measure the degree of correction with a Ronchi tester by placing the nulled zone on the edge of the mirror, marking where the screen is, then nulling the center and marking the position. Measuring the distance between the lines with a ruler will tell you if things are all right. Once again: remember to divide by 2 if the light source (whether it's a pinhole or the other side of the screen) moves with the testing Ronchi screen.

If the mirror is overcorrected, shorten your polishing stroke for a while. Long W strokes made with a sub-diameter lap or larger tend to overcorrect a sphere toward a parabola, while short W strokes with a 1-inch overhang tend to make the mirror more spherical. With a full-size lap, short strokes can even undercorrect a sphere. If you are polishing with a sub-diameter lap, you might even make the center a slightly undercorrected sphere while the edge becomes overcorrected. A very short stroke will make the edge take longer to polish out. You will find out, with time and experience, which stroke length or lengths work best for you.

B.10.3 Testing for Astigmatism

Astigmatism, or "cylinder" as it is often called, happens when one axis of the mirror has a shorter focus than the other. In the telescope, astigmatism shows up as star images that don't quite focus. When you perform the star test, the images on either side of focus are ellipses that focus to a cross shape, instead of circles that focus to a clean Airy disk.

At this stage the mirror should be free of astigmatism. If it shows astigmatism, something is wrong. In testing for this, you will find out why it took a long time for thin mirrors to become popular—and if things go well for you, you will learn why thin mirrors are now widely accepted.

The primary difficulty in testing thin mirrors for astigmatism is that when they are tested in a vertical position, they often bend. Thus, even a good mirror can look astigmatic. In the telescope, where most of the time the mirror is lying on its back in the flotation cell, thin mirrors perform very well.

To test for astigmatism indoors at the center of curvature, you examine a small illuminated pinhole with an eyepiece of approximately 9 mm focal length. Make pinholes by poking the smallest hole you can in a piece of aluminum foil or 0.001-inch thick brass shim stock. A small sewing needle works well. You will discover all kinds of ways to make smaller and more uniform pinholes by putting the brass on different things and poking it different ways with the needle. Strive for holes that are small, round, and have clean edges.

Place the pinhole in front of a light, preferably the bulb of your Foucault/ Ronchi tester, and examine at the image of the pinhole at the center of curvature with the eyepiece. Keep the distance between the pinhole and the eyepiece as small as you can or off-axis aberrations will confuse you. Do not allow stray light to enter the eyepiece or the image of the pinhole will be washed out, and make sure the eyepiece is square to the optical axis of the mirror.

A non-astigmatic image remains perfectly round on either side of, and through, focus. The idea is to examine the image of the pinhole very near focus where the test is most sensitive. Depending on the figure of the mirror, the image inside focus may be too soft for this test. Outside of focus you'll see a sharp bright ring on the edge of the image due to the edge focusing long. The bright ring should look perfectly round, with no trace of ellipticity.

Irregularities in the roundness of the pinhole sometimes masquerade as problems in the mirror. When the image is in focus, it shows the shape of the pinhole, so you can easily verify that the pinhole itself is round and smooth. If the pinhole is irregular, replace it with a pinhole that is perfectly round.

Air currents can disturb testing for astigmatism because they cause the image to slowly change shape. If you blow into the air path or stir the air with a small fan, the image will change more rapidly. If you don't see a round image, make sure your eyepiece is square on, then rotate it to make doubly sure the fault doesn't lie there. Tilt your head to the side to make sure the out-of-roundness is not due to astigmatism in your eyes. If the astigmatism is in the ocular or in your eye, rotating them will cause the out-of-roundness to rotate.

If the eyepiece and your eye past muster, make a note as to the shape and orientation of the problem and which side of focus you're on. Then take the mirror off the sling, *and without rotating it*, place it on the sling again. Make sure it hangs freely and evenly against the carpet. Inspect the image again. If the problem looks significantly different, then the mirror is not well supported on the test stand.

If the image still looks the same, this time rotate the mirror 45° in the sling and again observe the out-of-focus image. If you still see astigmatism but it has not rotated with the mirror, then the astigmatism you see is due to bending in the vertical position. You can probably get the astigmatism to come and go depending on just how the mirror rests in the sling. Develop a technique for putting the mirror in its sling to minimize the effect—and learn what it looks like and how it behaves. Since the astigmatism you observe will almost certainly be along the horizontal and vertical axes, you will be able to recognize and allow for it, and it will not significantly affect later zonal testing of the mirror.

With thin mirrors, the bending problem sometimes takes a form that is similar to but not quite the same as astigmatism. Called "potato chipping," it occurs when the bottom of the mirror bends but the top does not. You will see round images with a flat spot on the bottom of the outside-of-focus image, and the bottom area will have more light in it.

Even if you see a round image apparently free of astigmatism, rotate the mirror to make sure that astigmatism due to the support system has not cancelled astigmatism in the mirror.

The bad news starts if the out-of-roundness rotates reliably with the mirror. If the astigmatism is very slight or difficult to detect, there is a chance it will polish out. I do not recommend trying. Polishing out astigmatism by hand may prove fruitless and you can waste a lot of time and a great deal of energy. Make one final check before despairing. Allow the mirror to sit in the sling for at least 3 to 4 hours to come to thermal equilibrium, and then retest it. Some portholes become astigmatic while cooling.

If the mirror is astigmatic, the solution is to go back to #400 or 12-micron grit, whichever you used, and fine grind again. Scrupulously apply all the precautions I described above for avoiding astigmatism. Make sure the back is flat. If you are using a porthole and you did not grind the back flat, do so. Did you let the carpet pile get smashed down? If the tool was on the thin side, perhaps it was too thin. Examine your rotating techniques to insure you are rotating by random amounts, and not repeating positions.

Carefully determine and mark the axis of the astigmatism on the mirror. If the astigmatism returns after grinding and polishing, you can see if it lies in the same axis of the mirror as it did the first time. If it does, it could indicate bad glass, or that the back of the blank is not flat. Even if the tool is too thin, the astigmatism should be less on the second try.

You can, of course, check for this by putting the mirror in the telescope—just apply the same tests using a star. The diagonal adds another variable because it will introduce astigmatism if it is not flat. Homemade 9-point and 18-point flotation systems can also do strange things if they are not properly designed, constructed, and used.

The primary difficulty in testing for astigmatism in the telescope is air turbulence. This blurs the image and hides astigmatism if it is small. Uneven air temperatures in closed telescope tubes cause oblong star images, so be glad yours has an open one. Despite these drawbacks, there is no more practical test for astigmatism than observing a star through a high-power eyepiece on a night when the air is steady.

B.11 Figuring

Figuring consists of an equal mix of testing, careful thinking and deliberate polishing. Figuring is not action-oriented. There are many ways to determine what figure a mirror has, and just as many ways to approach figuring it. Thinking is an important part of figuring. As you become more familiar with testing and interpreting test results, it will become easier for you to formulate solutions to problems you will encounter.

On the whole, you will find parabolizing a large mirror simpler than you might have imagined it could be—providing that you have had adequate experience using optical tests before undertaking a large, thin mirror *and* providing that you start with a smooth, nearly spherical one. Given this state of affairs, figuring the mirror may simply consist of a series of tests and general correcting strokes. If you aren't so lucky—if you have trouble with zones—your strategy should be to take care of the mirror's worst problems first, to fix them as they occur, and to reduce the mirror to a smooth, undercorrected curve before you parabolize it.

B.11.1 Goals in Figuring

I recommend that you make your mirror as good as you can make it. It might make me look good to say, "nothing less than an excellent mirror should be tolerated," but that would be both foolish and arrogant. Only the most experienced opticians can hope to make such a mirror. The way you gain experience is by doing it, and you'll never do that if you get discouraged by a promise that only a perfect mirror will do. Mirror making is learned by doing—there is no other way. Work until you cannot detect any significant defects in the figure—and then call it done.

Amateur telescope makers often want me to say something quantitative about how good your mirror has to be—or rather how bad it can be—and still function in a telescope, but all my experience tells me that there is no real answer to this question. Therefore, I am not going to tell you something like "a ¼-wave mirror is good enough," or "a ¼-wave mirror is not good enough," because there are too many unknowns.

The largest unknown is whether a mirror alleged to be ¼-wave really is as good, or as bad, as the maker reports it to be. It is almost impossible for an amateur to make an accurate determination of the wavefront error of a mirror. It isn't that amateurs are in any way inadequate, it's just that an accurate wavefront analysis requires sophisticated interferometric test equipment that few people have access to. What amateur astronomers need is not an arbitrary quantitative standard, but a practical one; a qualitative standard based on performance.

B.11.2 Rating Telescope Mirrors

As a professional optician I judge optical work based on tests accurate to a very small fraction of a wavelength of light. As an amateur astronomer and telescope maker on the weekends, I divide telescope mirrors into five categories: excellent, very good, good, poor, and unusable. I rate telescope mirrors using the star test. I have tested hundreds of telescopes this way, and, with a certain amount of objectivity, I believe that the star test is a practical and effective method of judging telescopes.

Excellent is a mirror that appears to have no significant focus error at all. This represents a very small percentage of the telescopes I have tested, either amateur-made or commercial.

Very good is given to telescopes that display a small amount of focus error, but that error has little or no practical effect on the resolution. Perhaps 10% of active telescopes can claim this.

Good applies to a telescope if the focus error limits the optical performance only at high power. Most amateurs have a mirror that warrants this rating—and most never acquire the observing expertise necessary to feel limited by the problem. Not all, but most. From what I have seen, most of the discontent over the image quality of telescopes stems from poor alignment and/or turbulence that is mistaken for bad optics.

Poor goes to telescopes that are limited at all magnifications above low power. Telescopes of this caliber can be useful for deep-sky observing where the power is held to less than 15× per inch of aperture. For planetary observing and splitting double stars, the limitations are obvious.

Unusable is a rating I do not apply very often, and when I have, it has usually been the result of a bad mistake. One of the worst telescopes I ever saw was a 10-inch *f*/5 telescope from a long-defunct commercial supplier—and it turned out that the telescope had been accidentally shipped with a spherical mirror. Another was made by a young mirror-maker who used the Foucault test but calculated his zonal readings incorrectly, and made a mirror with four times the proper correction. Upon discovering the error, he refigured the mirror correctly.

The point is that a telescope in any of the first four ranks can be enjoyed and used for astronomy. It is the performance of the telescope that matters, and that is something that every observer must assess for himself or herself over a period time.

B.11.3 Do the Best You Can

Even the most inexperienced telescope maker who grasps the concept and has the determination to polish glass faithfully, working without test equipment more sophisticated than a 150-line Ronchi screen used on a star image at the focus of his telescope, can figure a 16-inch mirror that will rank in the poor to good class. This will give him a working telescope, albeit one with certain limitations, and the experience and skills necessary to do a better job the next time. However, it is my hope that the majority of mirror makers reading these words will have made a series of smaller mirrors and will apply the ideas presented here to obtain a well-figured mirror with no practical limitations.

B.12 Test Methods

Several tests exist for testing parabolic telescope primaries. Each test has its strong and weak points. In general, it is best to use several methods of testing so that you can cross-check figure errors.

The *auto-collimation test* employs an optical flat the same size or larger than the mirror to return light collimated by the mirror back to the mirror again. This test has many strong points, perhaps the strongest being that it is a null test. However, it is difficult for amateur telescope makers to obtain the use of an acceptable flat, so I will not dwell on this test.

The *Foucault test* has the opposite problem: it is quite easy to make a Foucault test rig. Virtually every book on telescope making has detailed instructions on how to build a tester and use it. The problem is that for a paraboloid, the Foucault test is not a null test. Instead, the paraboloid is seen as an error in the optical figure of a sphere. The test requires you to measure the errors in the sphere and then calculate the curve they represent. Fortunately, this is not an insurmountable problem. If the zonal measurements are made carefully and instructions are followed closely, it is possible to make excellent mirrors with the Foucault test.

The *star test* requires only the telescope itself and a bright star. It has the advantage that it is a null test for a paraboloid. If you know how to apply this test and have a night with good seeing, you can immediately evaluate a telescope mirror. Of course, it is not without its problems—the biggest of which is that air turbulence often makes it difficult to evaluate, blurring or obscuring problems in the mirror's figure. It also restricts testing to clear nights. Despite these problems, the star test is not a bad way to figure a mirror. I have done some of my best work using it.

My personal favorite for figuring telescope mirrors is the *Dall null test* and a more recent variation, the *Ross null test*. These are similar to the Foucault test except that a plano-convex lens made from optical crown glass is placed in front of a pinhole or slit. Light returning from the mirror is cut with a knife edge, just as in the Foucault test. The lens, called an "aspheric compensator," introduces aberrations into the outgoing light that cancel the errors a paraboloid would normally display at the center of curvature. If applied correctly, a good lens and test rig can compensate the paraboloid's aberrations to better than one-twentieth wave. The problem with both the Dall and Ross null tests is making the lens. Not all telescope makers are cut out for lens making, and the lens has to be a good one or the mirror will inherit its flaws. Furthermore, the spacing between the pinhole and the lens must be set within a couple thousandths of an inch, which requires access to a large set of micrometers or an accurate caliper.

There are many other tests available, most of them more sophisticated versions of the Foucault test. One that is particularly adapted to fast mirrors is the wire test, in which the returning light is cut at the center of curvature with a wire. I have never used it, but from what I have read, it appears to be a manageable one. However, these tests all too often rely on the skill of the person doing it, a skill which can only be acquired by long practice.

B.12.1 Testing the Sphere

A spherical figure is a section of a sphere. Its prime characteristic, from our point of view, as shown in **Figure B.2.A** is that it focuses all light emitted from the center of curvature directly back to the center of curvature with no deviation at all. So under the Foucault test, the knife edge cuts all rays simultaneously and the mirror appears to darken evenly. The Foucault test is a null test for the sphere, because you see no deviations from a uniform darkening.

All light returns evenly to the center of curvature of a spherical mirror. Since the surface of the mirror under this condition appears evenly lit, or what we call "nulled," the shape of a spherical mirror tested at its center of curvature can be expressed as flat or with a straight line (**Figure B.2.B**).

Any other figure or location of the knife will not darken evenly. On the basis of the departures from uniform darkening, you must learn to picture the mirror's curve in your mind. The darkenings provide evidence—indirect evidence—from which you will try to deduce the one particular shape from the family of possible

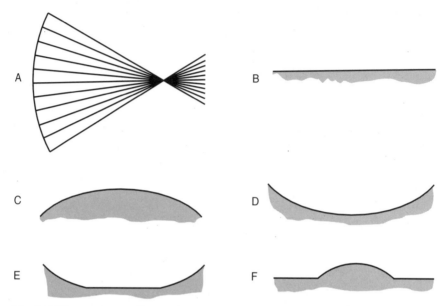

Fig. B.2 *When testing a sphere at the center of curvature, the position of the knife relative to the center of curvature determines the appearance of the mirror. At the center of curvature, the mirror is nulled, and appears evenly lit* (B). *If the knife is inside the center of curvature, the mirror appears relatively convex* (C); *and conversely, relatively concave when the knife is outside the center of curvature* (D). *Profiles* (E) *and* (F) *show a mirror with two zones at different knife edge positions.*

shapes for the mirror that will be easiest to fix with a polisher.

For example, the very same concave sphere can be expressed as too concave or too convex if you just think about it focusing "too long" or "too short." You must be careful here so as not to get confused by this telescope maker's jargon— what we mean is "relatively convex" and "relatively concave." The standard of comparison is another sphere whose radius is equal to the distance between the mirror and the knife edge. If we test a concave sphere with a radius of curvature of 50 inches at a distance of 49¾ inches from the mirror, or ¼ inches inside the actual point of focus, we can think of the mirror as having too long a focus, or being convex relative to a sphere of 49¾-inch radius (**Fig. B.2.C**).

Even though you know the mirror is concave, with respect to the shorter focus "reference sphere," the 50-inch radius mirror is relatively convex.

On the other hand, if the same 50-inch radius mirror were tested at a point 50¼ inches from the mirror, or ¼ inches outside focus, you would find it concave relative to the 50¼-inch radius reference sphere. Note that only the reference sphere has changed. Of course, the sphere under test is a deeper mirror, and really is concave relative to the reference sphere. Opticians express its shape as concave (**Fig. B.2.D**).

Remember that when we test this same mirror at its focus, it can be thought of as flat.

For an optician who must make a sphere to an exact radius, this is important. Even though he may have a ¹⁄₂₀-wave concave sphere, if the radius of the sphere is too long, it must be thought of as being too convex. For the amateur telescope maker, however, small errors in the radius of focal length of the mirror are not important. This freedom to change the radius, both conceptually and in reality, is extremely important when we have a problem in the figure and we are trying to fix it.

For example, suppose we have a mirror we wish to figure to a perfect sphere. Under test, we determine that the outer zones of the mirror are focusing too short. Therefore, the outer zones are high, which is concave with respect to the spherical curve in the center of the mirror which we have decided to treat as flat. See **Figure B.2.E**.

But this same curve can be expressed and treated quite differently. Instead of saying the edge zones are focusing short, and are therefore too high, we can move the knife inward to null of the outside zones. We can say that the center zones are focusing long, and they are too high or too convex. See **Figure B.2.F**.

This is the same mirror, but we have changed how we think about it and what radius we want as our reference. From a glassworker's point of view, it is much easier to polish down, or "reduce," the center than it is to reduce the entire edge. From this perspective, the time you spent thinking was worthwhile because it saved you a lot of polishing time.

B.12.2 The Sphere and the Paraboloid

Now consider a spherical mirror focusing parallel light from a star. No matter how you focus, a spherical mirror will not bring this light to a single point. The effect is called spherical aberration. Note in **Figure B.3.A** that the central regions of the mirror have the longest focus.

Testing a sphere with parallel light is not a null test. But parallel light is, of course, a null test for a paraboloid—the desired curve for a Newtonian telescope. A sphere is not represented as a straight line (i.e., "flat") when we use a test that nulls for a paraboloid, because now a paraboloid is seen as a straight line.

So, how does a sphere look when a paraboloid is thought of as a straight line? At a focus chosen midway between center and edge, with respect to a paraboloid, a sphere has a high center and a high edge. See **Figure B.3.B**.

Focusing near the edge of the sphere, or a shorter focus, the sphere departs from a paraboloid by having a progressively higher center as shown in **Figure B.3.C**.

Focusing near the center of the sphere, or a longer focus, the sphere departs from a paraboloid by having a progressively higher edge as shown in **Figure B.3.D**.

Starting from a sphere, you can see there are three basic ways to parabolize a spherical mirror. First, you can reduce both the center and the edge keeping the focus the same; second, you can reduce the center and shorten the final focus; and

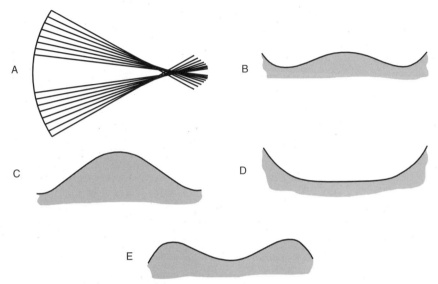

Fig. B.3 *Testing a sphere with parallel light* (A) *suggests how to correct the sphere into a paraboloid. Profile* (B) *shows the small set at the 70% zone, with a knife edge at center. Profile* (C) *shows the null at the outside zone, with the center high and* (D) *shows the edges high with the center nulled. Profile* (E) *shows how a paraboloid appears when tested at the null for the 70% zone.*

third, you can reduce the edge and lengthen the final focus.

Smaller mirrors are usually parabolized by the first method. Large mirrors, especially those with a large departure from a sphere, are parabolized using emphasis on the second method.

Let's return to the center of curvature test where a sphere gives a null. Relative to a sphere, a paraboloid gives the classic doughnut curve, **Figure B.3.E**, when focused for a reference sphere chosen to null the paraboloid 0.7071 of the way between its center and edge. This is the radius for which the edge and center are equal in area, and thus for the sphere that most equally fits the paraboloid.

It is possible to treat this topic in a much more mathematical way. A discussion of the curves using conic sections or as functions in x, y, and z axes can help you understand mathematically what's going on. But from a practical point of view, this information is already covered in many other books on mirror making.

B.12.3 Star Testing

Star testing is a null test for a paraboloid. When the optics are right, the image is right. But its limitations are obvious—you must test at night, at least in the final phase of testing. However, many of the early figuring steps can be done during the daytime using the glint of sunlight reflected from shiny electric power pole insulators or a glass tree ornament hung several hundred yards away.

The biggest problem is turbulence. Air currents stir up the star image and make testing very difficult at times. This is more disturbing in figuring a telescope mirror than in evaluating telescopes for practice at a star party. When you're figuring a mirror, you're probably working from your driveway; but when you're examining telescopes, you're probably at a star party at a site that has been chosen for its good seeing. Although you don't need perfect seeing for the star test, you must avoid really bad seeing.

For our purposes, the star test is applied to the out-of-focus images of a star formed by the mirror. Although there is plenty of information in the focused image, the most useful information about the overall figure shows up inside and outside focus. The reason is that the aberrations have longer or shorter focal lengths that cause light from a zone to concentrate, or reach its focus, inside or outside the main focus. What you'll be looking for with the star test, then, is the distribution of the light in the expanded near-focus disk of a star. **Figure B.4** shows how different aberrations appear inside, at, and outside the best focus.

Ideally, for a perfect telescope the inside-focus and outside-focus disks look exactly the same. In Dobsonians both appear round and evenly illuminated, with diffraction rings due to the wave nature of light and a dark region in the center due to the secondary obstruction. The outermost ring of the star image and the first ring around the diagonal hole in the center are brighter due to diffraction. The diffraction effects can be separated from the figuring defects because the diffraction effects are the same on both sides of focus and figure defects are different.

As you run from inside to outside focus with a less than perfect mirror, you can spot these differences, from not enough light in a zone to too much light. Light from a particular zone may bunch up on one side of focus and disperse unevenly on the other.

In order to make a proper diagnosis of the figure problems in the mirror, you have to understand how the rays from various zones behave as they go through focus. Then, when you actually observe the effects of the zones, you'll be able to estimate where the zone is and whether its radius is too long or too short.

When you look at **Figure B.5** compare a perfect mirror with one having spherical aberration. Notice that some rays of light bunch inside focus due to the short focus of the edge. Where those edge rays intersect, the light is more concentrated, and you would see brighter rings. Outside focus, the edge rays are spread out and you would see a dim area in the star disk.

B.12.3.1 Doing a Star Test

Now let's get down to actual testing. The first step is to collimate the telescope correctly. Post-it notes make dandy temporary center dots.

Next, view a fairly bright star, one of second magnitude or brighter. The glass surface of an unaluminized mirror reflects about 4% of the light that strikes it, so it will form an image as bright as a fully reflecting mirror of 20% of its aperture. To test an aluminized mirror, you'll want a star about sixth magnitude.

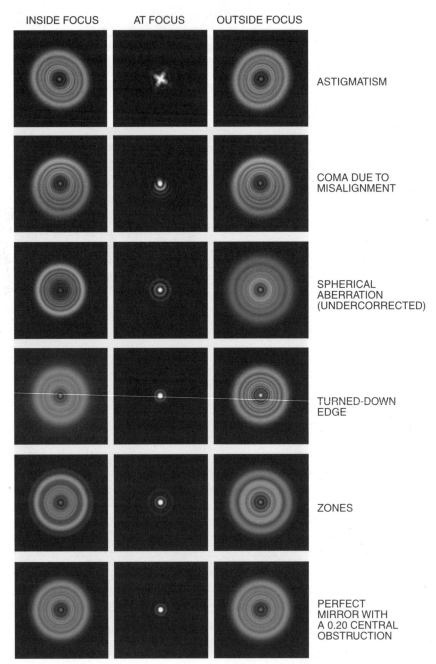

Fig. B.4 *Computer-simulated star test images, inside focus (left), focused (middle), and outside focus (right). From the top row: astigmatism, misalignment-induced coma, spherical aberration (undercorrection), turned-down edge and bad zones. The bottom row is a perfect mirror. All systems shown have a 20% central obstruction. Courtesy of H.R. Suiter.*

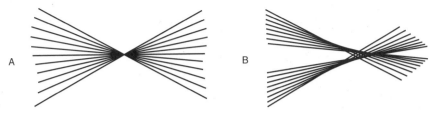

Fig. B.5 *The star test shows how light from the mirror passes through focus. In a perfect mirror* (A), *you see a uniform distribution of light in the out-of-focus disk. In a mirror with spherical aberration* (B), *the distribution of starlight changes as you move through focus. Diffraction adds rings to this simplified geometric picture.*

The magnification you use depends on how bad the mirror is and limiting factors like turbulence (see **Figure 4.5**). I find the most useful magnifications lie between 6× and 25× per inch of aperture. When you first test the mirror during polishing or in the early stages of figuring, you will probably see errors in the correction of the mirror with as little as 6× per inch of aperture. As the figure nears perfection, powers of 15× per inch and above become useful. Magnifications above 25× per inch are not often used on large mirrors because turbulence is likely to be a problem—and what you have to see can be perceived at 25× per inch of aperture anyway.

At magnifications less than 10× per inch of aperture, or with eyepieces longer than 25 mm focus, spherical aberration in your eye or in the ocular may appear as errors in the mirror. Above 10× per inch, this effect becomes negligible. It has been my experience that Kellner, Orthoscopic, and Plössl eyepieces with focal lengths under 25 mm do not introduce aberrations that affect the distribution of light in the star disk.

Your eyes may introduce problems if you have astigmatic vision. Nearsightedness and farsightedness do not matter unless severe enough to affect the afocal performance of the eyepiece, and most of other problems that may be present in the eye are unimportant because only small part of the pupil of the eye is involved when you test with a small exit pupil. Even with small exit pupils, severe astigmatism can remain a problem. You can counter it by wearing prescription glasses, although it may prove difficult to wear them while you look though a short-focus eyepiece. If you see astigmatism in the star test, check to see if it rotates when you rotate your head.

B.12.3.2 Interpreting the Star Test

Let's start with basic undercorrection, the most common type of spherical aberration. Remember, an undercorrected parabola has a short-focus edge and a long-focus center. Starting well inside focus and racking the eyepiece outward toward focus, you'll see the edge of the out-of-focus disk get brighter and bolder while the center starts losing light. Starting from outside focus and moving inward toward focus, you'll see the center fill with light and the edge become dim and poor-

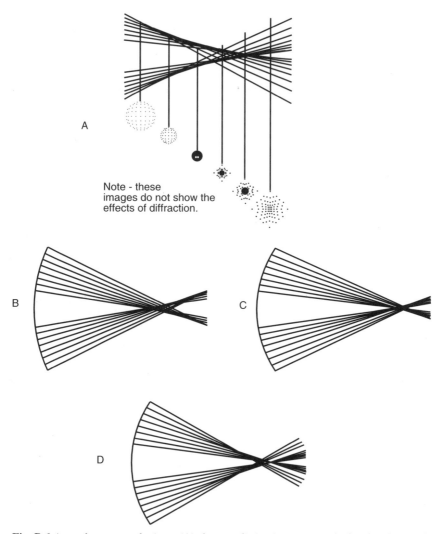

Fig. B.6 *An undercorrected mirror (A) shows a distinctive pattern of a bright edge inside focus and a bright center outside focus. An overcorrected (hyperboloical) mirror (B) shows the reverse sequence. Profiles C and D show mirrors with a defective outer zone. The turned-down edge (C) gives the "hairy edge" effect inside focus, while a turned-up edge (D) gives a "hairy edge" outside focus.*

ly defined. Very near focus or just outside it, the dark spot caused by the central obstruction will disappear entirely as the center focuses.

The computer-drawn ray-trace images in **Figure B.6.A** show how light from a 16-inch *f*/5 mirror corrected to 90% of a full paraboloid focuses over a range ±0.4 millimeter around the best focus. The degree of correction is, of course, proportional to how pronounced these effects are in the test.

An overcorrected mirror, one with a short focus center and a long focus edge, displays the same patterns but on the opposite side of focus. Outside focus, as shown in **Figure B.6.B,** energy builds up at the edge, and inside focus, energy concentrates in the center.

Overcorrection and undercorrection are easy to interpret. Even when they are very slight, they can be seen very near focus at high power. Temperature changes in the mirror may show up as correction errors, either under or over, depending on the direction of temperature change. Interpreting correction errors should be done only when the mirror is near thermal equilibrium, especially if the mirror is plate glass.

What's involved if the problem is not a smooth error in the correction? Suppose you have a parabolic mirror with a raised, or short focus, edge. If the only thing wrong is the short edge, and the rest of the mirror is paraboloidal, you'll see an effect at the edge that looks a lot like undercorrection. If you watch the behavior of the center part of the star image, though, you'll see that it focuses properly, which gives away that the problem is just a raised edge and the rest of the surface is okay.

A rolled or turned-down edge has a very distinct effect on the star image. The edge has a bad slope error so it will rapidly defocus as seen in **Figure B.6.C.** The outermost ring in the star image will be somewhat brighter outside focus, but the effect inside focus is dramatic. Around the edge, the disk just wastes away. This effect is sometimes called a "hairy" edge. An up-turned edge produces a "hairy edge" outside focus.

If the turned-down edge is on an undercorrected mirror, inside focus the usual bright outer ring will have a blur of diffuse light surrounding it. Outside focus, where the extreme edge would be weak due to the undercorrection of the mirror, the star disk will have a thin, bright outer ring.

If you defocus far enough, zones on the mirror show up as bright rings in corresponding parts of the defocused disk on one side of focus and as darker areas on the other side. Where a zone has a shorter focus, the bright ring will be inside focus. Long focus zones show a brighter ring outside focus. Defocusing less far allows the diffraction effects of different zones to mix, and the behavior becomes more difficult to untangle.

Don't be discouraged if all this is hard to visualize—if you remember the basics and understand how the test works, you'll recognize these things when you actually see them. One drawback of the star test is that even when you can tell a bad zone is there, it is not easy to know how wide it is. The solution is to place tape on the mirror or use stop-down rings. Several stop-down rings of different inside diameters will allow you to change the aperture of the mirror to isolate the zone.

A useful adjunct to the star test is a 150-line Ronchi screen. This allows you to see the boundaries of the zones as bumps or wiggles in the Ronchi bands. Used with stop-down rings or tape on the mirror, this Ronchi null test is helpful in marking zone boundaries. The Ronchi screen is also a valuable test for overall correction and smoothness of figure. Use it in conjunction with star testing as a check and backup for your interpretations. It is *always* a good idea to check the results of one test with another.

Turbulence affects the star test by stirring up the diffraction rings near focus. Bad turbulence usually gives the impression that the mirror is better than it really is since you will not be able to see zones or changes inside and outside focus. An experienced star tester can tell when this is happening, but it can fool almost anybody at any time, so watch out for it.

Warm air inside a closed tube can settle at the top and cause the stars to look something like tear drops. All too often this is blamed on the optics. Another effect I've seen is that air swirling inside the tube can have different densities at the center and edge, making the correction of a good mirror look wrong. With truss-tube telescopes, tube currents are not a problem, but you may see some interesting effects before the mirror box cools or when warm air from your body enters the light path.

If you're figuring your mirror with the star test, from time to time, check it at the center of curvature with the Foucault test or a Ronchi screen to make sure its surface is smooth. Figure smoothness is much more apparent in still indoor air.

When you think you've completed figuring, test the mirror again on a second night. You may find that exhaustion or the desire to complete the mirror has blinded you to obvious problems; or that the mirror, especially a plate glass one, was not allowed to cool long enough the first night. Remember that after a polishing session, a plate glass mirror needs as long as three hours to equilibrate. Pyrex needs at least an hour to cool, and should be given considerably longer before a critical evaluation.

Irregular or astigmatic star images were discussed above, as seen from center of curvature. The key to interpreting irregular images is whether or not the irregularity rotates with the mirror. Also check the diagonal mirror—a bad one can do some strange things to the shape of the star images.

B.12.4 Applying a Null Test

As mentioned before, I prefer the Dall and Ross null tests for figuring my own telescope mirrors. Their great advantage is that, like the star test, they are null tests, but you do not have to wait for a clear night, a star, a finished telescope, and good seeing. Besides, indoor tests offer much better control over air turbulence.

Their only real drawback is that you must make and mount a finely figured plano-convex lens. Although lens making is manageable enough, it's not a skill many telescope makers have acquired. You must also be able to measure the distance between the pinhole and the lens to 0.001 inch, a problem solved by access to a good vernier caliper. An article by David Stoltzmann and Marcus Hatch on performing the Dall null test is found in *Sky & Telescope*, September 1976, page 210. An article by Stoltzman and Peter Ceravolo on the Ross null test is found in *Telescope Making #39*, and software for calculating the necessary parameters is given by Doug George in *Telescope Making #45*.

The spacing for null tests is rather demanding, but not impossibly so. For example, to perform the Dall null test on a 25-inch *f*/6 parabolic mirror using a null lens with a focal length of 10 inches, the focal length of the lens must be known to better than 0.1 inch. An error of 0.005 inch in the spacing of the pin hole and lens is acceptable, as is an error of ±¼ inch in the placement of the knife edge.

In constructing a tester, I have found that a small slit works better than a pinhole. The slit allows more light, but it must be narrow or sensitivity is lost. In applying the test, I use both a knife edge and a Ronchi screen to analyze the aberrations in the returning beam. I find the information I get from both very useful.

The problem you will most likely run into with the Dall null test is collimation. Here's how to set it up. Start by placing the mirror so its center of curvature is over the test table. Using a Ronchi tester, find the exact location of the center of curvature on the test table and place a mark on the table; it will be necessary to place the knife edge within ¼ inch of this location. Now place the Dall null tester already setup for the mirror's correction on the table and aim it at the center of the mirror. Locate the return image from the mirror and move the tester laterally until the image is just beside the illuminated slit. Find the exact focus with a Ronchi screen and push the test unit forward or back until the focus lies over the original center of curvature mark you made on the table.

Now comes the hard part: adjusting the collimation of the tester until the Ronchi screen lines run regularly across the mirror. If the lines bend off to one side, rotate the tester left or right to remove the appearance. Correct lines that bow up or down by tilting the tester up or down. This step is just like collimating a very fast telescope.

It speeds testing if you can place the mirror on the test stand in exactly the same place each time so that the center of curvature line on the table does not have to be located each time. However, check it occasionally because figuring the mirror changes its radius.

Once you have the mirror accurately collimated, place a flashlight on the test table where it won't get moved (or put it in V-blocks so it can be lifted out and replaced exactly) so that it shines on the mirror. Clamp a receiving screen to the test table, and when the tester is aligned, mark the exact spot on the screen where the image of the filament formed by the mirror comes to focus. Every time you place the mirror on the test stand, align it so that the image of the flashlight filament falls on the mark on the screen. This places the mirror in its former position so the null tester will be very nearly collimated. Once the flashlight image is lined up, I can recollimate the tester in less than a minute.

B.12.5 Reading a Ronchi Screen

The Ronchi test is very handy, relatively simple to interpret, and a good cross-check on other tests especially if you have any doubt about what they mean. A Ronchi screen is a piece of glass (preferably), plastic, or film with alternate equal-

Fig. B.7 *This simple device can perform the Ronchi test, a test that is especially useful for checking for zones. When the Ronchi bands appear as smooth curves, the surface is free of zones; if the bands are kinky or have sharp changes in curvature, it's an indication of zoniness.*

ly wide opaque and transparent lines on its surface. The number of lines per inch usually lies between 50 and 500. A 150-lines-per-inch screen is suitable for most mirror-testing applications.

The Ronchi test is done by placing your eye to receive a beam from a small source of light while looking through a Ronchi screen, an arrangement much like the Foucault test. You will see dark "bands" on the mirror surface. The source of light can be a star for null testing in the telescope, or for indoor testing, a pinhole, slit, or light that has first passed through one side of the screen itself.

You will see the mirror surface illuminated as in the Foucault test, but with the addition of the dark bands running up and down it. As you move the screen toward focus from either side, you'll see fewer and broader lines on the mirror. The farther you move the screen from focus, the more lines you'll see.

The more accurately the mirror focuses light, the straighter the lines look. A perfect sphere tested at the center of curvature and a parabola tested in the telescope on a star both yield perfectly straight lines. However, if you test a parabolic mirror at the center of curvature, or a sphere at focus in a telescope, you will see the lines bent inward or bowed outward. The Ronchi test reads slope errors on the mirror which affect the focus of the different parts of the mirror and these, in turn, affect the straightness of the lines in the test.

To interpret the test, you only have to remember that the closer you are to the focus of any given zone, the further apart the lines get and the fewer lines you see.

You also must know whether you are inside focus or outside focus. This may seem like an oversimplification to those who are familiar with the test, but it is not. The test can become somewhat difficult to interpret when the problems in the figure are compound and you are forced to make many distinctions as to high and low zones.

Let's consider a few examples. If you're testing a sphere at center of curvature with the screen inside focus and the mirror has a hole (shorter radius region) in the center, you'll see straight lines everywhere on the mirror except in the defective center, where the lines bow outward. Outside focus, you'll see straight lines everywhere except in the center where the lines bow inward.

The reason the center lines bow away from the center inside focus is that the sphere has a hole in the center which makes the center zone focus short, and which means the screen is closer to the focus of the center than the rest of the mirror. The consequence of being closer to focus is that the lines are farther apart—exactly what you see when the lines spread apart in the center.

Outside focus, the screen is farther away from the center focus than the focus of the rest of the mirror, causing the lines to push closer together in the center. If the mirror had a hill (that is, long radius region) in the center rather than a hole, the whole thing would be reversed. Inside focus, the screen would be farther away from the focus of the center and the lines would bunch together there.

A mirror with a good figure except for a turned down edge, tested inside focus, shows straight lines across most of the mirror, but they bow inward at the edge. The defective edge focuses farther away from the screen than most of the mirror. Outside focus, where the edge is closer to the screen than the rest of the mirror, the lines bow away from the center of the mirror.

Interpreting zones midway across the mirror is difficult, but the same rules apply. Imagine a sphere with a high zone halfway to the edge. A high zone has a short focus on the side toward center and a long focus on the side toward the edge. A screen inside focus shows the inside of the zone is closer than the rest of the mirror, so the lines bow outward from the center. On the outside edge of the zone, the Ronchi screen will be farther away causing the lines to bow back toward the center. Of course, the opposite appearance is seen on the other side of focus.

One more example is worth mentioning since it can really surprise someone seeing it for the first time. Suppose you have a mirror that has zones with different focal lengths, but each zone is spherical. In the Ronchi test, the whole mirror shows straight lines, but the lines in different zones have different spacings. These zones are separated by a transition zone in which the lines jog sharply.

As the screen is moved closer to focus, the sensitivity of the Ronchi test increases. For null testing, place the screen to show one to five lines across the mirror. For mirrors with large departures from sphericity—for example, a paraboloid tested at the center of curvature—the lines bow so strongly that many lines will be needed.

Testing a parabolic mirror at the center of curvature shows a mirror with grossly bowed lines. Although you cannot easily determine the amount of correction from the bowing of the lines, you can judge whether the figure is smooth from how smooth the lines are. You can also judge whether the curve "looks right," that is, whether the curve of the lines is too strong in one zone with respect to the overall curve. These are skills you gain only by testing many mirrors.

Astigmatism also affects the Ronchi lines. If the astigmatism is bad, the lines will turn a little clockwise or counterclockwise when you move the Ronchi ruling through focus. If there is just one line on the mirror, it will be distorted into an S-shaped curve, and if there are many lines and the astigmatism is severe, they will all be distorted into S curves. If the Ronchi bands are aligned with the axis of the astigmatism, the effect becomes less noticeable.

You will see many other interesting optical phenomena when you test with a Ronchi screen. The effects are stronger in Ronchi screens with more lines per inch. Most of these are due to diffraction and interference. Be aware that such effects exist, and don't get confused when you discover them while applying the test.

B.13 The Art of Figuring

Figuring is an endeavor of strategy. The strategy comes into play in deciding how you will treat a certain problem. The trick is to examine all of the different contours a mirror may have with respect to different radii.

At the outset of figuring, the ideal starting point is a sphere. In the parabolizing stages of figuring, "undercorrected" or "overcorrected" are understood with respect to a paraboloid, not a sphere, so a sphere is an undercorrected paraboloid. To shape a sphere into a parabola, you can deepen the curve in the center with a correcting lap. If the strokes are applied with perfection, and care is taken not to overcorrect the mirror, it is possible, though not likely, to bring the mirror to a paraboloid with no other strategies. The same holds true if the mirror starts from any smoothly undercorrected figure.

Unfortunately, you are likely to find that the mirror is not an undercorrected parabola, and it isn't even a conic section. However, we will start by describing basic "parabolizing" strokes, that is, strokes that increase the correction of the mirror.

In figuring, you work with the mirror face up. Start by squirting polishing agent onto its center and smearing it around an area larger than your correcting lap. Place the lap on the mirror for a 5-minute wet press with 10 to 15 pounds on its back. Start every figuring session this way to insure a good fit between the polishing tool and glass.

B.13.1 High, Low, Long, Short

When you test a mirror by the methods usually available to amateurs—the Foucault, Ronchi, and star tests—you don't see the high and low zones on the mirror directly, but instead you see where they focus with respect to each other. Although the different focal points are caused by the highs and lows on the surface of the mirror, it is important to understand relationships between the highs and lows and the longs and shorts.

When the *edge* of a mirror has a longer focus than the rest of the mirror, it's called a *long edge* and is a *low* zone. When the *center* of a mirror has a longer focus than the rest of the mirror, it's called a *long center* and is considered a *high* zone.

Almost everybody with an interest in mirror making has to struggle with this, but it is impossible to emphasize too strongly. When you encounter a seemingly intractable problem and have put in hours of futile figuring, it is often necessary to re-explain to yourself why a mirror acts as it does by drawing light rays reflecting off curves representing mirror surfaces to see which way the light would be reflected by different errors.

Even though these concepts seem elementary, it is very important to bear in mind that whenever you test a mirror, you are measuring the slopes on the mirror's surface and not the highs and lows that cause the slopes.

Watch out for sneaky highs and lows. A quarter-wave high zone that reaches its peak slowly may look less severe than an eighth-wave zone that slopes up to its peak abnormally. Ultimately, it is the errors from proper focus that degrade the final image of a star or planet in your telescope.

B.13.2 Figuring Laps

Small mirrors are usually parabolized with a full-size lap. The long W stroke usually employed deepens the center and reduces the edge of an initial sphere until the mirror is paraboloidal. With a 6-inch *f*/8 mirror, parabolizing can take fewer than 100 strokes.

A large mirror departs far more strongly from a sphere than a small one— enough so that parabolizing with a full-size lap can be a lesson in frustration. While an 8-inch *f*/6 paraboloid and the best fitting sphere differ by about a half-wave at the 70% zone, a 16-inch *f*/5 differs by about 1½-waves. Because a full-size lap has a natural tendency to produce a spherical surface, it tends to work against you. Nevertheless, it is easy with a full-size lap to deform the figure of an 8-inch *f*/6 mirror smoothly by ½-wave. By contrast, it requires very aggressive strokes with a full-size lap to deform a 16-inch *f*/5 spherical mirror smoothly by 1½ waves.

When I first started making big mirrors, I parabolized them with full-size laps. It wasn't until I had great difficulty getting sufficient correction and a smooth curve on a 16-inch *f*/5 that I resorted to using an 8-inch lap. In no time it was parabolized—and really smooth. For the next mirror, a 16-inch *f*/6, I corrected directly from a sphere with an 8-inch lap. The mirror went from the sphere to a well-corrected paraboloid with no more than the 8-inch lap and a general correcting stroke. Since then, I have learned that almost everybody parabolizes big or fast mirrors with small laps.

Small laps are used for polishing in an aspheric correction and for figuring

Fig. B.8 *The star lap is used for correcting zones and for figuring mirrors with faster focal ratios. This is a 4-inch star lap; it would be used for working on mirrors between 16 and 24 inches diameter. Note that the "star" shape is grooved with non-centered arcs.*

zones in the mirror, as we will discuss later. Small laps come in many different types, but for our purposes, the two main shapes are the star lap and the scalloped lap. Both forms have "tapered" edges so that the correction being made blends with the rest of the mirror. Small round laps often leave the mirror with a rough surface texture.

For increasing or decreasing the correction of the mirror, a lap that is half the diameter of the mirror or less should be used. For a 16-inch *f*/5 or *f*/6 mirror, I would use an 8-inch scalloped lap for most of the correcting. For a faster 16-inch mirror, I would switch to an 8-inch five-point star lap because it would conform to the deeper curve more successfully. For a 25-inch mirror, a 10-inch scalloped or star lap gives good results. For mirrors in this size range under *f*/5, more deeply cut or smaller star laps are necessary. Your own experience and techniques will determine what you will need.

For figuring zones on the mirror, still smaller laps are needed. These laps are better off cut to a star shape, except for very small ones that can be scalloped or round for the extreme edge of the mirror—but more about this later. It is impossible to tell just how small a lap you will need for correcting zones, since that depends on the zone to be corrected, but I start with a 4-inch star lap. It can do the job on almost anything that an 8-inch or 10-inch lap is too large to handle.

I usually make small laps on glass mirror blanks or fairly thick aluminum disks. Pitch for these laps should not be too hard or they have a rough feel to them when working. I strongly recommend Adolf Miller #73 black pitch, the same material I use for full-size polishing laps. Groove the larger small laps, those of 8-inch and 10-inch diameter, with a standard square pattern, and groove smaller star laps with a couple of arcing curves.

Fig. B.9 *This 8-inch subdiameter polishing lap has scalloped edges so that its action tapers off smoothly. A lap like this one is used to add to or subtract from the mirror's overall correction. A lap this size should be used to figure a 16-inch to 20-inch mirror.*

B.13.3 Figuring Strokes

I could say a lot about figuring strokes, but it is really something that you have to learn for yourself from the direct experience of pushing the lap. However, I will pass on some general principles that hold for all figuring strokes. You will often be doing a short spell of work—just a few turns around the mirror. It is important to distribute the polishing action uniformly around the mirror, and to be consistent and even-handed. If your strokes vary as you progress around the mirror, irregular errors in the figure will result. Keep the stroke pressure, stroke length, stroke rate, and stroke position consistent.

It is also important to return to where you started without doing the starting point over. Even if you do several revolutions, be sure to stop your last stroke just before the place where the first stroke started. If, for example, you were to do two and one-half turns around the mirror, then one side of the mirror would receive 25% more correction than the other. Make sure that you do either two full turns or three full turns. Finally, start each figuring session at a new, randomly chosen orientation of the mirror so that over a period of time, the starting points will average out.

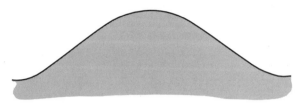

Fig. B.10 *Use the long straight stroke to reduce the high center of an undercorrected paraboloid.*

B.13.3.1 The Long, Straight, Central Stroke

The first stroke used for increasing the correction with large subdiameter laps is a straight stroke running nearly through the center of the mirror. You push the lap *nearly* through the center and continue until it's part way over the far edge, then you pull it back *nearly* through the center until it overhangs the edge near you, and then you push it back nearly through the center again—all while you walk slowly around the mirror. The place where the lap overhangs the edge should be a few inches over from the last stroke. I prefer to move over about one diameter of the pitch lap each stroke, thus making sure not to repeat an area of the mirror on my next time around.

This long, straight stroke works on the undercorrected parabola as if it were focused so that all the error was in the center (**Figure B.10**). On fast mirrors, this stroke tends to concentrate work in the center and let the edge lag behind, in effect raising the edge.

The long, central stroke normally doesn't go right through the center, but instead misses it by an inch or so. Either side will do, but stick to one side or the other and don't alternate. If the stroke goes directly over the center, that area tends to get overworked and the result is a hole in the middle. On the other hand, if the stroke passes too far off center, it will produce a flat or high center. You have a lot of control in this, so use it. To reduce a high middle, you may want to stroke right across the center. To bring up a central low while still adding correction to the mirror, stroke 3 to 4 inches off center. Once you get a feeling for how it works, this stroke proves to be quite versatile.

B.13.3.2 The W Stroke

The second major correcting stroke for large, subdiameter laps is a W stroke. It starts at one side of the mirror, with the lap hanging over the edge of the mirror 2 to 3 inches and proceeds forward and back across the mirror to the other side. Each complete W runs 4 to 5 forward strokes, with the lap slightly overhanging the edge of the mirror at the end of each stroke. This occurs as all the while you walk around the mirror (**Figure B.11**).

The W stroke reduces the edges somewhat while also reducing the center. For best effect, avoid stroking heavily over the 70% zone since this serves to re-

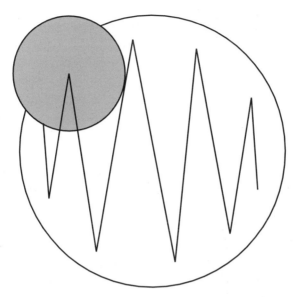

Fig. B.11 *The W stroke can reduce both the edge and the center simultaneously while leaving the 70% zone alone to reduce an overcorrected paraboloid. However, by varying the spacing of the strokes, the W stroke can be made to do many different things.*

duce the effect you're working for. However, do not skip the 70% zone, but work the W pattern to cross over it more quickly than other zones. Someone watching as you do this stroke might not even notice that the 70% zone is getting less work.

I use both the long central and W strokes when correcting a mirror by hand. How fast these strokes change the mirror's correction depends on dozens of variables unique to each telescope maker. I would start with a 15-minute session to get a gauge on how fast it works for you. Especially with a fast mirror, you will have to work harder and harder to reach full correction because as the mirror departs more strongly from a sphere, the strokes work less effectively.

B.13.3.3 Tangential Strokes

To decrease the correction, you must reduce the 70% zone. Remember that an overcorrected mirror has a high 70% zone relative to a paraboloid (**Figure B.12**).

If you have a full-size lap, polish with a short W stroke: this usually brings an overcorrected mirror down very smoothly. If your main polishing lap is subdiameter, be careful because doing a short W stroke can rapidly flatten the center while leaving the edge down, a problem you'd prefer to avoid.

A useful alternative is to use the pitch lap prepared for parabolizing the mirror with a tangential stroke right on the 70% zone. Starting with the lap centered on the 70% zone, push it forward and pull it back along a straight line while you walk slowly around the mirror. This stroke should remain wholly tangential—don't follow the

Fig. B.12 *An overcorrected paraboloid has a high 70% zone. You can reduce it using a short W stroke and a tangential stroke.*

zone with a curved stroke. The tangential stroke reduces the 70% zone and brings up the edge at the same time. If the mirror is strongly overcorrected, bringing up the edge will be a problem, and we'll discuss techniques for curing it below.

If, upon testing, you see that a bump is forming in the center, add a long center stroke to the tangential strokes. After every 3 or 4 tangential strokes, turn and run the lap right through the center of the mirror, then resume the tangential strokes.

B.13.3.4 Strokes for Small Polishers

For working zones with small subdiameter polishers, the usual polishing strokes are the tangential stroke and a variation called the tangential "tree" stroke or V stroke. The tangential stroke is applied using straight strokes centered on the zone you want reduced while walking slowly around the mirror. See **Figure B.13**.

Apply the stroke by pushing the lap forward and then pulling it back in a straight line while taking a step around the mirror. Depending on the amount of correction and your personal preferences, you can move less than 1 inch around the mirror with each stroke, or you can move several inches. The length of the tangential stroke will vary with what you wish to accomplish. Short strokes cut more narrowly and leave steep slopes, while long ones tend to blend adjacent zones more and leave less steep slopes at the edge of the correction. See **Figure B.14.**

The V stroke branches off to the left or right or both sides of a tangential stroke. This allows you to widen and blend the cutting action of the lap. Taking strokes off toward the center causes the lap to cut steeper on the edge side and less steeply on the center side. You can branch every stroke, every other one, or every third stroke—whatever the problem calls for. Just be sure not to miss sections of the mirror. Another stroke that serves to widen the action of a small lap is a W stroke centered on the zone.

I use these strokes with polishers from 1 inch to 16 inches diameter on optics from 16 inches to 36 inches diameter on a daily basis in my work. The results even after heavy handwork are amazingly smooth even when tested with sensitive interferometric equipment.

B.13.4 Correcting a Low Edge

A low edge or long edge, whether it is a half-inch wide or four inches wide, presents the same dilemma—you need to add glass. Since adding glass is impossible,

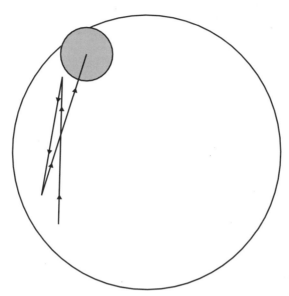

Fig. B.13 *The tangential stroke works directly on a high 70% zone. By varying the length of the stroke, you can focus the work on a narrow zone or blend it over a broad region.*

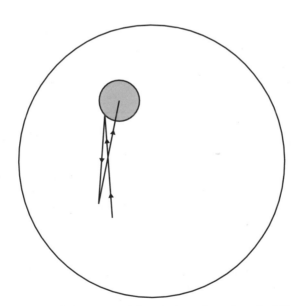

Fig. B.14 *Here, a relatively long tangential stroke is used to reduce the 40% zone and blend it into the surface. The stroke is "long" because it covers a wide arc relative to the center of the mirror.*

you are forced into either removing glass everywhere else on the mirror or blending the edge into the existing curve, thereby lengthening the overall focus. The second method is much less work.

If the edge is undercorrected with respect to the final figure, you can ignore a long edge. You simply treat it by applying strokes to increase correction without going all the way out to the edge until the center matches the edge. However, when the edge is overcorrected, or long, you must deal with it directly before parabolizing can proceed.

As an exercise, consider an 18-inch mirror about 50% parabolized with an edge zone 2 inches wide rolled-off with respect to a curve focused midway. One way to fix the edge is to use a 3-inch or 4-inch star lap to reduce the region just inside the edge. The stroke is a tangential one centered on the peak. Start with a well-pressed small star lap. Push the lap straight back and forth tangent to the center of the mirror while progressing 1 to 2 inches around the mirror between each stroke. By reducing the zone inside the edge, the zone becomes part of a sphere with a longer radius as shown in **Figure B.15**. After this fix, the mirror warrants a general correcting stroke that will also blend in any zoniness generated by the small lap on the edge.

Unless the edge is only slightly rolled, the simple tangential stroke often won't do the whole job. You must blend the reduction toward the center using a V stroke in conjunction with it. You start with a normal tangential stroke forward and back, then you add a stroke that veers off toward the center by 1, 2, 3, or more inches, depending on how much blending is needed. How often you add the side strokes and whether you make them every stroke or every third stroke, can be determined only by examining the problem on the mirror. The V stroke will cause the zone you are working on to slope less on the side that gets the V—just what's needed for the mirror in this example.

How much work will it take? This is determined by testing and evaluating the progress. Times typically range from one trip around the mirror for a very slight zone to 30 minutes of hard work followed by smoothing with a large polishing lap for gross zones.

With a badly rolled edge, you may encounter difficulty in getting the last inch or so of the edge to come into line. For this, a 2-inch scalloped lap may be needed, especially if the extreme edge is lagging behind. Apply the scalloped lap with a circular stroke, allowing the lap to overhang the edge of the mirror by ¼ inch to ½ inch as you progress around the mirror. Always keep the lap the same amount over the edge—and be aware that as this work contours the slope of the edge up, it also leaves a long zone a couple of inches wide inside the edge, which you will have to deal with later. Use the circular stroke sparingly, and use it early in figuring, since it can take considerable effort with your largest lap to smooth the damage caused by fixing the edge.

If you have a turned edge that is not perched on an undercorrected curve, the same techniques apply except you should use them at the expense of the center of the mirror. Work the edge as described above, but blend the curve much farther into

Fig. B.15 *The key to correcting problems is to visualize the surface in such a way that the error can be fixed by removing glass, then tailor a stroke to remove the higher zone. In this case, a zone inside the edge is reduced with a V stroke and then blended.*

the center. The resulting high center then can be worked with a subdiameter lap. After extensive work with a small figuring lap, a smoothing session with your polishing lap or a large subdiameter correcting lap will almost certainly be necessary.

B.13.5 Correcting a Turned-Down Edge

The much dreaded turned-down edge, or "TDE," of *Amateur Telescope Making 1* fame, is a long or rolled edge ¼ inch to ½ inch wide. You must deal with it as soon as you detect it, since the figure of the whole mirror may suffer in repairing it.

A full-size lap can sometimes eliminate the TDE if you use short strokes back and forth with no side motion. A lap the size of a silver dollar can also be used in much the same way a 3-inch or 4-inch star lap is used for the similar problem of a broader low edge. Hang the round lap over the edge around ¼-inch and proceed with circular strokes around the mirror. Remember, for a circular stroke, the lap must always hang off the edge ¼ inch. This will usually turn the edge up in a short time because even though turned edges slope sharply, they are often narrow zones and therefore not down very far. This technique leaves a low zone inside the edge that you must smooth with a larger lap—a problem many times more manageable than the rolled edge itself.

If the extreme edge proves intractable, do not ruin the figure of the mirror to save it. Work it down to ¼ inch wide and then forget about it — you can easily mask the extreme edge. In a working telescope, it is far more important to have a smooth overall figure than the extra ½ inch of aperture.

B.13.6 Correcting High Zones

High zones can be handled directly with the tangential stroke. The trick is to determine the exact shape of the zone and apply a stroke designed to remove it. If you get a high zone that rises steeply on the center side but drops slowly toward the edge, remove it with a tangential stroke on the crest of the zone with an added V stroke toward the edge on every other stroke. This will work on the crest while putting some extra action on the slope toward the edge where it needs a wider action.

In a case where the high zone is wider than your star lap, within reason, you can widen the action of your star lap by using V strokes centered on the zone but branching off to both left and right. Be careful not to do too much work and cut a narrow low zone in the center of a broad high zone because you did not spread the action of your small lap.

Central bumps are best treated with a W stroke on them while you circle the mirror. Of course the size of the lap and the length of the stroke depend on the width of the bump. Again, remember to spread the work. A common mistake is cutting a small hole in the center of a central bump. Be careful not to underestimate the diameter of the bump.

B.13.7 Correcting Low Zones

Low zones are treated by reducing the mirror in a general fashion on either side of the low zone, then smoothing the whole surface with a larger lap. This is very effective. For a small hole in the center, work down the sides of the hole until you've blended it into the rest of the mirror. By smoothing with a large lap, or "refacing the curve," a small hole in the center can vanish almost as if you had miraculously gained the ability to add glass.

B.14 A Few Final Words

Take routine care that the mirror hangs properly in its sling when you test it. Check regularly for astigmatism. Make sure the pile in the carpet under the mirror doesn't get flattened. Trim the facets on the laps. And keep your work area clean—you don't want to pick up a big scratch during the last stages of figuring.

With large mirrors, figuring sometimes goes smoothly and sometimes doesn't. Don't give up. But if things seem to get worse every time you work, make sure that you're interpreting the test results correctly. It is easy to get the high-low and long-short logic turned around and deepen a zone you were trying to eliminate. If in doubt, use several tests before doing more work. Think strategically—that's as important as the polishing.

At the outset, I suggested that the outer ¼ inch of the mirror would probably be lost to a turned edge. If this prediction has come true and your mirror tests out well otherwise, do not worry about it. Professional shops usually write off the extreme edge because it takes too much time and effort to make a mirror that is perfect to the edge.

Check your test results. If you rely on indoor tests like the Foucault and Ronchi tests all through figuring, put the unaluminized mirror into the telescope from time to time to triple-check its progress. Looking at a star image is an excellent reality check. If you have messed up in some big way, the terrible star images will let you know and you can fix the error.

When should you pronounce your mirror done? I'd suggest that you figure it until you cannot detect any significant problems. The reason for this is that for all but the most experienced telescope makers, once the mirror is aluminized and put into operation, the problems you see in the figure will turn out to be worse than you thought they were. It is best to eliminate all the errors you can detect even though a calculation may show a certain error will not be too severe. Besides, it will save you years of saying, "One of these days I need to refigure this darned mirror."

Fig. B.16 *The master himself, John Dobson, pushing glass at the 1994 Riverside Telescope Makers Convention. Dobson conducts mirror making classes throughout the world.*

Appendix C
Digital Setting Circles

Digital setting circles are a relatively new phenomenon, but probably only the first of numerous high-tech telescope accessories that will some day be part of amateur astronomy. You use digital setting circles to locate objects in the sky without recourse to a star map. This appendix tells you a little about what they are and how they work, and then tells you how to install digital circles on your truss-tube Dobsonian.

C.1 A Bit of History

Setting circles were standard issue on equatorial mounts for a hundred years. Because the celestial coordinate system of right ascension and declination corresponds to the axes of motion of the equatorial mounting, by attaching setting circles to the mounting, you could set the telescope on a celestial object by moving it until the readings on the setting circles matched the coordinates of the object. Setting circles are basically glorified protractors that measure the angular orientation of the telescope. Accurate setting circles work very well indeed, but many amateurs never used them, probably because they didn't know how.

Then came the Dobsonian revolution. Because the telescope no longer moved parallel to the standard sky coordinates, setting circles were useless. However, amateurs rapidly adopted the Telrad, a simple and inexpensive sighting device that made star-hopping and "point and shoot" observing quick and easy. Star-hopping is great, but doing it successfully requires either that you memorize the locations of hundreds of deep-sky objects or you wrestle with a flashlight and star atlas and make up hopping paths as you go. Lots of people found a third option: they look at the same three dozen objects over and over again because that's all they remember how to find.

Then came microprocessors and pocket calculators. Although he was not the first, amateur astronomer Bill Burton bought a $35 programmable pocket calculator from Radio Shack, attached a surveyor's compass and inclinometer to the axes of his carefully leveled 17½-inch Dobsonian, and persuaded John Kerns to write a program for the calculator to convert the right ascension and declination sky coordinates into the altitude and azimuth of the telescope. Since the surveyor's compass read out the azimuth where the telescope was pointing, and the inclinometer read out the altitude, Bill could find new objects easily.

To initialize the program, Bill had to enter his latitude, longitude, the date, and the time of day into the calculator. Thereafter, to find any deep-sky object, he could simply enter its right ascension, declination, and the current time plus 30 seconds—because that's how long it took the calculator to work out the new co-ordinates and for Bill to point his telescope.

Well, it worked and it worked very well. Furthermore, because Bill wrote an article about his technique for *Astronomy* magazine, thousands of people realized how easy it would someday become to find celestial objects with a big Dobsonian telescope. Thus the idea was planted.

It wasn't long before an electrical engineer named Rick McWilliams developed a preprogrammed "black box" computer for use with digital shaft encoders. When the units were introduced in 1989, they had a built-in database of stars and deep-sky objects and red LED readouts. McWilliams sold this product only through other manufacturers and retailers, and within a few years there were half a dozen firms competing to sell their own brand-name version of the McWilliams digital setting circle unit.

C.2 How Digital Circles Work

Digital setting circles are an amazing tool for the observer. The best thing is that, like everything else electronic, they get better every year. Rather than try to describe a list of features that grows and grows, let's just consider the parts of a basic, bare-bones digital setting circle system.

The key element is the computer unit. It houses a microcomputer, buttons to tell the computer what to do next, memory chips to store object locations, and interface chips that "talk" to the digital shaft encoders. It also houses a battery to run the system.

The observer uses the buttons to select the desired object. The computer looks up the object's location and figures out where it should be. Depending on the unit, you can manually steer your telescope until the display says that you have reached it, or in the newer units, let it run motors to drive the telescope to the object you want to see.

To align the system at the beginning of the night, all you must do is identify two or three reference stars and point the telescope at each of them in turn. The system keeps track of the time automatically, and once it has been aligned, displays the right ascension and declination of the telescope at all times.

The computer knows where the telescope is pointing from two digital shaft encoders mounted on the telescope. One encoder attaches to the center of the side bearing and the other to the center of the azimuth bearing. A metal arm holds the encoder body stationary while the shaft of the encoder rotates with the telescope bearing. The optical encoder transmits the measured angle of rotation to the indicator. Encoders used on telescopes today divide the circle into 4096, 8192, or

Fig. C.1 *Tom Dey was among the first to install alt-azimuth setting circles on his 17½ inch Dobsonian. Dey used a programmable calculator to transform celestial coordinates into alt-azimuth coordinates.*

Fig. C.2 *Once he knew the proper alt-azimuth coordinates, Dey could aim his telescope at the desired celestial object. His large setting circles were as accurate as today's digital shaft encoders.*

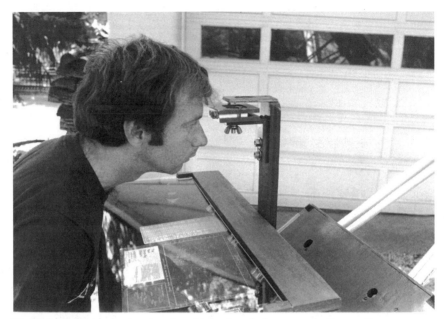

Fig. C.3 *Bill Burton reads azimuth on his setting circle system. It allows adjustment to remove the magnetic offset from north. All the hardware near the magnet is made of brass and power for illumination is separated using fiber optics. The altimeter is mounted on a strut. Photo by Robert Bunge.*

16,384 counts per rotation, which means that they divide a circle of 360° into 4096, 8192, or 16,384 parts. The resolution of the most common type of encoder, with 4096 counts, is about $\frac{1}{10}$°, which is precise enough to let you bring any object into the low-power field of view.

C.3 Accurate Circles Need an Accurate Telescope

Digital setting circles cannot be more accurate than the telescope they are mounted on. For digital setting circles to work right, the axes of the telescope must be at right angles to one another and the encoders must be mounted exactly on the axes of rotation. When these conditions are met, digital setting circles are accurate over the entire sky. On a poorly made Dobsonian, digital setting circles will give you acceptable accuracy over a small part of the sky, but not over a wide region.

These conditions mean that on a Dobsonian, the axis of rotation of the rocker (azimuth) and axis of rotation of the mirror box (altitude) must be perpendicular. The azimuth pivot bolt must be drilled dead center and perpendicular to the axis of the bolt. The altitude encoder must be dead-center in the rotational axis of the side bearing. Side-to-side slop in either axis will harm the accuracy of the system. Finally, the encoders must be supported tightly enough so they do not wiggle around but loosely enough so they do not get crunched.

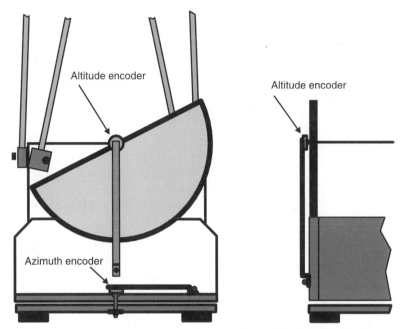

Fig. C.4 *Modern digital shaft encoders sense the position of the altitude and azimuth axes. The shaft of the encoder is attached to the axis of rotation and the body of the encoder held on a long flexible metal strip.*

However, don't despair. As long as you apply a little common sense when you install the shaft encoders, you can find lots of nifty stuff even with sloppily installed encoders on a sloppily made telescope. You'll have to realign on new stars when you swing the telescope more than about 45°, but they'll still speed up finding new objects. Installing the encoders accurately on an accurately made telescope gives you all the above and they will work accurately over the entire sky. We think the effort of installing them accurately is worth the reward.

What digital setting circles should you order? First, remember that most of the digital setting circles on the market are systems designed by McWilliams, so inside they are all pretty much the same. The differences come down to how much catalog memory a system has—500, 4,000, 16,000, or 65,000 deep-sky objects—and whether the computer knows how to calculate the positions of planets and asteroids. Top-of-the-line units allow you to set up an observing agenda: as soon as you locate one object, the computer is ready to guide you to the next object on your observing list.

So—which one should you get? Remember that because the electronics revolution will continue, any unit you buy will be "out-of-date" within a year, so pick a unit with reasonable specifications and a good price. Any unit you select today will provide a lifetime of observing service. If you think you'll exhaust a database with 8,000 objects any time soon, you'd better call NASA for a job interview.

Rocker base   Azimuth encoder   Support arm   Thumb screw

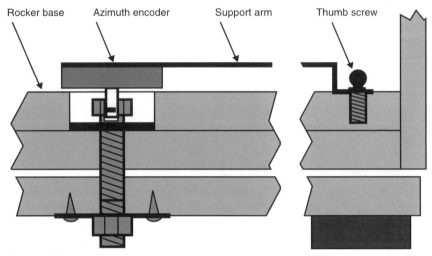

Fig. C.5 *This detail shows how the azimuth encoder shaft is held by the pivot bolt and the body, prevented from turning by a metal strip attached to the rocker box. The pivot bolt does not rotate because it is firmly held by the lock nut under the ground board.*

C.4 Installation

First of all, we assume you have built a *square* Dobsonian. "Square" is a carpenter's term: it means that you have paid attention to dimensions. This is, of course, an ideal that you cannot truly attain, but hopefully you were conscientious and got reasonably close. The mirror box and rocker sides are square, not 21⅝ inches on the left and 21½ inches on the right. The mirror box does not teeter in the rocker. The rocker sides are perpendicular to the bottom, not tilted at 89.5° or 90.5°. The side bearings are the same diameter, perfectly round, and have been mounted concentric to one another. The arcs cut into the rocker sides are concentric, and the Teflon pads all lie tangent to the same imaginary cylinder.

If you have failed to build your telescope truly "square," the setting circles will work over a 30° to 60° swath of sky, but become inaccurate when you swing from the eastern sky to the western sky. Not to worry: you simply "resynchronize" on a couple stars in the area you want to observe and the unit is back on target.

When you install the encoders, it is important that the encoder shafts be centered precisely on the axis of rotation of the bearing. If you mount the encoders crooked or off center, every time you turn the telescope, the motion will force the encoder shaft. Sooner or later, the encoders will fail. To prevent this, we recommend that you purchase the manufacturer's mounting hardware along with the setting circle system. Sure, it's fun to figure out your own mounting scheme, but it isn't worth ruining an encoder. Read the instructions that come with the encoders and follow them carefully.

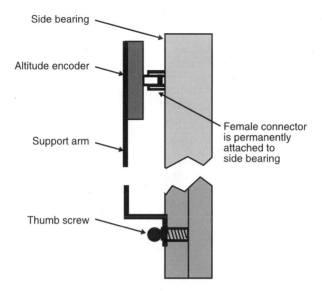

Fig. C.6 *The altitude encoder shaft turns with the side bearing while the support arm keeps the encoder body stationary.*

C.4.1 Mounting the Azimuth Encoder

Azimuth is the easier encoder to mount. The shaft of the encoder plugs into an oversized hole drilled in the azimuth pivot bolt in the rocker bottom. The pivot bolt is installed through the rocker bottom and is permanently screwed into an all-metal lock nut and plate that you attach to the underside of the ground board. These parts are supplied in the mounting kit from the manufacturer.

An O-ring on the shaft of the encoder provides a snug fit within the pivot bolt bore. Thus the encoder can be installed or removed with a simple press fit.

The hole you drill through the rocker bottom for the azimuth bolt must satisfy three conditions:

- It must be in the center of the rocker. Locate the bolt where pencil lines drawn from the corners of the rocker bottom intersect.

- It must be exactly perpendicular to the rocker bottom. Use a drill guide to make the hole perpendicular.

- It must be the same size as the bolt diameter. The rocker must not wobble on the bolt. If you drill the hole oversize, press fit a steel flat washer with the same hole size as the bolt diameter into a countersunk hole in the rocker bottom. The washer provides a metal-to-metal bearing that eliminates any chance for azimuth rocker wobble.

The encoders are about ½ inch thick. Make sure you have at least an inch of clearance between the mirror box and the top of the rocker bottom when the scope

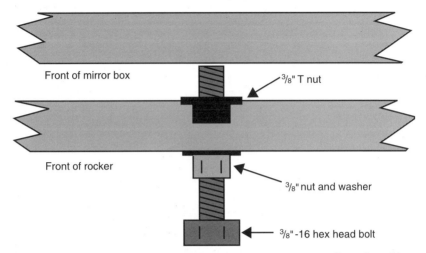

Fig. C.7 *To provide the zenith reference that some systems require, install an adjustable stop on the front board of the rocker. When you initialize the digital setting circles at the beginning of the night, the adjustable stop lets you aim the telescope at the zenith in a matter of seconds.*

is rotated in altitude. This leaves sufficient clearance if you don't push the encoder into the pivot bolt all the way. The last thing you want is for the tailgate or primary mirror to crush the azimuth encoder.

Permanently attached to the encoder is a mounting arm about 10 inches long. Attach the free end of the arm to the rocker bottom with two thumb screws so that you can remove the arm and the encoder for transport. Run the thumb screws into a couple of self-tapping threaded brass inserts installed in the rocker bottom.

The arm is wide so that the encoder is effectively prevented from rotating even a tiny fraction of a degree, but long and thin enough that the encoder can move up and down slightly. Never mount an encoder rigidly; if the rocker is lifted off the ground board and the pivot bolt moves up or down, the encoder would be damaged. By mounting it on a long, thin arm, the encoder is kept both accurate and safe.

C.4.2 Mounting the Altitude Encoder

The altitude encoder is more difficult to mount. The altitude encoder shaft must be mounted exactly at the center of rotation of the side bearing, and it must be in line with the axis of rotation. If the encoder is displaced from the center, it will not accurately report the rotational position of the telescope, and if the shaft is angled, the encoder could be damaged.

As we did for the azimuth encoder, the altitude encoder is mounted on a long arm. The mounting arm can flex slightly to allow lateral shifting of the mirror box within the rocker, but it cannot rotate and thus report a false position.

The upper end of the arm holds a female connector that attaches to the altitude encoder shaft. The female connector has an internal O-ring that slips snugly over the encoder shaft. (You can put the female connector on the side bearing and the encoder on the arm attached to the rocker, if you wish.)

Finding the center of rotation of the side bearing is tricky. Drill a close-fitting hole in a 2 by 4 pine board and press a pencil with a sharp point through the hole. Temporarily clamp this board to the side of the rocker. While you watch the pencil, have a friend move the tube of the telescope up and down. Adjust the arm until the point of the pencil remains at the exact center of rotation. Press the pencil point to mark this location. The encoder shaft must be centered at this point.

Connect the bottom end of the altitude support arm to the rocker side with a couple of thumb screws and threaded brass inserts, as you did for the azimuth encoder. After the encoder and the arm are mounted, bring them close together and check your work. Have your friend move the scope up and down while you observe the relationship of the female connector to the encoder shaft. If the male connector wanders around, it is not dead center and you must relocate it. When you have accurately located the pivot point and attached the encoder, the job is done. Slip the connector together and attach the bottom end of the arm with the thumb screws.

Leave both the altitude and azimuth encoder arms loose. Do not tighten the brass thumb screws on either support arm. The arms need to be able to move slightly as the scope is slewed or stress to the encoder shafts will cause errors or damage the encoders.

C.4.3 Mounting the Computer

Mounting the computer may seem like a small point, but it is not. When you observe with a telescope that is taller than you are, you will soon find that you really want the computer in two different locations. When your feet are on the ground and you are slewing the scope to a new object, you want the computer down by the mirror box or in your hand where you can see the digital readout and press the buttons.

However, once you have the telescope pointed close to the desired object, you want the computer with you at the eyepiece. You may want to read the display to find the magnitude of the object or its description, or you may want to center the object carefully in the field and update the computer so it will find the next object more accurately. All this must be done at the eyepiece.

The solution is to put the computer on a long cord and keep it mobile. When you are aiming the telescope, hang the computer on the mirror box. When you climb the ladder, take the computer with you and attach it to the focuser board where it is handy.

Place strips of self-adhesive Velcro on the back of the computer and at each of these locations. (All night long, people near your telescope will hear the distinctive "rrrip" of Velcro each time you climb or descend the ladder.) Or, if you like a telescope that moves too easily for detaching Velcro, attach small metal hooks and hang the computer where you need it.

C.5 General Advice

Digital setting circles are more delicate than the rest of your telescope. Remove the encoders when you transport the telescope. The azimuth encoder is especially vulnerable if the leading edge of the ground board catches on the floor of the trailer. It might not be harmed the first time, but sooner or later it will be. The altitude encoder sticks out on the side of the telescope, and is likely to get knocked off when you brush past some minor obstacle. If you are very lucky, no damage will be done.

To prevent either of these accidents from happening, remove the support arms and the encoders when you are done observing. Unplug these delicate components and stow them safely in a box. If you wait until the next morning to remove them, you will probably forget and—crunch—there goes another set of encoders.

Appendix D
Equatorial Platforms

Dobsonian telescopes have one major drawback: they do not track the stars. No matter how smoothly the telescope moves, the object moves and you must chase after it. This makes the careful study of fine detail difficult and makes seeing faint objects tricky, especially at high power.

Many observers like to star-hop around a page of *Uranometria 2000.0*, hunting out the galaxies and planetary nebulae. Without tracking, it's down the ladder to peek at the chart, over to the finder, up the ladder to the eyepiece—ah-ha—there it is! Then it's down the ladder for a different eyepiece, perhaps a quick peek at the chart to check on a faint field galaxy, then back up the ladder, fuss around recovering the object, plugging in the new eyepiece, then making countless gentle nudges to track the object. How nice it would be if the object would just stop drifting!

Tracking also opens the exciting potential of astrophotography and CCD imaging to the Dobsonian owner. There is so much light bouncing off your big mirror that long exposures and onerous guiding are just not necessary, at least not for the brighter objects. Wonderful pictures can be "snapped" in 5 to 10 minutes on fast color film; guiding during such a short period is not much of a chore. With CCDs, the central star in the Ring Nebula shows up in a 1-second exposure, and integrations of 60 seconds reveal stars four to five magnitudes below the visual limit.

One solution to the Dobsonian-builder's dilemma is the equatorial tracking platform. A French amateur astronomer named Adrien Poncet demonstrated, in 1977, a low platform that would, by virtue of a pivot point and two points sliding on a plane, rotate around an axis perpendicular to the plane. If this platform were motorized and the plane aligned to face the celestial pole, the motor counteracted the rotation of the Earth. For a telescope placed atop the platform, the stars would no longer appear to be moving.

Poncet's original design was restricted to small telescopes and cameras; the sliding contact would not slide if the platform carried more than a light load. However, by replacing Poncet's sliding contacts with ball bearings, amateurs realized lower friction and greater load-carrying capacity. In 1980, a group in England put a small observatory shed on an equatorial platform. In 1983, Kansas City amateur Tom Martinez was among the first, if not the first, to demonstrate a practical equa-

Fig. D.1 *Tom Martinez built one of the first equatorial platforms that was beefy enough to carry the weight of a large Dobsonian. However, because the rollers ride on inclined surfaces, heavy loads bend the platform slightly.*

torial platform sufficiently beefy to carry a large Dobsonian telescope without shaking.

In its modern incarnation, the equatorial platform is basically a motorized ground board. The telescope turns on this modified ground board, just as it always does. The only difference is that this particular ground board is moving equatorially at the sidereal rate—that is, it's tracking the sky. The price you pay for this wonderful convenience is small: another half-step that you must climb up the ladder.

Actually, there are quite a few different ways to build an equatorial platform, but all platforms have three components in common:

- the support plate,
- the rocking plate, and
- the drive system.

The support plate rests on the ground, the rocking plate sits on top of it, and motors in the drive system move the rocking plate. In different designs, the details of the pivot, ball bearings, segments of polar disks, inclined planes, gears and drive rollers vary. They may be mounted on the support or on the bottom of the rocking plate.

For example, the original Poncet design had a pivot and an inclined plane,

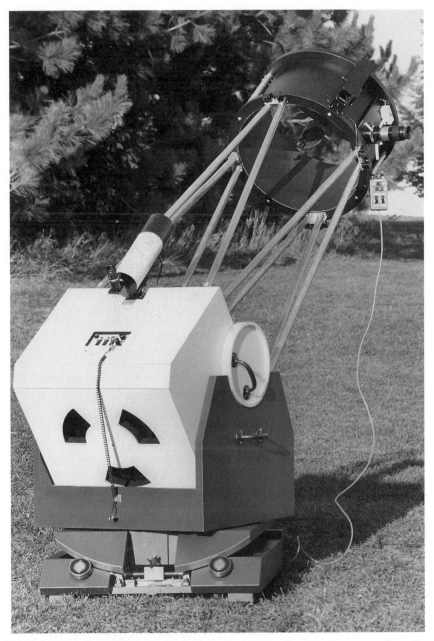

Fig. D.2 *In Pete Smitka's equatorial platform, stationary rollers support an arc of a cylindrical disk. This arrangement can carry heavier telescopes than its predecessors, but the load is still cantilevered on the rollers.*

Fig. D.3 *In Tom Osypowski's design, a conical bearing surface rides on horizontal rollers. This keeps the whole unit low and close to the ground. This design can carry heavy loads and still track precisely for an hour.*

but in 1988, another French amateur, Georges d'Autume, demonstrated a pivotless platform which rotates on bearings that support concentric cones mounted on the bottom of the rocking plate. Amateur astronomer Andy Saulietis formed a company, DIG, Inc., to manufacture his own variation of the pivotless platform.

Another amateur astronomer, Tom Osypowski, has spent many years bringing the variation on the original Poncet design proposed by Alan Gee to perfection. The Gee design has a pivot and two bearings mounted on the support plate, and a polar axis and a segment of polar disk mounted on the bottom of the rocking plate. Osypowski's design is capable of tracking large and heavy telescopes with high precision. The Gee-type platform employs the same mechanical principles as the horseshoe equatorial used on the 200-inch Hale telescope on Palomar mountain. The horseshoe is trimmed down to an arc attached to the bottom of the platform—just enough to give one hour of precise tracking. The rest of the structure is gone, replaced by the platform that carries the Dobsonian rocker of your telescope.

In Osypowski's platforms, the support plate sits on the ground, like the ground board in a regular Dobsonian. At the north end are two roller bearings inclined at the latitude angle, at the south end, a single bearing, the pivot, for the polar axis. Together, these bearings carry the rocking plate. The rocking plate swivels equatorially around the south bearing. The two north bearings support the small remaining part of the horseshoe attached to the bottom of the rocking plate.

Fig. D.4 *Tom Osypowski took this picture of M42/43 using one of his platforms, a 16-inch f/5 Newtonian, 5 minutes (unguided) on Ektar 1000 print film. Note the nice round star images.*

The rocking plate rides on the support plate. The curved sector attached to the rocking plate is inclined at the latitude angle; the sector is all that is left of the horseshoe. The rocking plate turns through an angle of 15°, corresponding to one hour of tracking time.

Osypowski uses a friction drive system. The remaining segment of the horseshoe has a radius of 25 inches or more, and it rests on two rollers. To achieve high precision tracking, two geared-down stepper motors used with their gear trains preloaded against each other turn one of the rollers that the horseshoe rests on. The stepper motors are driven by a small computer mounted in a small cavity on the side of the support plate. The computer drives the motors at different rates and accepts inputs from a hand-held control paddle.

To achieve tracking accurate enough for astrophotography, Osypowski places a small motor that can raise or lower the south Teflon pad. This moves the telescope a small distance in declination 20° on either side of the meridian, exactly what is needed for guiding a long-exposure photograph.

Constructing an equatorial platform is not a trivial project. Ideally, you should be a skilled woodworker with a well-equipped shop; at the very least, a table saw, router, power drill, and saber saw. While a platform is as difficult to construct as any equatorial mount, the rewards of motorized tracking make it all worthwhile. For anyone who has spent time viewing with a large Dobsonian at high power, it's somewhat of a shock at first to see the image remain motionless in the center of the field. Then, it begins to sink in: "Hey, now I can relax and really start to observe!"

Appendix E
Resources and Suppliers

This appendix addresses the question "Where do you get that?" The answer is that most of the parts used to construct your dream telescope can be found at the local hardware store and lumber yard. It is truly amazing how many things can be adapted to telescope making if you use your imagination. All too often, amateur laboriously hand-make parts that are readily available with standard, commercial, off-the-shelf industrial products.

The secret to finding the very best and cheapest industrial products for telescope making is simple: get obsessed. Invade countless hardware stores, question every clerk, make endless follow-up phone calls to run down the parts that you need.

The suppliers listed in this appendix make a fantastic array of useful gadgets and gizmos that become parts of the commercial Obsession telescope, and could easily become parts of your Dobsonian, too. However, remember that you are at a considerable disadvantage when you order some of these things. Distributors and wholesalers are used to taking an order for ten thousand of the doodad you want three of. You will almost hear the laughter in their voices as they incredulously repeat, "You want three?"

If possible, obtain a catalog ahead of time so you can decide exactly what you want and sound intelligent when you order. Many companies have minimum orders, don't sell retail at all, won't sell to anyone that hasn't passed their credit department or simply don't want to bother with your tiny order. Be prepared for this. Identify yourself and explain your project. Beg them. Offer to pay in advance or have them send it C.O.D.

Sometimes a simple request for a "sample" is all it takes. That's because they would rather give you three doodads for free than to go through all $15 worth of paperwork for a $2 order. However, the odds that this will get you what you want diminish in direct proportion to the success of this book! If all else fails, buy enough to meet their minimum order requirement and sell the extras to other telescope makers.

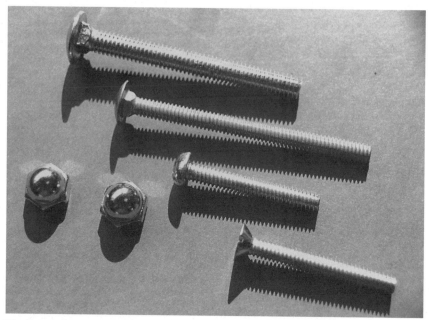

Fig. E.1 *Carriage bolts (top) have flat or round heads and a square neck that resists turning. Round-head and flat-head machine screws have slotted or Phillips heads. Acorn nuts are useful in constructing the mirror cell.*

E.1 Aluminum tubing

1¼-inch anodized aluminum tubing, 0.055-inch wall, 6-foot or 8-foot lengths.

• Ace Hardware stores. Cost is $15 to $20 each, six-foot length.

1½-inch non-anodized aluminum tubing, 0.049-inch wall, 6061-T6 alloy, 12-foot length.

• Macklanburg-Duncan, PO Box 25188 Oklahoma City, OK 73125. (800) 654-8454.

• Tube Sales Inc., Los Angeles, CA. (800) 565-8637. This company has a great variety of tube shapes and materials. If they don't have it, nobody does. Generally you'll pay from $2 to $3 per foot. Fantastic catalog of tubing in every shape and alloy imaginable.

E.2 Foam tube insulation

Ace Hardware stores or plumbing supply houses.

E.3 Double-sided tape

½-inch wide by 33 yard roll, item #CW-2570, $8. Other sizes available.

• Star Packaging, 616 S. 92 St., Milwaukee, WI 53214. (800) 634-0901.

• Ace Hardware stores

E.4 Specialty tools, knobs and latches

Bondhus Balldriver Hex Tools. The Balldriver is item #10705 CL, The L stands for long blade length of 5.5 inches; the cost is only $3. Be sure to get three button-head screws or socket-head screws from the hardware store first.

Self-tapping threaded inserts in brass, steel, and stainless. For use in plastics, metal, soft and hard woods. Coarse, fine and metric threads. A table in the Reid catalog gives the recommended pilot hole size.

Black knobs for split blocks with J threaded hole, ask for item #DK-55 Latches for "The Big Ones" ask for both item #DDL 802-G (latch) and #DDL 802-2 (strike)

• Reid Tool Supply, P.O. Box 179, Muskegon, MI 49443. (800) 253-0421. A great source of knobs, handles, and odd tools.

E.5 Threaded inserts

1¼-inch or 1½-inch with 1¼-20 thread, order numbers #3046-4 and #3047-4, respectively. The cost is about $0.65 each.

• Superior Components, 12006 Spaulding School Dr., Unit 104, Plainfield, IL 60544. (815) 254-2313.

• Plastiglide Manufacturing Corp, 19440 S. Dominguez Hills Dr., Rancho Dominquez, CA 90220. (310) 885-4500. Plastiglide makes the inserts sold by Superior Components. They have 25 catalogs and offer an abundance of neat stuff.

E.6 Levelers (collimation knobs)

Item #FC 3340, but they have a minimum order of 100 pieces. The cost is $0.85 each. See if they can give you the name of a distributor, or try asking for a "sample" of three.

• Buckeye Fasteners, 5250 W 164 St., Cleveland, OH 44142. (800) 437-1689.

• Ace Hardware stores.

E.7 Felt tabs and floor protector pads (mirror cell pads)

• Ace Hardware stores.

E.8 HVHC plywood (Hardwood Veneer Hardwood Core)

- AplePly brand plywood, State Industries, PO Box 7037, Eugene, OR 97401. (541) 688-7871.

- Baltic Birch brand plywood, Allied International, 200 Baker St., Suite 210, Concord, MA 01742. (508) 371-3399. Ask for a source near you.

E.9 Sliders or cord locks and sling webbing

Fastex Lock with Wheel, item #67794-E. A set of five is only $3.00.

- Campmor Supply, 810 Rt. 17, North Paramus, NJ 07652. (800) 226-7667. Ask for their catalog. It's one of the best camping supply mail order outfits. They have cold weather clothing, mini-flashlights, buckles, tarps, nylon fabric and Velcro by the yard, as well as many other things that amateur astronomers can use. Do not get their polyester webbing as it is much too elastic.

- The local junk yard. Take along scissors and cut out a seat belt from an old Ford. Makes a terrific mirror sling.

E.10 Heat ropes

- American Scientific, 3605 Howard St., Skokie, IL 60076. (847) 982-0874. Ask for their catalog. This is a mail-order warehouse with the oddest stuff. Surplus electronics, motors, fans, pulleys, gears, science things, military, magnets, optics and many strange items of dubious value all described in a clever 50-page catalog. A must for any telescope maker.

E.11 Heat paks

Chemical heating pads. Any good sporting goods store, especially those that supply hunters. If you have trouble locating them contact Grabber, the manufacturer.

- Grabber Inc., 205 Mason Cr., Concord, CA 94520. (510) 687-6606.

E.12 Black Ripstop Nylon (for the light shroud)

- JoAnn Fabrics, a national chain. Check your Consumer Yellow Pages. If you can't find a JoAnn Fabrics store, try Fabri-centers of America, 5555 Darrow Rd., Hudson, OH 44236. (216) 656-2600. It costs $6 per yard in 1997.

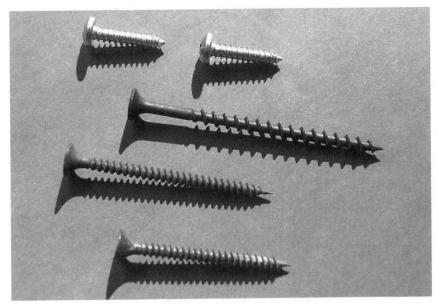

Fig. E.2 *Use screws for attaching parts to wood. At top are pan-head sheet metal screws you can use to attach pole seats to the rings of the secondary cage. Below are three sizes of flathead Phillips drywall screws used for the split-block pole fixtures.*

E.13 Kydex plastic for the light baffle

- Kleerdex Company, 100 Gaither Dr., Suite B, Mount Laurel, NJ 08054. (800) 325-3133. Available from 0.027-inch thickness on up. In the 0.027 thickness, a 4 by 8-foot sheet is $35. Call for the name of a distributor near you.

E.14 Etched virgin Teflon

- Midland Plastics, 5405 S Westridge Ct., New Berlin, WI 53151. (414) 938-7000. 0.093-inch thick Teflon is $35 per square foot. They have a huge inventory of plastics and their catalog is a good reference.

- AstroSystems, 5348 Ocotillo Ct., Johnston, CO 80534. (970) 587-5838.

E.15 #1782 Stardust Quarry Finish Formica

Formica Inc., 10155 Reading Rd., Cincinnati, OH 45246. (800) 524-0159. For a 4 by 8-foot sheet, the cost is $40.

Fig. E.3 *T nuts (left) place metal threads in plywood panels. At top center are brass threaded inserts that provide permanent threads in holes drilled in wood. All-metal locknuts (center) and plastic-insert locknuts act as non-loosening nuts on machine screws and bolts.*

E.16 #4552-50 Ebony Star countertop laminate

- Wilsonart Inc., 2400 Wilson Place, Temple, TX 76504. (800) 433-3222. For a 4 by 8-foot sheet, the cost is $70.

E.17 Glassboard (bead board or fiberglass reinforced panel)

- Sequencia Inc., 15900 Foltz Industrial Parkway, Strongsville, OH 44136. (800) 637-0095. For a 4 by 8-foot sheet, the cost is $30.

E.18 Self-tapping threaded wood inserts

- Woodworkers Supply, 1108 N. Glenn Rd., Casper, WY 82601. (800) 645-9292. Large catalog offering brass inserts, T-nuts, fasteners of all sorts, every wood working tool imaginable, finishes for wood, abrasives, adhesives. If you enjoy working with wood, get their catalog.
- McFeely's, PO Box 3, Lynchburg, VA 24505. (800) 443-7937. Great fastener catalog including decorative screw caps, specialized tools, threaded inserts. Highly recommended.
- Ace Hardware Stores.

E.19 Cam levers

- Quik-n-Easy Parts, PO Box 874, Monrovia, CA 91017. (818) 358-0562. Order the lever, item #U111, and the nut, item #U112 together. The cost is $3 per set.

E.20 12-Volt cooling fans

4½-inch fan, a real blower, for about $15.

- Skycraft Parts, 2245 W Fairbanks Ave.; Winter Park, FL 32789. (407) 628-5634, is another candy store for amateur telescope makers. If you're ever passing through the Orlando area on the way to Disney World or the Winter Star Party, stop and browse.

- Radio Shack and American Scientific also sell cooling fans under the name "muffin fans."

E.21 Ladders (for big telescopes)

- McMaster-Carr Supply Co., Chicago IL. In the Chicago area (630) 833-0300; in the Los Angeles area, (310) 692-5911; in the New York area, (908) 329-3200. McMaster-Carr sells every imaginable ladder from little two-step ones to big, double-access industrial types. We recommend the extra heavy duty ladder, #7957T16, listed at $200 in catalog #94. McMaster-Carr is the largest mail order industrial supply company in the world. Their catalog is over 2,200 pages long and contains just about everything. If you would like to "page through" the ultimate hardware store get this catalog.

E.22 Small parts

- Small Parts Inc., PO Box 4650 Miami Lakes, FL 33014. (305) 557-8222. As the name implies this company sells small parts, useful and unusual items, stainless steel fasteners, clamps, gears, tubing, Teflon and other plastics, and tiny drills, just to name a few. Great for high-tech telescopes makers as well as Dobsonian buffs.

E.23 Epoxy resins

- Defender Industries, 42 Great Neck Rd., Waterford, CT 06385. (800) 435-7180. Epoxy and fiberglass resins by the quart on up to 5 gallons, Styrofoam, great varnishes, and a multitude of odd stuff for the marine industry. If you're interested in making ultralight foam-core telescope parts, get their 200-page catalog.

Fig. E.4 *Threaded inserts are the best way to attach end brackets to the thin-wall aluminum truss poles in your telescope. Once inserted, threaded inserts will not pull out. Also pictured are T nuts in lower left corner and threaded brass inserts for wheel barrow handles, upper right corner.*

E.24 Loading ramps, pneumatic wheels for wheelbarrow handles, and marine trailer jacks

- Northern Hydraulics, 2800 S. Cross Dr., W. Brunsville, MN 55306. (800) 222-5381. This discount hardware store has a 135-page catalog with a full line of trailer jacks, hand dollies, pneumatic wheels, castors, electric fans, kits to make your own loading ramps, winches, tools, trailer kits.

E.25 Black aperture f/ratio labels

- Tom Macomber, PO Box 39, Fremont, IN 46737. (219) 495-9991. Custom made self-adhesive black labels. Great for labeling your scope's focuser board with the aperture, focal ratio or any other name you want to put on it. Makes it look professional. (Tom is a professional sign maker and enthusiastic astronomer too!)

E.26 Plastic grommets for light shroud

- Coghlan's 8 Snap n Tap made in 121 Irene St., Winnipeg, Manitoba Canada R3T 4C7. (204) 284-9550.
- Check camping supply stores and hardware stores.

E.27 Duct tape

Duct tape is your friend. Get it at your local hardware store. Great for impromptu repairs, holding counterweights, charts, or anything that won't stay put. Get a roll and never go observing without it.

E.28 Collimation tools

- Tectron, 2111 Whitfield Park Ave. Sarasota, FL 34243. (941) 758-9890.
- AstroSystems, 5348 Ocotillo Ct., Johnston, CO 80534. (970) 587-5838.

E.29 Commercial Components for Telescope Making

E.29.1 Optics, focusers, digital setting circles, eyepieces, Telrads

- See ads in *Sky & Telescope* or *Astronomy* magazines.

E.29.2 Spiders and secondary mirror holders

- Novak & Co, Box 69W, Ladysmith, WI 54848. (715) 532-5102.
- AstroSystems, 5348 Ocotillo Ct., Johnston, CO 80534. (970) 587-5838.
- ProtoStar, PO Box 258, Worthington, OH 43085. (614) 785-0245.

E.29.3 Cast aluminum pole seats, clamping wedges, side bearings

- AstroSystems, 5348 Ocotillo Ct., Johnston, CO 80534. (970) 587-5838.
- Tectron, 2111 Whitfield Park Ave. Sarasota, FL 34243. (941) 758-9890.

E.30 Mirror Making Supplies

- **Pyrex glass:** United Lens, 259 Worchester St., Southbridge, MA 01550. (617) 765-5421 or United Lens, 2781 East Regal Park Dr., Anaheim, CA 92806. (714) 630-6035.
- **Kerr dental plaster:** Kerr Co., Romulus, MI. (800) 537-7333. Order Vel-Mix Stone Pink, for making plaster tools. This stuff hardens like

stone! Ask for names of dental suppliers in your area. Kerr has 400 distributors nationwide. It costs about $12 for 25 pounds.

- **Microgrit abrasives:** Micro Abrasive Corp., 720 S. Hampton Rd., Westfield, MA 01086. (413) 562-3641. Ask for names of distributors.

- **Microgrit abrasives:** Fusco Abrasive Systems, 17899 S. Susanna Rd., Compton, CA 90221. (310) 637-3427. Order Adolf Miller pitch #73. It comes in 1 kilogram tubes.

- **Microgrit abrasives:** Edmund Scientific Company, 101 E. Gloucester Pike Barrington, NJ 08007. (609) 573-6259.

E.31 Mirrors for Large Dobsonians

The mirror is the most important component in your large Dobsonian. We strongly recommend that you read current astronomical magazines for optical shops that make large mirrors, check with amateur telescope making discussion groups on the Internet, and attend star parties to compare the quality of the mirrors in working telescopes.

- Galaxy Optics, P.O. Box 2045, Buena Vista, CO 81211, 719-395-8242.

- Nova Optical, P.O. Box 80062, Cornish, UT 84308, 801-258-5699.

- Pegasus Optics, RR 5-Box 502, Huntsville, Arkansas 72740, 501-738-1650.

- Star Instruments, P.O. Box 597, Flagstaff AZ 86002, 520-774-9177.

- Telescope Engineering Co., 747 Sheridan Blvd. #4 B, Lakewood, CO 80214, 303-274-7944.

Afterword

Since John Dobson constructed his first telescope, the Dobsonian has undergone a truly remarkable evolution. The telescopes that we describe in this book are outstanding performers, the beneficiaries of the best thinking of a generation of amateur telescope makers. Anyone who builds a telescope based on the ideas set forth in this book will certainly be pleased with the result. However, we recognize that the evolution of the Dobsonian has not yet ceased and will presumably continue in the future. Inevitably, amateur astronomers will refine their telescopes and experiment to discover new ways to build better ones. Exotic materials may prove better than Pyrex for mirrors, all-dielectric high-reflectance coatings may become the norm, and plastics with lower friction and better stick-slip properties than anything we have today could replace our present Stardust-and-Teflon bearings. However, there are five areas that we believe amateur telescope makers should watch with special attention: larger apertures, new optical systems, egg-crate mirror blanks, improved mirror cell designs and computer driven telescope mountings.

Afterword 1: Larger Apertures

Within a few years of their conception, Dobson's large telescopes broke through the pre-Dobsonian "16-inch barrier" and reached the unheard-of aperture of 24 inches. When the Dobsonian lost its solid tube, the peak apertures built climbed to 30 inches, then 36, and finally reached what some telescope makers consider a natural limit in the vicinity of 40 or 41 inches aperture. At this size, many factors conspire: the difficulty of finding large blanks, the time and cost of grinding the mirror, the growing problems of weight and length and the difficulties involved in setting up and using the finished telescope.

One potential road leading to larger mirrors is faster mirrors. We may see amateur or commercial opticians attempt mirrors with focal ratios of $f/3.5$ or even $f/3.0$ — focal ratios superseded decades ago in large professional telescopes. Relatively few amateurs will attempt to crash through the "40-inch barrier," because such fast mirrors are difficult to make, but the reward of owning a 50-inch Dobsonian may be sufficient inducement. For innovative instruments at the large-aperture end of the telescope continuum, keep an eye on amateur telescope makers who learn how to make fast mirrors.

462

Afterword 2: New Optical Systems

The desire to observe with a large-aperture telescope and still keep both feet on the ground has inspired many amateurs to seek innovative optical designs. For example, the folded Newtonian built in the late 1980s by the Los Angeles Astronomical Society employs a 10-inch flat to bring the eyepiece of this 31-inch altazimuth telescope down to eye level. Amateur astronomers have traditionally preferred to avoid the complexities of figuring, mounting, and collimating Cassegrain or Gregorian secondary mirrors — yet the payoff in convenient eyepiece location may outweigh the difficulties. Imagine looking through a 40mm Panoptic eyepiece located at the Nasmyth focus of a 40-inch f/8 Cassegrain, enjoying celestial sights with a magnification of 200 and an exit pupil of 5 millimeters! Despite the difficulties, there is reason to believe that not all future Dobsonians will be Newtonians.

Afterword 3: Egg-Crate Mirrors

Today's big Dobsonians have thin mirrors primarily because thick mirrors weigh too much for a portable telescope. However, these thin mirrors are quite rubbery. Unless carefully supported in a flotation mirror cell, the mirror bends under its own weight. However, by making a mirror from a properly designed internally hollow "egg-crate" blank, it is possible to make mirrors that are both lighter and more rigid than today's mirrors. One amateur astronomer who has made a telescope with an egg-crate mirror is John Vogt. He ground, polished, and figured his pioneering 30-inch f/4 primary mirror on a blank manufactured by the Hex-Tek Corporation. The mirror is 4 inches thick yet it weighs only 40 pounds, so it is very stiff compared to a typical "thin" mirror. The front and back faces of the mirror are ½-inch thick, and the interior of the blank is a cellular structure with ribs ⅜-inch thick. Holes in the internal cells allow air circulation, so that the mirror structure cools rapidly.

As a pioneering worker with egg-crate optics, John encountered significant difficulties in figuring the mirror because the rib structure tended to "print though" the thin front plate of the blank. He ultimately overcame the difficulties to produce a mirror that performs well in his telescope. Although it may be a long shot, egg-crate blanks for lighter, stiffer mirrors are another area worth keeping an eye on.

Afterword 4: Improved Mirror Cell Designs

Flotation mirror cells date back to the 1830s and 40s, when the great English amateurs Lord Rosse and William Lassell were building large telescopes. Lassell used levers to apply a restoring force to the rear of his mirrors. Lassell's greatest triumph was the famous 48-inch reflector installed on the island of Malta in 1861. Likewise, Lord Rosse supported the mirror of the 72-inch reflector at Burr Castle on a 81-point cell and that of his 36-inch looks remarkably similar to the 27-point cell described in this book. For American telescope makers, however, an article by John H. Hindle titled "Mechanical flotation of mirrors" published in 1933 in *Amateur Telescope Making*, Book 1, became the source most frequently consulted by telescope makers. In the article, Hindle states: "The mirror must rest on a number of self-adjusting supports, geometrically symmetrical ... each taking an equal share of the weight." This was generally understood to mean that the mirror should be divided into zones of equal area, and later, into areas of equal weight.

For almost fifty years, this formulation stood unchallenged until amateur astronomer Richard Schwartz began to model mirror cells using the inexpensive student version of a sophisticated finite element analysis software package. He soon discovered that the areas, weights, and support forces did not need to be equal, and then he found that engineering studies made in 1982 for the 10-meter Keck Telescope (see Jerry Nelson, Jacob Lubliner and Terry Mast "Telescope mirror supports: plate deflections on point supports" in *SPIE Proceedings* Volume 332) supported his findings.

Schwartz soon discovered that he could design cells that theoretically outperform those cells based on Hindle's formulation, and that he could design excellent cells that required fewer points than Hindle's. Although flotation cells based on Hindle's formulation work quite well, better ones are possible. As this book goes to press, the new designs have not been built and are therefore untested, but improved mirror cell designs clearly represent an area that interested telescope makers should watch closely.

Afterword 5: Computer-Controlled Mountings

Controlling telescopes by computer is hardly a new idea. Since the 1930s, virtually every major professional telescope has incorporated some degree of automatic control, although in the older instruments, the controller unit was an analog computer. Widespread use of digital computers followed the introduction of DEC's PDP-8, an "inexpensive" minicomputer with a 100kHz CPU and 4 kilobytes of random-access memory. Twenty years later, few professional telescopes remained uncomputerized.

Computer-controlled Dobsonian telescopes entered amateur astronomy in the late 1980s. An outstanding early example was David Gedahlia's 10-inch *f*/4.5 Dobsonian driven by an Atari 800XL computer, shown at the 1987 Riverside Telescope Makers Conference. With the Atari driving altitude and azimuth stepper-motors, the telescope would move automatically to coordinates entered on the computer's keyboard. David was a third-year engineering student when he built this telescope.

Then, in 1989, *Sky and Telescope* published a brief article by a Japanese amateur named Toshimi Taki describing an efficient method for aligning an alt-azimuth telescope mounting by sighting just two stars. Taki's software used an efficient matrix method to perform the coordinate conversions, thus removing a significant mathematical obstacle that had frustrated many would-be digital-control-engineer/amateur-astronomers. More recently, amateur astronomer Mel Bartels pointed the way to full computer-control, sharing his ideas on the Internet with amateur astronomers around the world. In Mel's prototype, a home-built 20-inch Dobsonian, a laptop computer drives stepper motors for slewing to a fraction of a degree and tracking accurate to about 1 arcsecond, which is entirely adequate for direct photography and CCD imaging.

Since the entire project — including the mechanical design, the electronic circuits, and the software — has been "published" on the World Wide Web, Mel has received advice and feedback from electronics and software experts that has enabled him to refine and improve the basic design. It is clear that computer control is an area of future Dobsonian development that bears close watching. Even if the technology is a bit too involved for most amateur telescope makers in the late 1990s, that situation is likely to change as faster, less expensive computers and microcontrollers become available.

As individuals, we amateur telescope makers may not advance the art of telescope making very much. Collectively, however, we have totally changed amateur astronomy in the last two decades. This book is necessarily a snapshot in time, portraying a particular type of telescope in its current state of development. Although no one can predict exactly how, I am certain that the Dobsonian will continue to evolve, and that all of us will benefit from its further evolution.

Richard Berry
Lyons, Oregon

The Dobsonian Bibliography

Telescope Making Books

Berry, Richard. *Build Your Own Telescope*, details the construction of five astronomical telescopes from 4-inches to 10-inches aperture. Willmann-Bell, Richmond, VA, 1994.

Ingalls, Albert G. *Amateur Telescope Making* Books 1, 2, and 3, this three-volume compendium reaches back to the roots of amateur telescope making in America, and covers every aspect of the subject. Willmann-Bell, Richmond, VA, 1996.

Macintosh, Allan. *Advaced Telescope Making Techniques, Volume 1: Optics, Volume 2: Mechanical,* feature selected articles from the Matsutov Circulars which were published for about 20 years starting in 1957. Willmann-Bell, Richmond, VA, 1986.

Rutten, Harrie and Martin van Venrooij. *Telescope Optics: Evaluation and Design*, provides a comprehensive overview of telescope optics for the amateur astronomer. Willmann-Bell, Richmond, VA, 1988.

Suiter, Harold Richard. *Star Testing Astronomical Telescopes, A Manual for Optical Evaluation and Adjustment*, the best (and only) must-read for testing the telescopes. Willmann-Bell, Richmond, VA, 1994.

Texereau, Jean. *How to Make a Telescope*, Second English Edition, the bible for amateur astronomers who wish to grind and polish a high-quality telescope mirror. Willmann-Bell, Richmond, VA, 1984.

Trueblood, Mark and Russell Merle Genet, *Telescope Control*, is the second edition of *Microcomputer Control of Telescopes* (1984), and deals with every aspect associated with computerizing a telescope. Willmann-Bell, Richmond, VA, 1997.

Dobsonian Books and Booklets

Berry, Richard, editor. *How to Build a Dobsonian Telescope*, a booklet compilation from *Telescope Making* magazine, Astromedia, 1980.

Clark, Tom. *The Modern Dobsonian*, a useful booklet from the owner of Tectron Telescopes. Tectron Telescopes, Sarasota, FL, 1992.

Cunningham, Randy. *Truss Tube Telescopes*, a handy booklet that systematizes the design and construction of a large Dobsonian. Available from Astrosystems, 1993.

Dobson, John. *How and Why to Make a User-Friendly Sidewalk Telescope*, how to build a Dobsonian telescope of the classical design by Dobson himself. Everything in the Universe, Oakland, CA, 1991.

Overholt, Steven. *Lightweight Giants: Affordable Astronomy at Last*, this booklet offers a plentitude of tips for reducing the weight of large telescopes. Owl Books, San Juan Capistrano, CA, 1991.

Periodical Literature (in chronological order)

McDonald, Lee. "Deep-sky objects in a 24-inch reflector," an enthusiastic description of observing with Dobson's largest telescope, with only the slightest description of the telescope itself. (See also "A September conference of amateurs at Oakland," in the same issue.) *Sky and Telescope*, January 1973.

Peters, William, and Robert Pike. "The size of the Newtonian diagonal," gives formulae for computing the size and vignetting losses from secondary mirrors. *Sky and Telescope*, March 1977.

Smith, Stephen. "Compact 6-inch refractor," about refractors on simple alt-azimuth mountings. *Telescope Making #2*, Winter 1978.

Berry, Richard. "Elegant design in tubing and framework," about truss tube telescopes. *Telescope Making #4*, Summer 1979.

Berry, Richard. "Large portable aperture and ease of use," describes telescopes built by Doug Berger and Earl Watts. *Telescope Making #4*, Summer 1979.

Berry, Richard. "Ultimate simplicity with a 'big box' telescope," a design that foreshadows the Dobsonian. *Telescope Making #4*, Summer 1979.

Kestner, Robert. "It's stability that counts," on the key ideas behind the Dobsonian. *Telescope Making #4*, Summer 1979.

Kestner, Robert. "The Dobsonian telescope," the first detailed instructions for building a classic Dobsonian. *Telescope Making #5*, Fall 1979.

Sinnot, Roger. "Spinoffs of the Poncet mounting," a collection of ideas gleaned from the readership. *Sky and Telescope*, February 1980.

Berry, Richard. "A 20" Newtonian on an alt-azimuth mount," about the construction of a Dobsonian. *Telescope Making #7*, Spring 1980.

Sinnot, Roger. "Further notes on the Poncet platform," focuses on larger, load-bearing equatorial tables. *Sky and Telescope*, March 1980.

Dobson, John. "Have telescope, will travel," about the public observing program of the San Francisco Sidewalk Astronomers. *Sky and Telescope*, April 1980.

Mayer, Ben. "Astronomy on the sidewalk," an interview with John Dobson. *Astronomy*, April 1980.

Berry, Richard. "Variations on a Dobsonian theme," a description of telescopes by Dennis diCicco, Gerald Sibert, George Scotten, and others, in

How to Build a Dobsonian Telescope, Astromedia, 1980.

Berry, Richard. "How to control friction in a Dobsonian telescope," gives the derivation of Dobsonian friction equations. *Telescope Making #8*, Summer 1980.

Kestner, Robert, and Richard Berry. "How to build a Dobsonian telescope—I," three-part series on building a basic 10-inch Dobsonian telescope. *Astronomy*, June, 1980; "How to build a Dobsonian telescope—II," *Astronomy*, July, 1980; "How to build a Dobsonian telescope—III," *Astronomy*, August, 1980.

Berry, Richard. "The tale behind the telescope," in a brief editorial, introduces the Dobsonian concept to *Astronomy's* readership. *Astronomy*, August 1980.

Berry, Richard. "Newtonian telescopes," has everything you want to know about the Newtonian optical system. *Telescope Making #9*, Fall 1980.

Lasher, Richard W. "The Poncet mounting with a difference: it's for low latitudes," about modifying the Poncet. *Telescope Making #9*, Fall 1980.

Hamler, Walter. "Bowie's Dobsonian," letter describing the Bowie Astronomical Society's 14" club Dobsonian. *Telescope Making #10*, Winter 1980.

Sahs, Roger. "Mass production," letter describing a 6" f/8 Dobsonian design for teenagers. *Telescope Making #10*, Winter 1980.

Enright, Leo. "A revolution in amateur astronomy," how the "aperture explosion" is improving observing for amateurs. *Astronomy*, March 1981.

Kestner, Robert. "Grinding, polishing, and figuring thin telescope mirrors: part I - grinding," part 1 of 3 concerning grinding. *Telescope Making #12*, Summer 1981.

Kestner, Robert. "Grinding, polishing, and figuring thin telescope mirrors: part 2 - polishing," part 2 of 3. *Telescope Making #13*, Fall 1981.

Kysor, Stephen. "Using setting circles on a Dobsonian telescope," uses a TRS-80 pocket computer, protractor, and a roofer's level. *Telescope Making #13*, Fall 1981.

Berry, Richard. "Dual axis equatorial platforms," demonstrates how a platform can track in declination. *Telescope Making #14*, Winter 1981.

Dodson, Steve. "A Poncet/Dobson hybrid mounting for a 22" reflector," an alt-azimuth mounting with built-in tracking. *Telescope Making #14*, Winter 1981.

Houskeeper, Dick. "Astrophotography on a shoestring budget," describes an early Poncet platform. *Telescope Making #14*, Winter 1981.

Berry, Richard. "Equatorial platforms," describes how equatorial platforms work. *Astronomy*, January 1982.

Dey, Tom. "Dobsonian developments," a detailed account of using a TI-59 pocket calculator with Dobsonian setting circles. *Telescope Making #15*, Spring 1981.

Berry, Richard. "How to plan a Newtonian reflector," explains how to size and space the optical components of a Newtonian telescope. *Astronomy*, March 1982.

Burke, Paul. "Alt-azimuth setting circles," a detailed article on computing star positions for Dobsonian setting circles. *Astronomy*, April 1982.

Kestner, Robert. "Grinding, polishing, and figuring thin telescope mirrors: part 3 - figuring," part 3 of 3. *Telescope Making #16*, Summer 1982.

Hamberg, Ivar. "An extremely portable 17.5" Dobsonian," a significant breakthrough in truss-tube design. *Telescope Making #17*, Fall 1982.

Edberg, Steve. "How to determine the size of diagonals, draw tubes, and dew caps" gives equations for designing a telescope free of vignetting. *Telescope Making #18*, Winter 1982.

Berry, Richard. "The telescope revolution," a report on the 1982 Riverside, Astrofest, and Stellafane conventions, and a prediction of amateur astronomy's future. *Astronomy*, December 1982.

Kysor, Stephen. "Making an observing table and cover for a 10" reflector," also shows how to put precision alt-azimuth setting circles on a Dobsonian. *Telescope Making #19*, Spring 1983.

Martinez, Tom. "A State-of-the-art Poncet equatorial platform for large reflectors," show to build an equatorial platform able to carry a heavy load. *Telescope Making #19*, Spring 1983.

Pursell, Wally. "Pursell's platform: a southern alternative to the Poncet," describes an alternative geometry for equatorial platforms. *Telesoope Making #19*, Spring 1983.

Suiter, Dick, and Tom Burns. "An easy design for a truly portable 17.5" reflector," describes the "Boppian" mounting, an alternative way to build an open-tube telescope. *Telescope Making #19*, Spring 1983.

Cain, Lee. "Designing and constructing a 13" Dobsonian binocular," about the first "big" binocular telescope. *Telescope Making #20*, Summer 1883.

Dobson, John. "Dobson on Dobsonians," from a talk given at the 1983 Riverside Telescope Makers Conference. *Telescope Making #20*, Summer 1983.

Maxwell, Jason. "Teflon telescope bearings," examines the performance of Teflon/Formica bearings. *Telescope Making #21*, Winter 1983.

Allen, D. Eric. "Updating Newtonian collimation," gives procedures for collimating Newtonians with centered and offset secondary mirrors. *Telescope Making #22*, Spring 1984.

Cain, Lee. "Lee Cain's transportable Dobsonian binocular," describes the design and construction of a 17.5" truss-tube Dobsonian binocular. *Telescope Making #24*, Summer 1984.

Osypowski, Tom. "A 16" Dobsonian on a Pursell platform," details a load-bearing low-latitude equatorial platform. *Telescope Making #24*, Fall 1984.

Dodson, Steve. "A 22-inch equatorial Dobsonian," describes a hybrid three-axis mounting. *Sky and Telescope*, August 1984.

Shumaker, Bryan. "Computing under the open sky," basic alt-azimuth conversion for setting Dobsonian telescopes. *Sky and Telescope*, August 1984.

Smith, Steve. "Conical Poncet platform," this short letter describes a significant breakthrough in equatorial platforms. *Telescope Making #25*, Winter 1984.

Chandler, Dave. "Flotation cell design," includes a program for calculating 9-point and 18-point cells. *Telescope Making #26*, Summer 1985.

Multiple authors. "Sharing Dobsonian technology," nine authors offer hints on building Dobsonian telescopes. *Telescope Making #26*, Summer 1985.

Berry, Richard. "Highlights of the 1987 Stellafane convention," featuring John Dobson as the evening speaker. *Telescope Making #31*, Winter 1987.

d'Autume, Georges. "Equatorial tables without a pivot," describes a class of equatorial tables with conical bearings. *Sky and Telescope*, September 1988.

Kriege, David. "Dobsonian evolution: the construction of Obsession 1," details an optimized truss-tube telescope. *Telescope Making #35*, Winter 1987.

Ravneberg, Ronald. "Dobsonian evolution: refining the basic design," loaded with tips on building "new" Dobsonians. *Telescope Making #35*, Winter 1987.

Smitka, Peter. "Dobsonian evolution: the telescope is just one part of an observing system," shows that "system" thinking applies to telescope construction. *Telescope Making #35*, Winter 1987.

Albrecht, Richard. "The design of telescope structures, 1," a two-part article on building rigid telescopes; *Sky and Telescope*, January 1989; "The design of telescope structures, 2," *Sky and Telescope*, February 1989.

Kriege, David. "The making of Obsession 2," with details on building a 25" truss-tube Dobsonian. *Telescope Making #37*, Summer 1989.

O'Meara, Stephen J. "John Dobson: a man with a mission," transcript of a interview on the goals of the Sidewalk Astronomers. *Sky and Telescope*, November 1989.

Berry, Richard. "Inside the well-baffled Newtonian," covers the basic techniques for fixing stray light. *Deep Sky 25*, Winter 1988.

Clark, Tom. "Supertune your telescope for the very best performance," provides ground rules for building a highly effective telescope. *Telescope Making #39*, Winter 1989.

Ceravolo, Peter. "High-finesse figuring," tips for making telescope mirrors from a professional. *Telescope Making #39*, Winter 1989.

Stoltzmann, David and Peter Ceravolo. "Ross null test," a null method for figuring large, high-quality telescope mirrors. *Telescope Making #39*, Winter 1989. (See also, George, Doug. "Ross null software," *Telescope Making #45*, Summer 1991.)

Saulietis, Andy, and Al Kelly in "Top ten telescope ideas of 1989," describes a 32" telescope on a d'Autume platform. *Sky and Telescope*, December 1989.

Berry, Richard. "The d'Autumne equatorial platform," shows how Andy Johnson built a three-cone equatorial platform. *Telescope Making #40*, Spring 1990.

Kriege, David. "An observing ladder for large telescopes," describes constructing this vital observing accessory. *Telescope Making #41*, Summer 1990.

Kelley, William. "Free-floating Dobsonians," friction reduction strategies for Dobsonian bearings. *Sky and Telescope*, July 1990.

Clark, Tom and Jeannie. "Building a light-weight 30-inch Dobsonian," with useful information on balancing telescopes. *Telescope Making #44*, Spring 1991.

Kriege, David. "Low-tech R&D and off-the-shelf parts for super-big telescopes," an invaluable guide to telescope parts suppliers. *Telescope Making #44*, Spring 1991.

Sinnot, Roger. "Focus and collimation: how critical?," describes how the focus, focal ratio, and collimation, affect the performance of a telescope. *Sky and Telescope*, May 1991.

Diffrient, Roy. "Flexure in a Serrurier Truss," an engineering analysis of the truss tube in amateur telescopes. *Sky and Telescope*, February 1994.

Taki, Toshimi. "Mirror support: 3 or 9 points?," discusses mirror flotation systems. *Sky and Telescope*, September 1994.

Reference Works

----. *Machinery's Handbook*, 1990 Edition #23, an engineering reference book with complete formulae and tables of the properties of materials. Industrial Press, New York, 1990.

----. *The Aluminum Tube Book*, a catalog with the dimensions of nearly every type of commercial tube sold in North America. TubeSales, 1997.

----. W*ood Handbook: Wood as an Engineering Material*, a comprehensive manual for anyone who uses wood or wood products in construction. Forest Products Laboratory, U.S. Department of Agriculture, 1987.

Index